A PRACTICAL APPROACH TO DIGITAL ELECTRONICS

ALAN C. DIXON

Broome Community College

JAMES L. ANTONAKOS

Broome Community College

Prentice Hall

Upper Saddle River, New Jersey
Columbus, Ohio

Library of Congress Cataloging-in-Publication Data

Dixon, Alan C.
    A practical approach to digital electronics / Alan C. Dixon, James L. Antonakos.
       p.   cm.
    ISBN 0-13-727595-1
    1. Digital electronics.   I. Antonakos, James L.   II. Title.
TK7868.D5D595   2000
621.381—dc21
                                                             99-26192
                                                               CIP

Cover photo: FPG International
Editor: Scott Sambucci
Production Editor: Stephen C. Robb
Production Supervision: York Production Services
Design Coordinator: Karrie M. Converse-Jones
Cover Designer: Thomas Mack
Production Manager: Patricia A. Tonneman
Illustrations: York Graphic Services, Inc.
Marketing Manager: Ben Leonard

This book was set in Times Roman, Trajan, and Universe by York Graphic Services, Inc. and was printed and bound by Von Hoffman Press, Inc. The cover was printed by Von Hoffman Press, Inc.

©2000 by Prentice-Hall, Inc.
Pearson Education
Upper Saddle River, New Jersey 07458

Printed in the United States of America

10 9 8 7 6 5 4 3 2 1

ISBN: 0-13-727595-1

Prentice-Hall International (UK) Limited, *London*
Prentice-Hall of Australia Pty. Limited, *Sydney*
Prentice-Hall Canada, Inc., *Toronto*
Prentice-Hall Hispanoamericana, S. A., *Mexico*
Prentice-Hall of India Private Limited, *New Delhi*
Prentice-Hall of Japan, Inc., *Tokyo*
Prentice-Hall (Singapore) Pte. Ltd., *Singapore*
Editora Prentice-Hall do Brasil, Ltda., *Rio de Janeiro*

To Our Parents

It is easy to find evidence of digital systems all around us. Just look in your home, and you could list many different kinds of digital equipment:

- Telephone
- Microwave
- Television
- Clock
- Children's toys
- Music CD player
- Personal computer

We are clearly in the digital age.

What is a digital circuit? What is it made of? How is it designed and analyzed? This book answers these questions in detail, providing students with the skills necessary to work comfortably in a digital environment. Students in two- and four-year programs in Electrical Engineering, Electrical Engineering Technology, and Computer Science will be able to use this book to learn and apply the digital techniques necessary in today's market.

OUTLINE OF COVERAGE

The material in the book is divided into chapters that slowly build from the most basic material (number systems, basic logic gates) to advanced digital topics (analog-to-digital interfacing, microprocessors). Where applicable, circuit files for Electronic Workbench™ are provided to allow students to examine the operation of a working application. In addition, many C/C++ programs are included that let students get results to a question quickly and correctly. These programs perform number conversions from decimal to hexadecimal (and many others), solve Karnaugh maps, and analyze Boolean equations.

The chapters are organized as follows:

Chapter 1, Number Systems: The details of the binary, decimal, and hexadecimal number systems are covered. Techniques to convert from one base to another are explained.

Chapter 2, Binary Arithmetic: Reader can learn how to add, subtract, multiply, and divide in binary. Signed and unsigned numbers are presented, as well as BCD numbers.

Chapter 3, Logic Families: The major logic families and their operating characteristics are covered. TTL and CMOS are emphasized, and students are introduced to VLSI and ASIC technology.

Chapter 4, Basic Logic Gates: This chapter sets the stage for all of the following chapters. Everything associated with the basic logic gates is covered: truth tables, Boolean equations, propagation delay, loading DeMorgan's Theorem, and troubleshooting.

Chapter 5, Combinational Logic Circuit Design: In this chapter students can learn how to determine the hardware cost of a digital circuit and how to reduce the circuit through techniques such as factoring and the use of Karnaugh maps. PALs and GALs are introduced to illustrate how multichip circuits can be further reduced.

Chapter 6, Flip-Flops: Several flip-flop types are covered, as are the differences between level-sensitive inputs and edge-sensitive inputs. Typical applications for flip-flops, such as switch debouncing, are also provided.

Chapter 7, Counters: The material on flip-flops is expanded, showing students how to construct ripple counters for binary and BCD operation. Standard counter packages are also presented, as are timing diagrams.

Chapter 8, Synchronous Logic Circuit Design: This chapter explains the method used to design synchronous logic circuitry. Excitation tables, state diagrams, and state machine design are covered in detail.

Chapter 9, Circuit Design Using Programmable Logic: The use of PALs and GALs is explored. The programming process is explained, from equation to fuse map. The benefits of using programmable logic are also covered.

Chapter 10, Complex Logic Functions: In this chapter, a large subset of the TTL family is highlighted. Many of the most common logic functions are covered, from decoders and encoders, to comparators, arithmetic functions, and A/D and D/A conversion.

Chapter 11, Memories: One of the most fundamental building blocks of microprocessor-based systems is the memory chip. In this chapter students are introduced to the typical RAM and EPROM devices used in computers. The differences between static and dynamic RAM are discussed, as are special memories such as EEPROM and cache.

Chapter 12, Digital Data Transmission: All of the details of parallel and serial data communication are presented. Students can learn how an ASCII character is transmitted from one system to another. Techniques for digital data compression are outlined, and the operation of MODEMs and LANs is covered.

Chapter 13, Troubleshooting Techniques: Working with high-speed, microprocessor-based digital systems is the subject of this chapter. Methods for finding bad components, improper timing relationships, and the use of logic analyzers are discussed.

Chapter 14, Organization of Computers: This last chapter shows students the next level of digital architecture: the microprocessor. The operation of the personal computer is discussed from the hardware perspective. The design of an actual working single-board computer is presented.

Many different types of troubleshooting tips and helpful facts are sprinkled throughout the text. Look for the following icon in the margin to find these hints for quick reference.

The C and C++ programming languages illustrate important concepts and applications in a feature titled **C/C++ Helpers,** an example of which is shown here:

C/C++ HELPER

The following C program converts a user supplied input number into its corresponding 8-bit binary equivalent. This is the program DECTOBIN on the companion CD-ROM.

```c
//Decimal to Binary Conversion
#include <stdio.h>

main ()
{
    unsigned char number;
    int pattern;

    printf("Enter a decimal value between 0 and 255 --> ");
    scanf("%d",&number);
    printf("The decimal number %d equals ",number);
    for(pattern = 0x80; pattern != 0; pattern >>= 1)
        (pattern & number) ? printf("1 ") : printf ("0 ");
    printf ("binary\n");
}
```

A sample execution is as follows:

```
C> DECTOBIN
Enter a decimal value between 0 and 255 --> 143
The decimal number 143 equals 1 0 0 0 1 1 1 1 binary
```

Use DECTOBIN to help you check the results of your own conversions.
In addition to DECTOBIN, the programs DECTOHEX, BINTODEC, and HEXTODEC are also included on the companion CD-ROM.

All of the source and executable files for the C/C++ Helpers are included on the companion disk.

Furthermore, Electronics Workbench™ (EWB) examples are used throughout the text to provide students with a ready-to-run circuit that illustrates one or more main points from the current chapter. All EWB files are included on the companion disk. Look for this icon in the margin to find an EWB example quickly:

Detailed appendixes summarize important material and provide a quick reference to it. Answers to selected odd-numbered questions are also provided.

THE COMPANION CD-ROM

The companion CD-ROM packaged with this text contains several items:

◆ Approximately 50 figures from the text rendered in Electronics Workbench™ Version 5.x software. Users with access to this software program can open these files in this format.

◆ A demonstration version of Electronics Workbench™. For users without a copy of the Electronics Workbench™ software, this demo version allows the user to view 15 of the rendered circuits mentioned above.

◆ The Student Version of Electronics Workbench Version 5.x software. This software is made available for the user by directly contacting Interactive Image Technologies and purchasing a passcode from them. Any questions or concerns regarding Electronics Workbench™ software and its capabilities should be directed to Interactive Image Technologies, the producers of this software.

◆ C/C++ programs that generate truth tables, solve Karnaugh maps, and perform number conversions.

A detailed description of the Companion CD-ROM is provided by the README.COM program.

Finally, a complete package of ancillaries is available for this book, including:

◆ **VHDL Supplement.**   An introduction to High-Level Hardware Design using VHDL is available upon request from the publisher.

◆ **Laboratory Manual.**   This manual (ISBN 0-13-833823-X) consists of laboratory experiments that the authors use in their own classes. Special attention was paid to make each lab adaptable to various lab settings. All experiments can be performed in a laboratory environment or using Electronics Workbench™ software.

◆ **Laboratory Solutions Manual.**

◆ **Instructor's Resource Manual.**   This manual includes solutions to selected text problems, teaching tips, and sample syllabi.

◆ **PowerPoint® Slides.**   Figures from the text, as well as additional illustrations prepared by the authors, are designed to help instructors with classroom/lecture presentations.

◆ **Test Item File.**   This ancillary features test questions and sample tests provided by the authors.

◆ **PH Custom Test.**   This computerized testing system allows instructors to build tests from an electronic database of questions. This program also has on-line testing capabilities.

◆ **Companion   Website — www.prenhall.com/dixon**   The companion website complements the text as an on-line study guide. Review questions help students to understand the topics presented in each chapter. In addition, students will find links to many sites in the field of electronics. Using the Syllabus Builder feature, instructors can post on the Internet syllabi specifically designed for their classes.

ACKNOWLEDGMENTS

We thank the following reviewers for their insightful suggestions: Mauro Caputi, Hofstra University; Julio R. Garcia, San Jose State University; Tony Hearn, Community College of Philadelphia; and Byron Paul, Bismark State College.

We also thank the students who helped us develop this material and who offered suggestions and improvements for our lecture notes. Many thanks to our copy editor, Cindy Lanning, and to Kirsten Kauffman at York Production Services, who managed the book through production. Finally, we thank our editor, Scott Sambucci, and his assistant, Marcie Wademan, for their help during the development of this project.

Alan C. Dixon
dixon_a@sunybroome.edu
http://www.sunybroome.edu/~dixon_a

James L. Antonakos
antonakos_j@sunybroome.edu
http://www.sunybroome.edu/~antonakos_j

BRIEF CONTENTS

CONTENTS

CHAPTER 6: FLIP-FLOPS   136

CHAPTER 7: COUNTERS   162

CHAPTER 11: MEMORIES   253

CHAPTER 12: DIGITAL DATA TRANSMISSION   277

NUMBER SYSTEMS

INSTRUCTIONAL OBJECTIVES

All chapters begin with chapter objectives and self-evaluation questions, to help you get more from the book. Read this material as you start a chapter and then reread it when you have finished. If you cannot answer any of the questions, reread the section that applies. Your reading of this chapter should enable you to:

1. Count in all four bases described.
2. Use the standard and shortcut methods to convert a decimal number to other bases.
3. Convert hexadecimal pairs into bytes for use in microcomputer programming.

4. Be familiar with powers of 2, 8, and 16.
5. Tell the differences between nibbles, bytes, words, and long words.
6. See how numbers compare from one base or radix to another.
7. Understand the difference between analog and digital signals.
8. Find the least significant bit (LSB) and the most significant bit (MSB) in a binary number.

SELF-EVALUATION QUESTIONS

Keep the following questions in mind and try to answer them when you have completed the chapter:

1. What four bases are described in this chapter, and what are the legal characters used in each?
2. How can all number systems be represented by the radix method?
3. What is the difference between a digital and an analog signal?
4. How do hexadecimal pairs relate to a micro-processor program?

5. What is a register used for?
6. What is the largest number that can be stored in a byte, a word, or a long word?
7. How does counting in hexadecimal differ from counting in decimal?
8. Can a number like 0.7 (7/10) be accurately represented in binary, octal, or hexadecimal? (Answer: No. Why not?)

1.1 INTRODUCTION

Before we begin to learn the four number systems important to digital logic and micro-computers, let us take a brief look at the history of computers.

In 1642 the French mathematician-philosopher Blaise Pascal invented the first mechanical adding machine, which consists of eight 10-digit wheels connected to advance each other by the rules of addition. In 1670 the German mathematician Gottfried Wilhelm von Leibniz improved Pascal's design by adding multiplication gears.

In the nineteenth century in Great Britain Charles Babbage designed complicated machines that could have solved complex mathematical problems. The machines were never built, however, because they were too complex for the technology of the times.

Alan M. Turing, a twentieth-century mathematician, introduced the idea of a general-purpose computer in 1937 when he proposed a simple hypothetical machine, since named the universal Turing machine. A general-purpose computer is one that can be programmed to handle a variety of diverse problems.

The first electronic computer appeared in 1946 at the University of Pennsylvania. The computer was called ENIAC, which is the acronym for *E*lectronic *N*umerical *I*ntegrator *A*nd *C*omputer. It contained 18,000 vacuum tubes and occupied 15,000 square feet of floor space. ENIAC was able to perform 5000 additions or between 360 and 500 multiplications per minute.

With the invention of the transistor and later on the integrated circuit, computers shrunk in size, and now powerful devices can sit on the top of a desk. Some calculators are even small enough to fit inside a wristwatch.

The one link from Pascal's adding machine through ENIAC to today's microchip computers is found not in the machine's internal circuitry, but in its external communication. After all, computers are number-oriented machines, and in the next section we will see exactly why numbers are so important.

1.2 DECIMAL, BINARY, OCTAL, AND HEXADECIMAL

Today's microcomputers deal in four common number systems: decimal, binary, octal, and hexadecimal. Although most people are very much at ease with decimals, the other three number systems can seem awkward at first. In this chapter we examine all four systems in detail and learn how they operate in digital computers. After you have had some practice, these number systems will become powerful tools for designing and using all types of digital circuits.

The circuitry on the inside of any microcomputer must deal with the binary number system only. On the outside, however, the users of microcomputers deal with all four number systems or bases. Table 1.1 shows a summary of the four systems.

It can be easily shown why a microcomputer must deal in binary on the inside. Modern microcomputer chips have thousands of transistors packed into an area the size of your thumbnail. Each transistor can be thought of as a switch that is either completely turned on or completely turned off. These two states, on and off, are used to represent the two

TABLE 1.1 **Summary of Number Systems**

NUMBER SYSTEMS	BASE OR RADIX*
Decimal	10
Binary	2
Octal	8
Hexadecimal	16

*The words *base* and *radix* are used interchangeably to refer to the base of the number system.

characters of the binary number system, one (1) and zero (0). These two states may also be thought of as high/low, true/false, yes/no—or on/off. A circuit that has only two states and can be represented by a pair of voltage levels is said to be a *digital* circuit. All microcomputer chips and integrated circuits that make up a microcomputer are digital circuits. Electronic circuits that consist of many voltage levels and are not limited to two different voltages are considered to be nondigital. These circuits are used to amplify or produce music or voice or complex waveforms, and they have an infinite number of voltage levels. Such circuits are called *analog* circuits. Simple digital and analog signals are shown in Figure 1.1.

There are two worlds to consider then, the digital and the analog. This text is mainly concerned with digital signals and circuitry. To combine the two forms of circuitry, a converter or conversion process is necessary. To connect digital circuitry to analog circuitry, a digital-to-analog converter is used. Likewise, to connect an analog circuit to a digital one, an analog-to-digital converter is used. These conversions are illustrated in Figure 1.2. The subject of D-to-A and A-to-D converters is discussed at length in a later chapter. First we must study the four number systems that we are going to use in our work with digital circuitry and computers.

Decimal

The *decimal* number system has 10 legal characters or symbols. We are accustomed to calling them digits and are aware that they begin with zero and end with nine. Decimal numbers are made up of the characters zero to nine. We are also familiar with the

FIGURE 1.1 **Simple Digital (*a*) and Analog (*b*) Signals**

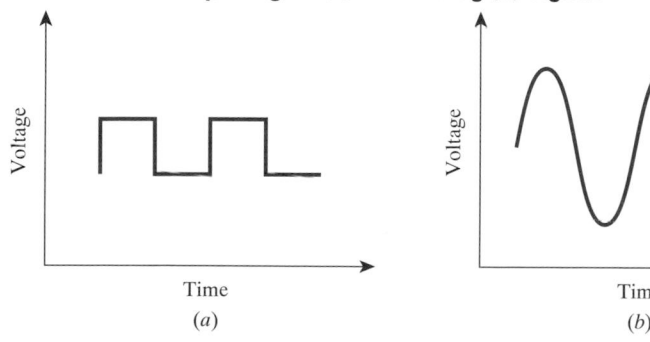

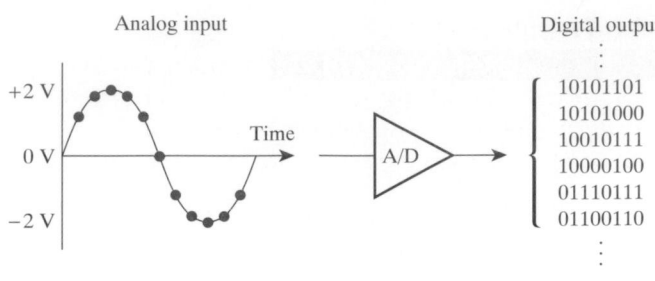

(*a*) A/D conversion

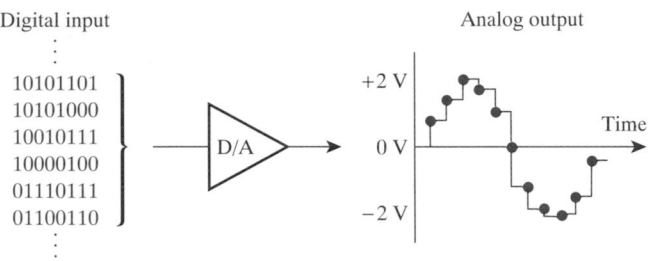

(*b*) D/A conversion

FIGURE 1.2   A/D and D/A Conversions

weight or importance that we ascribe to the position each digit has within a number. The units, tens, hundreds, and thousands have the weights of 1, 10, 100, and 1000, respectively.

Let's consider the number 491.68. This decimal quantity contains 4 hundreds, 9 tens, 1 unit, 6 tenths, and 8 one-hundredths. Where do these weights come from? A weight can be easily shown to be the base of a number system raised to a power. Here is a simple table detailing the weights involved in our example number.

$$4 \quad 9 \quad 1 \quad . \quad 6 \quad 8$$
$$10^2 \quad 10^1 \quad 10^0 \quad . \quad 10^{-1} \quad 10^{-2}$$

This is the basis for all our number systems. The weight of a position is related to the base you are operating in, raised to a power. This also holds when we deal with the binary number system.

Binary Number System

The **binary** or base 2 number system operates in a similar manner. In a binary number there are only two legal characters or symbols: zero (0) and one (1). Hence, binary numbers consist entirely of zeros and ones. Each binary character (i.e., each "0" or "1") is called a **bit**. Groups of bits are stored together in digital devices called *registers*.

A sample binary number might look like this: 10101100.1101 B (or 10101100.1101_2), where the B denotes the binary number system (8-bit binary numbers are used in Figure 1.2).

This serves to avoid confusing binary numbers with decimal or hexadecimal ones, which may be written 491.68, 491.68 D or 491.68 T (D means decimal, T means ten). We will use this method, following a number with a letter when dealing with all four bases. We evaluate the binary number by first determining the weight of each binary position or bit. Once again, the weight is a function of the base raised to a power.

EXAMPLE 1.1

Find the decimal equivalent of the binary number 10101100.1101 B.

SOLUTION

The number is converted to base 10 by using the following method:

Input bits →	1	0	1	0	1	1	0	0	.	1	1	0	1
Binary weights →	2^7	2^6	2^5	2^4	2^3	2^2	2^1	2^0	.	2^{-1}	2^{-2}	2^{-3}	2^{-4}
Decimal weights →	128	64	32	16	8	4	2	1	.	$\frac{1}{2}$	$\frac{1}{4}$	$\frac{1}{8}$	$\frac{1}{16}$

The left-most bit of a number is called the *most significant bit (MSB),* since it has the most weight. The left-most bit of our example has a total weight of 128. The right-most bit is called the *least significant bit (LSB)* because it has the least weight. To calculate the decimal number, you now simply multiply the input bit by its associated weight. Thus we have

$$
\begin{array}{rcl}
1 \times 128 & = & 128 \\
+ \ 0 \times 64 & = & 0 \\
+ \ 1 \times 32 & = & 32 \\
+ \ 0 \times 16 & = & 0 \\
+ \ 1 \times 8 & = & 8 \\
+ \ 1 \times 4 & = & 4 \\
+ \ 0 \times 2 & = & 0 \\
+ \ 0 \times 1 & = & 0 \\
+ \ 1 \times 0.5 & = & 0.5 \\
+ \ 1 \times 0.25 & = & 0.25 \\
+ \ 0 \times 0.125 & = & 0 \\
+ \ 1 \times 0.0625 & = & \underline{0.0625} \\
\text{Total} & = & 172.8125
\end{array}
$$

Put another way, simply add the weights of the positions whose bit value is a one.

The powers of 2 are very important, and you are encouraged to learn them as soon as possible. You will need to know the range from 2^0 through 2^{20} (very important in microcomputer work) and also the numbers 2^{-1} through $2^{-8.}$ Use Tables 1.2 and 1.3 to familiarize yourself with these numbers, which will appear in your work. To help acquaint you further with the binary number system, let's evaluate another sample number.

TABLE 1.2 Powers of 2

n	2^n	2^{-n}
0	1	1
1	2	0.5
2	4	0.25
3	8	0.125
4	16	0.0625
5	32	0.03125
6	64	0.015625
7	128	0.0078125
8	256	0.00390625
9	512	0.001953125
10	1,024	0.0009765625
11	2,048	0.00048828125
12	4,096	0.000244140625
13	8,192	0.0001220703125
14	16,384	0.00006103515625
15	32,768	0.000030517578125
16	65,536	0.0000152587890625
17	131,072	0.00000762939453125
18	262,144	0.000003814697265625
19	524,288	0.0000019073486328125
20	1,048,576	0.00000095367431640625

TABLE 1.3 Basic Data Sizes

1 *Nibble* = 4 bits	
1 *Byte* = 2 nibbles = 8 bits	
1 *Word* = 2 bytes = 16 bits	
1 *Long word* = 4 bytes = 32 bits	
Nibble	1101
Byte	11011101
Word*	1001101001101010
Long word*	10011010011110110111001101011100

*Typical names for 16-bit and 32-bit data sizes

···

EXAMPLE 1.2

Convert the number 1001011011.101101 B into base 10.

SOLUTION

As in Example 1.1, we sum only the weights of the positions that have the value "1"; thus we have

$$
\begin{array}{r}
512 \\
+ \quad 64 \\
+ \quad 16 \\
+ \quad 8 \\
+ \quad 2 \\
+ \quad 1 \\
+ \quad 0.5 \\
+ \quad 0.125 \\
+ \quad 0.0625 \\
+ \quad \underline{0.015625} \\
603.703125
\end{array}
$$

···

We have seen that in a binary number each digit is called a bit. Other data sizes (nibble, byte, etc.) are listed, with examples, in Table 1.3. The number of bits in a binary number determines the maximum size of the quantity to be represented. For instance, the largest number that can be represented with four bits is 1111 B. This represents the quantity 15 in decimal. Table 1.4 shows the largest number for some representative numbers of bits. This largest number can be computed from the expression $2^n - 1$, where n is the number of bits.

TABLE 1.4 Largest Value Possible with n Bits

n	LARGEST DECIMAL VALUE
1	1
2	3
3	7
4	15
5	31
6	63
7	127
8	255
10	1023
12	4095
16	65535
32	4294967295

Octal

Another number system that finds limited use in computer systems is *octal* or base 8. Octal numbers require eight legal characters, and we use the digits 0 through 7. An octal number would never contain the digits 8 or 9. An octal number will be followed by the letter O or Q to distinguish it from other bases. A Q is preferred, since an O might be mistaken for a zero and the number interpreted as a base 10 number. A sample octal number is 1423.41 Q (or 1423.41_8). The weights of the positions are again related to the base raised to a power.

··

EXAMPLE 1.3

Evaluate the following octal number: 1423.41 Q.

SOLUTION

Unlike binary, where we used powers of 2 to evaluate the weights, we will now use powers of eight.

$$
\begin{array}{ccccccccc}
1 & 4 & 2 & 3 & . & 4 & & 1 & Q \\
8^3 & 8^2 & 8^1 & 8^0 & . & 8^{-1} & & 8^{-2} & \\
512 & 64 & 8 & 1 & . & .125 & & .015625 &
\end{array}
$$

Now, by multiplying the first and third rows and taking the sum, we have

$$
\begin{array}{lll}
1 \times 512 & = 512 \\
+\ 4 \times 64 & = 256 \\
+\ 2 \times 8 & = 16 \\
+\ 3 \times 1 & = 3 \\
+\ 4 \times 0.125 & = 0.5 \\
+\ 1 \times 0.015625 & = \underline{0.015625} \\
\text{Total} & = 787.515625
\end{array}
$$

··

EXAMPLE 1.4

Convert the binary number 11011101.11011 B to decimal form.

$$
\begin{array}{lll}
1 \times 128 & = & 128 \\
+\ 1 \times 64 & = & 64 \\
+\ 1 \times 16 & = & 16 \\
+\ 1 \times 8 & = & 8 \\
+\ 1 \times 4 & = & 4 \\
+\ 1 \times 1 & = & 1 \\
+\ 1 \times \frac{1}{2} & = & 0.5 \\
+\ 1 \times \frac{1}{4} & = & 0.25 \\
+\ 1 \times \frac{1}{16} & = & 0.0625 \\
+\ 1 \times \frac{1}{32} & = & \underline{0.03125} \\
\text{Total} & = & 221.84375
\end{array}
$$

EXAMPLE 1.5

Convert the octal number 3407.521 Q to decimal form.

$$3 \times 512 = 1536$$
$$4 \times 64 \ = \ 256$$
$$7 \times 1 \ \ = \ \ 7$$
$$5 \times \tfrac{1}{8} \ \ = \ \ 0.625$$
$$2 \times \tfrac{1}{64} \ = \ \ 0.03125$$
$$1 \times \tfrac{1}{512} \ = \ \ \underline{0.001953125}$$
$$\text{Total} \ = 1799.658203125$$

Hexadecimal

The *hexadecimal* or base 16 number is very important and enjoys widespread use in computers. Hexadecimal (sometimes called just "hex") requires 16 legal characters. Since we only have 10 digits, we use not only zero to nine, but also the first six letters of the alphabet, A through F. In hexadecimal we write the letter A to represent 10 decimal. Table 1.5 shows the relationship between decimal, binary, and hexadecimal.

A typical hexadecimal number *may* contain both digits and letters as follows: 2AC6.4BH. Notice the H to indicate that we are dealing with a hexadecimal number. (We need the H to ensure that a decimal number, which is followed by a D, will not be mistaken for a

TABLE 1.5 Relationship Between Decimal, Binary, and Hexadecimal Number Systems

DECIMAL VALUE	4-BIT BINARY	HEXADECIMAL SYMBOL
0	0000	0
1	0001	1
2	0010	2
3	0011	3
4	0100	4
5	0101	5
6	0110	6
7	0111	7
8	1000	8
9	1001	9
10	1010	A
11	1011	B
12	1100	C
13	1101	D
14	1110	E
15	1111	F

hexadecimal number, where D is a legal character.) As in the three other number systems, the weight of each position is related to the base.

. . . .

EXAMPLE 1.6

Calculate the decimal value of the hexadecimal number 2AC6.4B H.

SOLUTION

As in Example 1.1, we make a small table of weights and take the sum of the products in the columns.

$$
\begin{array}{cccccccc}
2 & A & C & 6 & . & 4 & B & H \\
16^3 & 16^2 & 16^1 & 16^0 & . & 16^{-1} & 16^{-2} & \\
4096 & 256 & 16 & 1 & . & 0.0625 & 0.00390625 &
\end{array}
$$

$$
\begin{aligned}
2 \times 4096 &= 8{,}192 \\
+\ 10 \times 256 &= 2{,}560 \\
+\ 12 \times 16 &= 192 \\
+\ 6 \times 1 &= 6 \\
+\ 4 \times 0.0625 &= 0.25 \\
+\ 11 \times 0.00390625 &= \underline{0.04296875} \\
\text{Total} &= 10{,}950.29296875
\end{aligned}
$$

Notice that in the conversion, A becomes 10, C becomes 12, and B becomes 11.

. . . .

We frequently use hex numbers to four places, and it is useful to know that the largest number in hex to four places left of the decimal point is 0FFFF H (0FFFF H = 65535 D = $2^{16} - 1$). Notice the zero (0) in the "hex" number. This is common when dealing with hex numbers. If the hexadecimal number begins with a letter, we customarily precede it with a zero. When writing computer programs that utilize hexadecimal numbers, using the leading zero helps distinguish hexadecimal numbers (such as 0BEE) from variable names (such as BEE), which may not start with a numeric digit. Both the following are correct examples of hexadecimal notation: 0A412.4B H or 426.1B H.

Let us look at another example of a hexadecimal-to-decimal conversion.

. . . .

EXAMPLE 1.7

Convert the following hexadecimal number to decimal:

$$4A29.C3\ H$$

$$
\begin{aligned}
4 \times 4{,}096 &= 16{,}384 \\
10 \times 256 &= 2{,}560 \\
2 \times 16 &= 32 \\
9 \times 1 &= 9 \\
12 \times \tfrac{1}{16} &= 0.75 \\
3 \times \tfrac{1}{256} &= \underline{0.01171875} \\
\text{Total} &= 18{,}985.76171875
\end{aligned}
$$

. . . .

TABLE 1.6   Different Ways of Representing Numbers

Hexadecimal	3F7C.44B3 H = $(3F7C.44B3)_{16}$
Decimal	489.24 D = $(489.24)_{10}$ = 489.24
Binary	1101.11 B = $(1101.11)_2$
Octal	4531.45 Q = $(4531.45)_8$

An alternate method of representing numbers from different number systems is illustrated in Table 1.6, where the base of the number is placed at the end of the number being considered.

For instance, 47.9 D will become $(47.9)_{10}$ and 3F7 H will become $(3F7)_{16}$. Both methods work to convey the idea of a different number system.

··

1.3   SIMILARITIES BETWEEN BASES

Many bases are possible, and we are discussing only those used most commonly in computer systems. Regardless of the base (or radix), the weight assigned to a position is always the radix raised to a power as follows:

$$R^n \ldots R^3 R^2 R^1 R^0 . R^{-1} R^{-2} R^{-3} \ldots R^{-m}$$

To review then, in base N the weights of the positions are as follows:

Base (N)	R^4	R^3	R^2	R^1	R^0	.	R^{-1}	R^{-2}	R^{-3}
$N = 10$	10,000	1,000	100	10	1	.	0.1	0.01	0.001
$N = 2$	16	8	4	2	1	.	$\frac{1}{2}$	$\frac{1}{4}$	$\frac{1}{8}$
$N = 8$	4,096	512	64	8	1	.	$\frac{1}{8}$	$\frac{1}{64}$	$\frac{1}{512}$
$N = 16$	65,536	4,096	256	16	1	.	$\frac{1}{16}$	$\frac{1}{256}$	$\frac{1}{4096}$

These known weighting factors make it easy (with practice) to convert any binary, octal, or hexadecimal number to our familiar decimal number system. Section 1.5 shows how to convert in the other direction—that is, from decimal to another base.

1.4   COUNTING IN DIFFERENT BASES

We know that computers require many memory locations, which are numbered from zero up. We can count or number these locations in any of the bases we are studying. Therefore, we need a good understanding of how to count in the various number systems.

Table 1.7 offers a comparison of a continuous count for you to study. Memory locations are frequently numbered only in hexadecimal to four hex positions. A full memory range is from 0000 H to 0FFFF H. Some sequential locations are as follows:

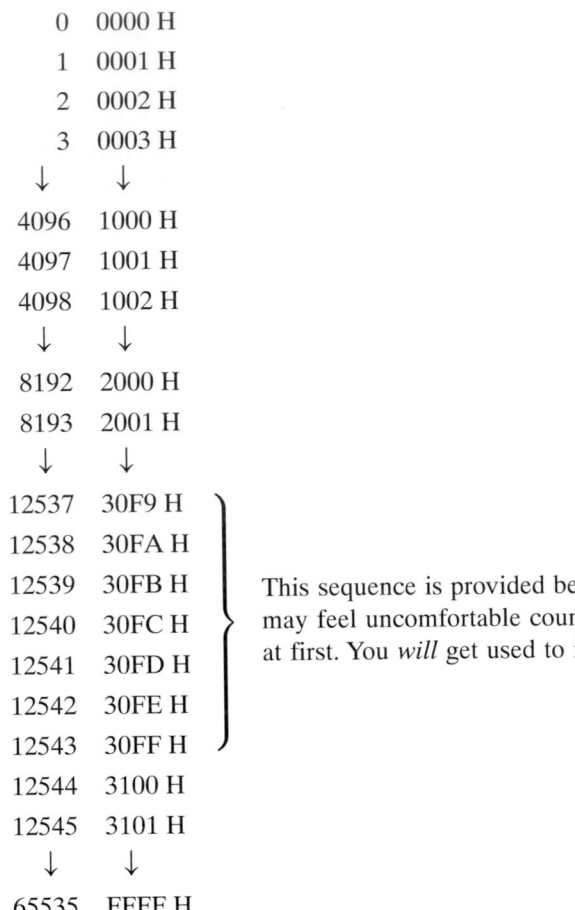

0	0000 H
1	0001 H
2	0002 H
3	0003 H
↓	↓
4096	1000 H
4097	1001 H
4098	1002 H
↓	↓
8192	2000 H
8193	2001 H
↓	↓
12537	30F9 H
12538	30FA H
12539	30FB H
12540	30FC H
12541	30FD H
12542	30FE H
12543	30FF H
12544	3100 H
12545	3101 H
↓	↓
65535	FFFF H

This sequence is provided because you may feel uncomfortable counting in hex at first. You *will* get used to it!

It is useful to refer to the number of a memory location as its **address.** The address of each memory location is unique and is a sequential number beginning with zero and usually given in hexadecimal. Since we begin counting memory locations at address zero, the address of the fifth location is 4 decimal, or 0004 H. Similarly, the address of the 12,540th location is 12539 decimal, or 30FB H. It is always one less because our count starts at zero and not one. In other words, location 1 is address zero. You see now why it is important to be able to count in the different bases.

Another aspect of counting in the binary number system involves determining *how many* bits are needed to represent a count. For instance, the number 7 requires three bits (111 B) to be represented in binary. The number 8 requires four bits (1000 B). Consider the numbers 700 or 25,000. How many bits are required to represent these numbers?

In general, any decimal number we can imagine can be found by raising the base 2 to a specific exponent. We saw some examples of this in Table 1.4. The exponent (power

TABLE 1.7 Comparison of a Continuous Count

DECIMAL	BINARY	OCTAL	HEXADECIMAL
0	00000 B	0 Q	0 H
1	00001 B	1 Q	1 H
2	00010 B	2 Q	2 H
3	00011 B	3 Q	3 H
4	00100 B	4 Q	4 H
5	00101 B	5 Q	5 H
6	00110 B	6 Q	6 H
7	00111 B	7 Q	7 H
8	01000 B	10 Q	8 H
9	01001 B	11 Q	9 H
10	01010 B	12 Q	A H
11	01011 B	13 Q	B H
12	01100 B	14 Q	C H
13	01101 B	15 Q	D H
14	01110 B	16 Q	E H
15	01111 B	17 Q	F H
16	10000 B	20 Q	10 H
17	10001 B	21 Q	11 H
18	10010 B	22 Q	12 H
19	10011 B	23 Q	13 H
20	10100 B	24 Q	14 H

TABLE 1.8 Bits Required to Represent a Number

N	$1 + \log_2 N$	BITS REQUIRED
1	1	1
3	2.58	2
7	3.8	3
10	4.32	4
25	5.64	5
40	6.32	6
100	7.64	7
700	10.45	10
1500	11.55	11
25000	15.6	15

of 2) actually tells us how many bits we need. To determine the exponent, we take the base-2 logarithm of the original value. Table 1.8 shows a few examples. Notice that we throw away the fractional part of the logarithm, since we can use only entire bits and not portions of them.

Knowing how many bits are required to represent a number is a valuable technique to have in your toolbox when solving a problem or starting a digital design.

1.5 CONVERTING BETWEEN BASES

We will now learn how to convert decimal numbers to any of the other three bases in use today. There are two separate methods, one for converting the integer part of a number and one for converting the fractional part. The integer part uses a divide-down procedure while the fractional method uses a multiply procedure. These two methods convert a decimal number to any base.

$$\underset{\swarrow}{\underline{293}} \cdot \underset{\searrow}{\underline{625}}$$

Integer part   Fractional part

Converting Decimal Integers to Another Base

To convert a decimal number (integer part only) to another base, we divide by the base we are converting to, and find the number of times that this division is possible. Any remainders will be the converted result in the new base.

· ·

EXAMPLE 1.8

Convert the decimal number 293 to base 2.

SOLUTION

To convert, we simply divide by 2 and write the remainders in a separate column, which will be the converted answer.

2|293	. binary point
146	1 (i.e., remainder of 1)
73	0 (i.e., remainder of 0)
36	1
18	0
9	0
4	1
2	0
1	0
0	1

When we get to zero, we are finished with the dividing, so now we read the remainder digits from the bottom up to the decimal point. Doing this we have 100100101 B. The result is always carefully checked by the method we know from Section 1.2: 256 + 32 + 4 + 1 = 293. It checks!

This method works when converting decimal integers to any base. Let us now convert the same number to octal:

$$
\begin{array}{r|l}
8\underline{\smash{|293}} & . \\
36 & 5 \\
4 & 4 \\
0 & 4
\end{array}
$$

The result is 445 Q and we also check the result:

$$
\begin{array}{rl}
5 \times 1 = & 5 \\
+ 4 \times 8 = & 32 \\
+ 4 \times 64 = & \underline{256} \\
& 293 \quad \text{Check!}
\end{array}
$$

Let's convert 293 into hexadecimal:

$$
\begin{array}{r|l}
16\underline{\smash{|293}} & . \\
18 & 5 \\
1 & 2 \\
0 & 1
\end{array}
$$

This result is 125 H, which we check quickly (by the same method):

$$
\begin{array}{rl}
5 \times 1 = & 5 \\
+ 2 \times 16 = & 32 \\
+ 1 \times 256 = & \underline{256} \\
& 293 \quad \text{Correct again!}
\end{array}
$$

··

Let us now learn how to convert the fractional part of a decimal number to another base.

Converting Decimal Nonintegers to Another Base

To convert fractional numbers to any other base, a multiply procedure is used. We simply multiply the number to be converted by the new base, with one limit: Only multiply what is to the right of the decimal point! Numbers appearing to the left of the decimal point are part of the result.

EXAMPLE 1.9

Convert 0.625 decimal to base 2.

SOLUTION

To solve, we simply multiply all numbers to the right of the decimal point by 2 until the number immediately to the right of the decimal is zero. Then we read the digits on the *left* of the decimal point from the top down. Doing this, we have

$$
\begin{array}{r}
0.375 \\
\times\ 2 \\
\hline
0.750 \\
0.750 \\
\times\ 2 \\
\hline
1.500 \\
0.500 \\
\times\ 2 \\
\hline
1.000
\end{array}
$$

Now we are done because only zeros are on the right of the decimal point. Reading from the top down (opposite of the integer routine), we get .011 B = 0.375 decimal. We also check our result to be sure:

$$
\begin{array}{ll}
0 \times 0.5 & = 0.0 \\
+\ 1 \times 0.25 & = 0.25 \\
+\ 1 \times 0.125 & = \underline{0.125} \\
& \quad 0.375 \quad \text{Check!}
\end{array}
$$

The conversion of fractions is not always exact, so it will be useful to do another example.

EXAMPLE 1.10

Convert the number 0.2 to binary.

SOLUTION

Some numbers, like 0.2, will not convert exactly. Rather, they continue forever with a particular binary pattern of ones and zeros.

$$
\begin{array}{r}
0.2 \\
\hline
\times\ 2 = 0.4 \\
\times\ 2 = 0.8 \\
\times\ 2 = 1.6 \\
\times\ 2 = 1.2 \\
\times\ 2 = 0.4 \\
\times\ 2 = 0.8 \\
\times\ 2 = 1.6 \\
\times\ 2 = 1.2 \\
\times\ 2 = 0.4
\end{array}
$$

The result is .0011001100$\overline{11}$ B: The sequence 0011 will continue forever. (*Note:* The repeating portion of such figures is indicated with an overline.) Taking the first six ones and their weights, we have:

$$
\begin{aligned}
&1 \times .125 \\
+ \ &1 \times .0625 \\
+ \ &1 \times .0078125 \\
+ \ &1 \times .00390625 \\
+ \ &1 \times .00048828125 \\
+ \ &1 \times \underline{.000244140625} \\
& .199951171875
\end{aligned}
$$

This is not quite 0.2, nor will it ever be. To simplify matters, if we take only the first six 1s (we selected six 1s for convenience here, but this choice does not represent any standard and you must determine your own requirements for accuracy) we will find that we are closely approximating 0.2.

In Chapter 2 we discuss how decimal numbers may be more accurately represented in binary codes. The error between 0.2 and 0.199951171 can be lessened, but not eliminated, by going to more places. This is one problem with binary numbers that must be overcome. It is usually solved by adding a slight "fuzz," or rounding many digits down the line. In other words, once the result is 0.19999999, a small part is added to this (0.00000001) to force the result to 0.2, which is the correct answer.

The fractional conversion from decimal to another base can also work for octal, so let us convert both example numbers 0.625 and 0.2.

EXAMPLE 1.11

Convert 0.625 and 0.2 to octal.

SOLUTION

0.625	0.2
$\times\,8$	$\times\,8$
5.00	1.6
0.00	4.8
	6.4
	3.2
	1.6
	4.8

Remember, we only multiply what is to the right of the decimal point. By this example you can see that 0.625 decimal = 0.5 Q and that 0.2 decimal once again does not convert exactly, but repeats: $.1463\overline{1463}$ Q. If we use our familiar method of checking our results, we will get

$$
\begin{array}{ll}
1 \times .125 & = .125 \\
+ 4 \times .015625 & = .0625 \\
+ 6 \times .001953125 & = .01171875 \\
+ 3 \times .00024414 & = \underline{.000732421} \\
& .199951171 \quad \text{Close!}
\end{array}
$$

$5 \times 0.125 = .625$ Check!

In doing fractional conversions that never stop, we will find that the answer always closes in on the expected result. It does this from below; that is, the result gets closer and closer to .2 (in this case), but never exceeds .2.

Let us try another example.

EXAMPLE 1.12

Convert 0.625 to binary.

SOLUTION

$$
\begin{array}{r}
0.625 \\
\underline{\times \ 2} \\
1.25 \\
0.50 \\
1.00
\end{array}
$$

The result is .101 B. Therefore, 0.625 D = 0.101 B = 0.5 Q.

$$
\begin{array}{ll}
1 \times .5 & = .5 \\
+ 0 \times .25 & = .0 \\
+ 1 \times .125 & = \underline{.125} \\
& .625 \quad \text{Correct!}
\end{array}
$$

EXAMPLE 1.13

Try a hexadecimal conversion with the two numbers used in Example 1.11.

SOLUTION

$$
\begin{array}{rr}
0.625 & 0.2 \\
\underline{\times \ 16} & \underline{\times \ 16} \\
10.00 & 3.2 \\
& 3.2 \\
& 3.2 \\
& 3.2
\end{array}
$$

We convert the 10 to an A, obtaining our hexadecimal answer 0.625 decimal = 0.A H. Once again we see, too, that 0.2 converts into a repeating number, namely 0.3333̄ H. To check:

$$3 \times .0625 = .1875$$

A × .0625 = 10 × .0625 = .625!

$$+ 3 \times .00390625 = .01171875$$
$$+ 3 \times .00024414 = \underline{.000732421}$$
$$.199951171 \quad \text{Close!}$$

Shortcut Between Bases

A relationship exists between bases 2, 8, and 16 that is very useful in making conversions from binary to or from octal and hexadecimal. This relationship comes from the fact that $2^3 = 8$ and $2^4 = 16$. To convert 46 Q to binary, we use three binary bits for each octal digit. We then use a 4-2-1 weighting scheme—the 421 code, as it is called.

EXAMPLE 1.14

Convert 46 Q into binary.

SOLUTION

By making use of the 421 code, we have, simply:

```
421   421
100   110.B

 4     6      Therefore 46 Q = 100110 B.
```

The 4 and 6 are each encoded using the 421 code.

EXAMPLE 1.15

Convert 763.421 Q into binary.

SOLUTION

Making use of the 421 code again, we have

```
421   421   421  .  421   421   421
111   110   011  .  100   010   001B

 7     6     3   .   4     2     1
```

There is also an easy way to go between binary and hexadecimal numbers. To convert 4AC3H to binary, an 8421 code or weighting scheme is used. In this case, four binary bits are allowed for each hexadecimal symbol. Knowing this, let us look at another example.

EXAMPLE 1.16

Convert 4AC3 H and 2864.D2C3 H into binary.

SOLUTION

The characters are encoded using the 8421 code as follows:

```
8421  8421  8421  8421  .  8421  8421  8421  8421
0100  1010  1100  0011  B
  4     A     C    3 H
0010  1000  0110  0100  .  1101  0010  1100  0011 B
  2     8     6     4    .   D     2     C    3 H
```

These shortcuts are very useful in working with numbers in various computer systems. Let us try one more example, converting first to base 16 and using the shortcut to obtain the binary and octal results.

EXAMPLE 1.17

Convert 293.41 decimal to binary, octal, and hexadecimal.

SOLUTION

First do the integer part:

```
16|293      .
      18    5
       1    2
       0    1
```

Then the fractional part:

$$
\begin{array}{r}
.41 \\
\times\ 16 \\
\hline
6.56 \\
8.96 \\
15.36 \\
5.76 \\
\end{array}
$$

The result is then 125.68F5 H. Let's check the result:

$$
\begin{array}{rcl}
5 \times 1 & = & 5 \\
+\ 2 \times 16 & = & 32 \\
+\ 1 \times 256 & = & \underline{256} \\
& & 293
\end{array}
$$

$$6 \times .0625 \quad\quad = .375$$
$$+ 8 \times .00390625 \quad = .03125$$
$$+ F \times .00024414 \quad = .0036621$$
$$+ 5 \times .000015258 = \underline{.000076293}$$
$$.409988393 \quad\quad \text{Close!}$$

We can now use the shortcut to convert over to binary.

125.68F5 H

1 0010 0101 . 0110 1000 1111 0101 B

And now we convert into octal by grouping the binary bits into triplets.

100 100 101 . 011 010 001 111 010 B

445.32172 Q

....

1.6 USES OF HEX AND BINARY IN COMPUTERS

The circuitry inside a computer must deal in the binary number system because transistors operating either fully on or fully off can be used to represent a binary "1" or "0." While bases 2, 8, and 16 are used in computers, the predominant mode is binary (inside the machine) and hexadecimal (on the outside, whereby the user specifies the binary). A person could easily slip an extra 0 or 1 into a string of binary bits, but one is less likely to drop a hexadecimal character; hence the use of both numbering systems. Octal is used in some applications, but it has been largely supplanted by hexadecimal. Frequently, when programming a microcomputer it is necessary to rapidly convert from hex pairs to 8 bits of binary. For example, a programmer will need to convert 0C3 H to 11000011 B. Several examples are now shown.

....

EXAMPLE 1.18

Convert the following hex pairs into binary:
(a) 03E H   (b) 2F H   (c) 76 H   (d) 0CD H

SOLUTION

We simply use the 8421 code to convert separate characters, and then group them into 8-bit numbers.

(a) 03E H = 0011 1110 B   (c) 76 H = 0111 0110 B
(b) 2F H = 0010 1111 B   (d) 0CD H = 1100 1101 B

....

Frequently a program appears as a hexadecimal string, and it is left to the programmer to divide the string into hex pairs and then convert quickly into binary.

EXAMPLE 1.19

Convert the following hexadecimal string into binary numbers (8 bits each): 3EFFD300C20010AFCD102176.

SOLUTION

To solve, we take two characters at a time and convert into binary.

$$
\begin{aligned}
3E &- 0011\quad 1110 \\
FF &- 1111\quad 1111 \\
D3 &- 1101\quad 0011 \\
00 &- 0000\quad 0000 \\
C2 &- 1100\quad 0010 \\
00 &- 0000\quad 0000 \\
10 &- 0001\quad 0000 \\
AF &- 1010\quad 1111 \\
CD &- 1100\quad 1101 \\
10 &- 0001\quad 0000 \\
21 &- 0010\quad 0001 \\
76 &- 0111\quad 0110
\end{aligned}
$$

Practice using hexadecimal pair–binary conversion and become expert at it. This is a common requirement for microcomputer programming.

1.7 TROUBLESHOOTING TECHNIQUES

Even though we will eventually be working with hardware during our study of digital electronics, we can begin using the computer as an educational tool right now. All of the conversion techniques presented in this chapter can be coded into computer programs that can be used to solve the problems for us, or check our results. Often, while troubleshooting a computer program or analyzing a circuit, a number is encountered whose value does not immediately come to mind, such as 3E8H. What is the decimal equivalent, you may wonder. This is why having a few conversion programs available during troubleshooting could save you some time.

In this section we encounter our first *C/C++ Helper*, a C or C++ program written to show you how the chapter material is applied in useful ways. You are encouraged to run the program (a copy resides on the companion CD-ROM), modify it, or even write your own. As you can see, the C/C++ Helper provides the program, a sample execution, and a list of related programs. Make good use of the provided programs. They allow you to do as many examples as you wish, and the "instructor" never gets tired.

·····C/C++ HELPER·····

The following C program converts a user-supplied input number into its corresponding 8-bit binary equivalent. This is the program DECTO-BIN on the companion CD-ROM.

```
//Decimal to Binary Conversion
#include <stdio.h>
main()
{
    unsigned char number;
    int pattern;

    printf("Enter a decimal value between 0 and 225 --> ");
    scanf("%d", &number);
    printf("The decimal number %d equals ",number);
    for(pattern = 0x80; pattern != 0; pattern >>= 1)
        (pattern & number) ? printf("1 ") : printf("0 ");
    printf("binary\n");
}
```

A sample execution is as follows:

```
C> DECTOBIN
Enter a decimal value between 0 and 255 --> 143
The decimal number 143 equals 1 0 0 0 1 1 1 1 binary
```

Use DECTOBIN to help you check the results of your own conversions.
    In addition to DECTOBIN, the programs DECTOHEX, BINTODEC, and HEXTODEC are also included on the companion CD-ROM.

SUMMARY

In this chapter we learned about the four most common bases used in microcomputers: decimal (base 10), binary (2), octal (8), and hexadecimal (16). To convert the integer part of a number to a different base, a divide-down procedure is used. To convert the fractional part of a number, a multiply procedure is used. Numbers are easily converted between the bases 2, 8, and 16 by using the 421 and 8421 codes.

STUDY QUESTIONS

1. Convert each binary number to a decimal number.
   (a) 1101.101 B
   (b) 110111.110 B
   (c) 1101111010.11 B
   (d) .0001110101 B
   (e) 110111010011 B
   (f) 10110111.111 B
   (g) 11111111.00101 B
   (h) 1101101111.100101 B

2. What is the MSB of each number in question 1? What is the LSB?

3. Convert each octal number to a decimal number.
   (a) 26.32 Q
   (b) .001634 Q
   (c) 46721.43 Q
   (d) 5472.623 Q

4. Convert each hexadecimal number to a decimal number.
   (a) 6AC.4B H
   (b) .46ACB H
   (c) 4CFA.B2 H
   (d) 2631.42 H

5. In a microcomputer memory with addresses beginning at address 0, what is the hexa-decimal address of each of the following locations?
   (a) The 2467 Dth location
   (b) The 65,000 Dth location
   (c) The 4000 Dth location
   (d) The 8191 Dth location

6. In the same microcomputer, what is the decimal location of each of the following ad-dresses (location zero is the "first" location)?
   (a) 2000 H
   (b) 200 H
   (c) 200D H
   (d) 0CC00 H

7. Convert the following integers to bases 2, 8, and 16.
   (a) 75 D
   (b) 289 D
   (c) 4071 D
   (d) 28431 D

8. Convert the following decimals to bases 2, 8, and 16.
   (a) .2 D
   (b) .1 D
   (c) .0461 D
   (d) .879 D

9. Convert the following "hex" pairs into 8-bit binary.
   (a) 0FE H   (e) 0C3 H
   (b) 55 H   (f) 0CD H
   (c) 0AA H   (g) 94 H
   (d) 0DB H

10. What are the MSB and LSB in each hex pair from question 9?

11. Convert the following microcomputer program to byte form (8-bit binary):
    210040DB40FE02C2034076.

12. Convert the following octal numbers to binary, hexadecimal, and decimal values.
    (a) 263.41 Q   (e) 777.77 Q
    (b) 46.235 Q   (f) .1 Q
    (c) 20634 Q   (g) 22.22 Q
    (d) .00431 Q   (h) 234.4 Q

13. Convert the following decimal numbers to bases 2, 8, and 16.
    (a) 268.49   (d) 999.99
    (b) 34019.2   (e) 2.414
    (c) 222.41   (f) .000141

14. What is the largest number (decimal) that can be represented if a 20-bit binary number is all 1s? What is the hex equivalent?

15. What is the largest number (decimal) that can be represented if a 24-bit binary number is all 1s? What is the hex equivalent?

16. How many bits are required to represent each number?
    (a) 6
    (b) 18
    (c) 27
    (d) 320
    (e) 186,000

17. Convert the following numbers to binary.
    (a) 4768219.4083
    (b) 20A.4CD H
    (c) 372.43 Q

18. Convert the following numbers to decimal.
    (a) 11011010111011.10011 B
    (b) 473F.4C2 H
    (c) 333.47 Q

19. Convert the following numbers to octal.
    (a) 2A3.41 H
    (b) 1100010011010.11111101 B
    (c) 3476.004

20. Fill in the following table:

N	2^N	8^N	16^N
0			
1			
2			
3			
4			
5		****	****
6		****	****
7		****	****
8		****	****

21. Compute the following:

$2^{20} =$	$8^3 =$
$2^{16} =$	$8^4 =$
$2^8 =$	$16^2 =$
$16^3 =$	$16^4 =$

22. Convert to 8-bit binary:

A7 H =	21 H =
D3 H =	E7 H =
C9 H =	DB H =
76 H =	F8 H =

23. A number is written simply as 110. What are its possible interpretations (values), using bases 2, 8, 10, and 16?

24. Convert the integer part of your age into binary, octal, and hexadecimal.

25. How many bits are needed to represent all the uppercase and lowercase characters of the alphabet, the digits 0 through 9, and an additional 20 punctuation symbols?

2

BINARY ARITHMETIC

When finished with this chapter, you should be able to:

1. Directly add or subtract complex binary numbers.
2. Multiply binary numbers by using the shift and accumulate method.
3. Add and subtract BCD numbers.
4. Determine the sign of a binary number.
5. Represent negative numbers using binary.
6. Correct the result of a BCD add operation.
7. Have an understanding of adding multiple binary numbers.
8. Use the carry bit and the auxiliary carry bit in math operations.

Keep the following questions in mind and try to answer them when you have completed the chapter.

1. What are the simple rules for binary addition?
2. How does the "control word" control the result in the accumulator in a binary multiplication?
3. Why do BCD numbers have to be adjusted after addition or subtraction?
4. What are the carry bit and the auxiliary carry bit?
5. When performing BCD subtraction, how is the number corrected when the result will be negative?
6. In the addition of hexadecimal numbers, how much does a carry represent?
7. Why is 2's complement arithmetic used?
8. What are the four variations of binary addition?

2.1 INTRODUCTION

In Chapter 1 we became familiar with binary numbers and their importance in digital circuitry and microcomputers. Now we will see how to perform addition, subtraction, multiplication, and division with binary numbers. In the following chapters, we will see how these arithmetic techniques are performed in a digital circuit.

2.2 SIMPLE ADDITION IN BINARY

As we learn to add in binary it is helpful to remember how we add numbers in decimal.

$$
\begin{array}{r}
1 \quad \leftarrow \text{Carry into tens column} \\
8 \\
+\ 9 \\
\hline
17
\end{array}
$$

In the addition of 8 and 9, the total is 7 for the units column with a 1 to carry into the tens column.

The carry into the next column is common in binary addition as well. A few rules of binary addition are now in order. Figure 2.1 gives the details of simple binary addition. These simple rules are all we need to know to add two binary numbers together. One or more rules will be used in each addition. For example, when adding 2 + 3, three different rules are used:

$$
\begin{array}{ll}
1 \qquad \leftarrow \text{Carry bits} & \\
010\ \text{B} & (2) \\
+\ 011\ \text{B} & +\ (3) \\
\hline
101\ \text{B} & (5)
\end{array}
$$

FIGURE 2.1   Simple Binary Addition Rules

$$
\begin{array}{ccc}
0 & 0 & 1 \\
+0 & +1 & +0 \\
\hline
0 & 1 & 1
\end{array}
$$

(*a*) No carry out to next column

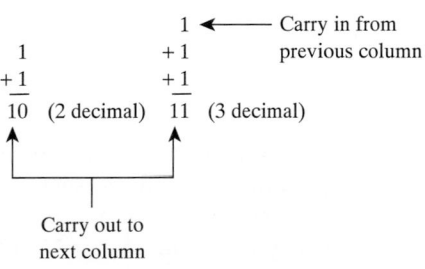

$$
\begin{array}{ll}
& 1 \longleftarrow \text{Carry in from} \\
1 & +1 \qquad\quad \text{previous column} \\
+1 & +1 \\
\hline
10\ \ (2\ \text{decimal}) & 11\ \ (3\ \text{decimal})
\end{array}
$$

Carry out to
next column

(*b*) Carry out to next column

2.3 COMPLEX ADDITION IN BINARY

The larger the binary number, the more complex the addition can become. It is important to watch any carries and place them in the correct column or columns.

....

EXAMPLE 2.1

Let's add 47 and 29 in binary.

SOLUTION

$$
\begin{array}{rr}
 & 111111 \quad \leftarrow \text{Carry bits} \\
47 & 101111 \text{ B} \\
+\ 29 & +\ 011101 \text{ B} \\
\hline
76 & 1001100 \text{ B}
\end{array}
$$

Begin with the LSB and work towards the MSB, writing down carries as they are generated.

....

Let us try another example.

....

EXAMPLE 2.2

Add 29 + 37.

SOLUTION

$$
\begin{array}{rr}
 & 1111\ \ 1 \\
29 & 011101 \text{ B} \\
+\ 37 & +\ \ 100101 \text{ B} \\
\hline
66 & 1000010 \text{ B}
\end{array}
$$

To check our result:

$$
\begin{array}{rcl}
1 \times 2 & = & 2 \\
+\ 1 \times 64 & = & \underline{64} \\
 & & 66 \quad \text{Check!}
\end{array}
$$

....

In a computer, the addition of binary numbers is performed in a special area called the arithmetic logic unit (ALU). The ALU is capable of performing additions, subtractions, and many other operations. The results are almost always placed in a general-purpose register called the **accumulator.** The accumulator may be an 8-, 16-, or 32-bit register. For this reason it is common to add 8-bit numbers together. Example 2.3 shows the addition of hex pairs.

····C/C++ HELPER·····

The BINADD program shown here asks the user for two numbers between 0 and 255 and shows how they are added together in binary.
Both input numbers are converted into binary as the addition is performed.

```c
//8-bit Binary Addition
#include <stdio.h>
main()
{
    unsigned int num1, num2, n1, n2;
    unsigned char n1bits[8], n2bits[8];
    unsigned char carries[9], sums[8];
    int k,carry,sum;
    printf("Enter the first number (0 to 255) ---> ");
    scanf("%d",&num1);
    n1 = num1;
    printf("Enter the second number (0 to 255) ---> ");
    scanf("%d",&num2);
    n2 = num2;
    carry = 0;
    for (k = 0; k < 8; k++)
    {
        n1bits[k] = num1 % 2;
        num1 /= 2;
        n2bits[k] = num2 % 2;
        num2 /= 2;
        sum = n1bits[k] + n2bits[k] + carry;
        if (sum > 1)
        {
            sum -= 2;
            carry = 1;
        }
        else
            carry = 0;
        sums[k] = sum;
        carries[k+1] = carry;
    }
    printf("The results are:\n\n");
    printf("\t\t");
```

(continued)

```
          for(k = 0; k < 8; k++)
             carries[8-k] ? printf("1 ") : printf("  ");
          printf("\n\t\t   ");
          for(k = 0; k < 8; k++)
             n1bits[7-k] ? printf("1 ") : printf("0 ");
          printf(" = %d\n\t\t+ ",n1);
          for(k = 0; k < 8; k++)
             n2bits[7-k] ? printf("1 ") : printf("0 ");
          printf(" = %d\n\t\t-----------\n\t\t",n2);
          carries[8] ? printf("1 ") : printf("  ");
          for(k = 0; k < 8; k++)
             sums[7-k] ? printf("1 ") : printf("0 ");
          printf(" = %d\n",n1+n2);
      }
```

Here is what BINADD does when executed:

```
C> BINADD
Enter the first number (0 to 255) ---> 110
Enter the second number (0 to 255) ---> 44
The results are:

          1 1   1 1
          0 1 1 0 1 1 1 0 = 110
        + 0 0 1 0 1 1 0 0 =  44
          -------------------
          1 0 0 1 1 0 1 0 = 154
```

The individual carries are shown to assist in checking your own hand
solution.
    BINADD can be easily extended into larger bit sizes by changing
the `for()` loop parameters and the array declarations.

EXAMPLE 2.3

Add 2C H and 48 H in both binary and hexadecimal.

SOLUTION

1	1	← Carry bits
2C H	00101100 B	
+ 48 H	01001000 B	
74 H	01110100 B	

In the hex addition of C + 8 we must convert C to 12 and call this the addition of 12 | 8 = 20. Then, since this is base 16, we subtract 16 from 20 and find that C + 8 is 4 with a 1 to carry to the next column. Let's try another example.

••

EXAMPLE 2.4

Add C3 H and B7 H in both binary and hexadecimal.

SOLUTION

$$
\begin{array}{llll}
1 & \quad 1 \quad\; 111 & \leftarrow \text{Carry bits} \\
\text{C3 H} & \;\; 11000011 \text{ B} \\
+\; \text{B7 H} & \;\; \underline{10110111 \text{ B}} \\
\hline
\text{17A H} & 101111010 \text{ B}
\end{array}
$$

In the hex addition, 3 + 7 = 10 or A, C + B becomes 12 + 11 = 23, and 23 − 16 = 7. Therefore, C + B is 7 with 1 to carry.

The final result of adding C3 H and B7 H is 17A H. The binary equivalent, which is really what appears in a computer, is also shown.

••

In Example 2.4 the result ends up in a carry in a ninth binary position, which is outside the 8-bit range of the input numbers. This indicates that the result of an addition may require more bits than either input number.

2.4 SIGNED AND UNSIGNED NUMBERS

In Chapter 1 we saw that the largest number that could be represented with n bits was $2^N - 1$. In general, this type of number is called an ***unsigned number.*** Unsigned numbers have a range that begins at zero and does *not* go negative. Only positive values are allowed. Thus, for four bits we get a range of numbers from 0 to 15.

The same four bits may also be used to represent a ***signed number*** range as well, including both positive *and* negative numbers. The actual range is from −8 to +7. Table 2.1 shows the way each 4-bit pattern is interpreted. Note that each 4-bit pattern that begins with a one has a corresponding negative value when interpreted as a signed number. This is an accepted convention when dealing with binary numbers, and results from our interpretation of a ***sign bit.*** The sign bit takes the place of the MSB when we are using signed numbers. This is necessary, since we have no other way of storing a minus sign with our binary number. Figure 2.2 illustrates the differences between signed and unsigned numbers.

In general, the largest positive value possible with n bits is

$$2^{N-1} - 1$$

and the most negative value is

$$-2^{N-1}$$

TABLE 2.1 Signed and Unsigned Range of Numbers

4-BIT BINARY	SIGNED VALUE	UNSIGNED VALUE
0000	0	0
0001	1	1
0010	2	2
0011	3	3
0100	4	4
0101	5	5
0110	6	6
0111	7	7
1000	−8	8
1001	−7	9
1010	−6	10
1011	−5	11
1100	−4	12
1101	−3	13
1110	−2	14
1111	−1	15

When a binary number is interpreted as an unsigned number, as in Figure 2.2(a), the MSB has an associated weight that may be included in the result. When interpreted as a signed number, the MSB becomes the sign bit (0 means positive, 1 means negative), and no numerical weight is associated with it. If the sign bit is zero, the remaining bits retain their original weights as we are used to seeing. If the sign bit is a one, we do not assign weights to the remaining bits. Instead, a technique called *2's complement* is used to determine the negative result. We will examine the 2's complement technique in detail in the next section, and see how it applies to binary subtraction. Keep in mind that it is important to know, in advance, which type of number is being represented in binary.

FIGURE 2.2 Signed and Unsigned Formats

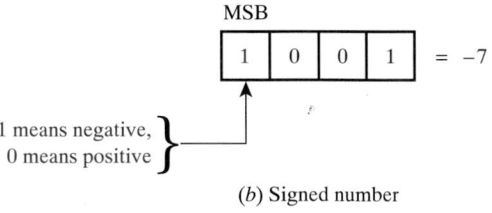

(a) Unsigned number

(b) Signed number

2.5 SIMPLE BINARY SUBTRACTION

In subtraction with decimal numbers we have become accustomed to "borrowing" from another column to find a result. This is not easily done on a computer, and so computer subtraction is not done in the way that is intuitively familiar to us. The computer subtracts by using the *complement* of a number. The complement of a binary number (*1's complement*) is found by inverting each bit (changing each zero to a one and each one to a zero).

One's Complement

Consider the binary number for 28: 28 D = 11100 B. The 1's complement (0s become 1s, 1s become 0s) of 11100 B is 00011 B. In working with complements, it is very important to hold an equal number of places for numbers being subtracted. That is:

$$
\begin{array}{r}
101101 \\
- \ 000110 \\
\hline
\end{array}
$$

Each number uses six bits in its representation.

∙∙∙

EXAMPLE 2.5

Find the 1's complement in binary of 47 H, 0B3 H, and 76 H.

SOLUTION

$$
\begin{array}{lll}
47 \text{ H} = 01000111 \text{ B}, & 1\text{'s complement} = 10111000 \text{ B} \\
0B3 \text{ H} = 10110011 \text{ B}, & 1\text{'s complement} = 01001100 \text{ B} \\
76 \text{ H} = 01110110 \text{ B}, & 1\text{'s complement} = 10001001 \text{ B}
\end{array}
$$

∙∙∙

Subtraction in binary is accomplished by adding the 1's complement of the number that is to be subtracted to the second number.

∙∙∙

EXAMPLE 2.6

Subtract 23 H from 81 H in binary.

SOLUTION

First, convert the hexadecimal values into binary:

$$
\begin{array}{cc}
81 \text{ H} & 10000001 \text{ B} \\
- \ 23 \text{ H} & - \ 00100011 \text{ B} \\
\hline
? & ?
\end{array}
$$

Rewrite the problem so that the $-$ 00100011 B is replaced with its 1's complement, then add.

$$
\begin{array}{r}
1 \qquad\qquad\qquad \leftarrow \text{Carry bits} \\
10000001\ \text{B} \qquad\qquad\qquad \\
+\ 11011100\ \text{B} \leftarrow \text{1's complement of 00100011 B} \\
\hline
\end{array}
$$

End-around carry ┌ 101011101 B
                 └────────→1
                 ────────────
                 01011110 B → 5E H

The correct answer must be 5E H. So far we have 15D H. The "1" at the left-most end of the binary result indicates a positive result and is a reminder to add 1 to 5D H to obtain the correct result, namely, 5E H.

Actually, the "1" can be added to the least significant bit of the result as shown. Adding the "1" from the carry into the LSB position is called an "end-around carry." Let's try another example.

....

EXAMPLE 2.7

Subtract 27 D from 72 D.

SOLUTION

$$
\begin{array}{rr}
72 & \quad 01001000\ \text{B} \\
-\ 27 & -\ 00011011\ \text{B} \\
\hline
45 &
\end{array}
$$

First, be sure both binary numbers are of the same length. We rewrite the problem using the 1's complement of 27 in binary.

$$
\begin{array}{r}
1 \qquad\qquad\qquad \leftarrow \text{Carry bits} \\
1001000\ \text{B} \qquad\qquad\qquad \\
+\ 1100100\ \text{B} \qquad\qquad\qquad \\
\hline
\end{array}
$$

End-around carry ┌ 10101100 B
                 └────────→ 1
                 ────────────
                 0101101 B → 45 D

Check it for yourself. It's 45! The "1" in the eighth position indicates two things: first, that the result is positive and, second, that an end-around carry is necessary.
....

Two's Complement

The "1" can be added to the 1's complement before the add operation takes place. In this case, the complement is called the 2's complement.

∙∙

EXAMPLE 2.8

Find the 2's complement of 28 in binary. Use 8 bits in your representation.

SOLUTION

$$28\ D = 00011100\ B, \qquad 1\text{'s complement} = 11100011\ B$$

Add 1 to find the 2's complement.

$$
\begin{array}{r}
11100011\ B \\
+ \qquad 1 \\
\hline
11100100\ B
\end{array}
$$

∙∙

Figure 2.3 shows how the 2's complement actually changes the *sign* of a number.

∙∙

EXAMPLE 2.9

Subtract 47 D from 123 D using 2's complement binary arithmetic.

$$
\begin{array}{rr}
123 & 01111011\ B \\
-\ 47 & -\ 00101111\ B \\
\hline
76 & ?
\end{array}
$$

FIGURE 2.3   Changing the Sign of a Number

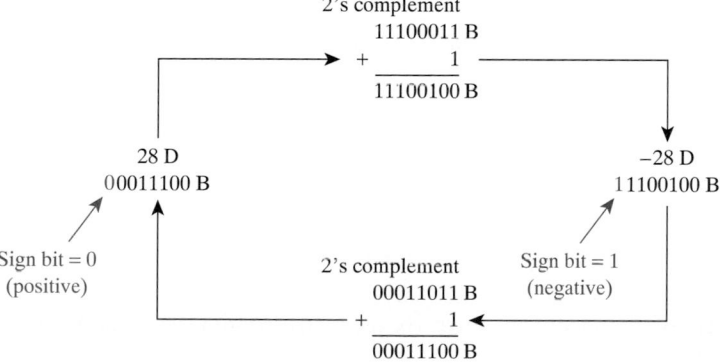

SOLUTION

First find the 1's complement of 47 D, then add 1.

$$00101111 \text{ B}$$
$$\underline{11010000 \text{ B}} \leftarrow 1\text{'s complement}$$
$$11010001 \text{ B} \leftarrow 2\text{'s complement}$$

Now perform the addition.

$$1\ 111\quad 11\quad \leftarrow \text{Carry bits}$$
$$01111011 \text{ B}\quad (123)$$
$$+\ \ \underline{11010001 \text{ B}}\quad (-47)$$
$$①01001100 \quad \to 76 \text{ D}$$

$$\text{ignored} \nearrow \quad \underbrace{\qquad} \\ \text{Answer}$$

When using 2's complement arithmetic, the resultant carry has a meaning different from its meaning in 1's complement arithmetic. A "1" means that the answer is positive and a "0" means that it is negative. In the event of a zero, the answer is recomplemented to obtain the correct result. The zero in the carry position then also means that the result is negative. In a computer, the use of 2's complement arithmetic generally saves a step in the addition process.

Let us reverse the numbers of Example 2.9 to see how this may work.

EXAMPLE 2.10

Subtract 123 D from 47 D using 2's complement arithmetic.

$$
\begin{array}{rr}
47 & 00101111 \text{ B} \\
-\ 123 & -\ 01111011 \text{ B} \\
\hline
-76 & ?
\end{array}
$$

SOLUTION

$$00101111 \text{ B}$$
$$+\ \underline{10000101 \text{ B}} \quad (2\text{'s complement})$$
$$⓪10110100 \text{ B} \quad \textit{Note: The carry is 0.}$$

Since the carry is zero, the result is negative, and recomplementing causes the number to appear in proper form:

$$01001011 \text{ B}$$
$$+\ \underline{\qquad\qquad 1}$$
$$01001100 \text{ B} \to 76 \text{ D}$$

Since the 2's complement of the result is positive 76, the actual result (10110100 B) must be -76.

Let us look at one more example of signed subtraction.

• •

EXAMPLE 2.11

Use signed 2's complement notation to find the result of 68 D − 49 D.

SOLUTION

$$
\begin{array}{r}
68\ D\ =\ 01000100\ B \\
\underline{-\ 49\ D\ =\ 00110001\ B} \\
19\ D \qquad ?
\end{array}
$$

Find the 1's complement of −49.

$$
\begin{array}{lr}
& 11001110\ B \\
\text{Add 1} & +\qquad\quad 1 \\
\hline
\text{2's complement} & 11001111\ B
\end{array}
$$

Now perform the addition.

$$
\begin{array}{r}
01000100\ B \\
\underline{+\ 11001111\ B} \\
①00010011\ B
\end{array}
$$

The correct answer is 00010011 B or 19 D. The carry bit indicates a positive result.

• •

Unsigned Two's Complement Arithmetic

A number that ignores the MSB as a sign bit is called unsigned. Positive 8-bit numbers may then be in the range 0–225.

$$
\begin{array}{l}
0\ =\ 00000000\ B\ =\ 0\ H \\
1\ =\ 00000001\ B\ =\ 1\ H \\
\qquad\qquad \downarrow \\
127\ D\ =\ 01111111\ B\ =\ 7F\ H \\
128\ D\ =\ 10000000\ B\ =\ 80\ H \\
\qquad\qquad \downarrow \\
255\ D\ =\ 11111111\ B\ =\ 0FF\ H
\end{array}
$$

Subtraction is still performed using 2's complement notation, but all 8 bits are available for use.

EXAMPLE 2.12

Using unsigned 2's complement arithmetic, find the result of 112 D − 49 D.

SOLUTION

$$
\begin{array}{rll}
112\ D = & 01110000\ B = 70\ H \\
-\ 49\ D = & -\ 00110001\ B = 31\ H \\
\hline
63\ D & ? \qquad = 3F\ H
\end{array}
$$

$$
\begin{array}{lr}
\text{1's complement of } 49 = & 11001110\ B \\
\text{Add 1} & +\qquad 1 \\
\hline
\text{2's complement} & 11001111\ B
\end{array}
$$

$$
\begin{array}{r}
01110000\ B \\
+\ 11001111\ B \\
\hline
100111111\ B \\
\uparrow
\end{array}
$$

Carry bit

In a microcomputer the carry out of the eighth bit is called the **carry bit.** A "1" indicates a positive result. The answer is $+63$ D or $+3F$ H or $+\ 00111111$ B. If the result is negative, the carry bit is a zero and the result must be complemented to obtain the correct answer.

EXAMPLE 2.13

Subtract 112 D from 49 D.

SOLUTION

$$
\begin{array}{rll}
49\ D = & 00110001\ B = & 31\ H \\
-\ 112\ D = & -\ 01110000\ B = & 70\ H \\
\hline
-63\ D & ? & -3F\ H
\end{array}
$$

Find the 2's complement of −112 D.

$$
\begin{array}{lr}
\text{1's complement} & 10001111\ B \\
\text{Add 1} & +\qquad 1 \\
\hline
\text{2's complement} & 10010000\ B
\end{array}
$$

$$
\begin{array}{lr}
49\ D & 00110001\ B \\
\text{2's complement of } -112\ D & +\ 10010000\ B \\
\hline
\text{Carry} = 0 \longrightarrow & 011000001\ B
\end{array}
$$

To obtain the correct answer, we note that the carry bit is a zero. The result is a negative number. Take the 2's complement and you have the correct result!

$$
\begin{array}{lr}
& 11000001\ B \\
\text{1's complement} & 00111110\ B \\
\text{Add 1} & +\qquad 1 \\
\hline
(-) & 00111111\ B = -3F\ H
\end{array}
$$

The correct answer is -00111111 B or $-3F$ H or -63 D.

If this seems complicated, remember that all arithmetic on a computer or in a digital circuit must be done using binary numbers and must follow a routine easily adhered to by digital circuitry. We have discussed three types of subtraction:

1. 1's complement.
2. Signed 2's complement.
3. Unsigned 2's complement.

The use of 2's complement is preferred and is often extended to be used in 16- and 32-bit accumulators. An 8-bit accumulator can operate on numbers ranging from $+127$ to -128. A 16-bit accumulator can operate on numbers ranging from $+32767$ to -32768. Thirty-two bit numbers have a range of over $+/-2$ *billion.*

....

EXAMPLE 2.14

Subtract 255 D from 47 D.

SOLUTION

Convert both numbers to binary.

$$47\text{ D} = \quad 00101111\text{ B} = 2\text{F H}$$
$$\underline{-\ 255\text{ D} = \ -\ 11111111\text{ B} = \ -0\text{FF H}}$$
$$-208\text{ D} \qquad\qquad ?$$

Find the 1's complement of 255 D.

1's complement	00000000 B
Add 1	$+\qquad 1$
2's complement	00000001 B

Now perform the addition.

47 D	00101111 B
2's complement of -255 D	$+\ 00000001$ B
Carry bit $= 0$	000110000 B

The answer is negative, and we must find the 2's complement to get the correct result.

	00110000 B
1's complement	11001111 B
Add 1	$+\qquad 1$
2's complement	11010000 B

The answer is -11010000 B $= -0$D0 H $= -208$ D.

....

2.6 BINARY MULTIPLICATION

There are several methods to accomplish multiplication. Some are less accurate than others as a result of rounding errors and the nature of representing decimal numbers in binary.

One way to perform a multiplication that gives a satisfactory result is to do successive additions the correct number of times. For instance, to multiply 5 × 17, we add 17 to itself a total of five times. This method works well for integers that are also small numbers.

. . . .

EXAMPLE 2.15

Find 5 × 17 using 8-bit binary additions.

SOLUTION

5 × 17 = 85.

$$
\begin{array}{r}
1 \quad 1 \quad\quad\quad\\
00010001 \text{ B}\\
00010001 \text{ B}\\
00010001 \text{ B}\\
00010001 \text{ B}\\
+\ 00010001 \text{ B}\\
\hline
01010101 \text{ B} \rightarrow 85 \text{ D}
\end{array}
$$

This type of solution is limited to numbers that are integers, or an integer times a fractional. A different method must be used when both numbers are fractional.

. . . .

Accurate Multiplication

We have already described the concept of a register as storage for binary numbers. Even though a register may not be long enough in reality, it can be considered to have storage capacity for a very large number of bits if a few tricks are used.

Such a register can be made to shift or rotate the bits to the left or to the right. This is usually done by the application of a clock pulse, a fast binary signal used to sequence or control digital circuitry. It takes one clock pulse to shift the entire binary string stored in the register. Figure 2.4 shows such a register with the binary number 11011.011 B stored in it. This is the equivalent of 27.375 D.

FIGURE 2.4   Shift Register

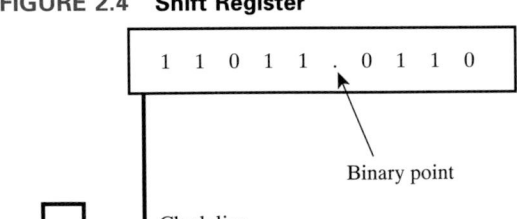

Binary point

Clock line

If we cause the register to perform a shift left by the application of a single clock pulse, the bits in the register move one place left with respect to the binary point. The new number looks like this:

$$\boxed{110110.110}$$

Checking the new result, we find a larger number, 54.75 D. This number is exactly twice the size of the original number. From this we can immediately and correctly conclude that each time a string of bits (binary number) is shifted to the left, it doubles. Furthermore, each shift to the right will halve the number. Shifting to the left again will give us a second doubling of the number, or four times the original.

The highly accurate multiplication by computer of binary numbers is based on the use of a shift register, which enables us to shift bits left and right, and an accumulator.

Integer Multiplication

We first learn how to multiply a number by an integer using a shift register and an accumulator.

We will multiply 18.75 by the integer 12 for a result of 225. We enter 18.75 into the shift register and use 12 as the *control word*, to tell us exactly what operations to perform. We use the accumulator, initialized to zero, as a place in which to add numbers. First let's convert the problem to binary.

$$
\begin{array}{rl}
18.75 & 10010.11 \text{ B} \leftarrow \text{Shift register number} \\
\times\ \ 12 & \times\ \ 1100.\ \ \text{B} \leftarrow \text{Control word} \\
\hline
225.0 &
\end{array}
$$

We enter 10010.11 B into the shift register. We will interpret the control word (1100 B) in the following way: A "0" will mean to shift left; a "1" will mean to add what is in the shift register to the accumulator, then shift left again.

Shift Register

$$\boxed{10010.110000 \text{ B}}$$

We have entered 10010.11 B into the shift register.

The control word is actually a set of instructions to be read, beginning with the least significant bit. The instructions are as follows:

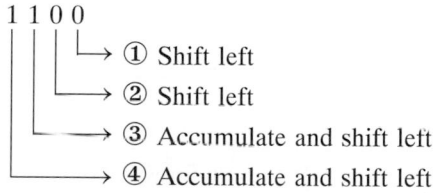

We will follow the instructions in that order.

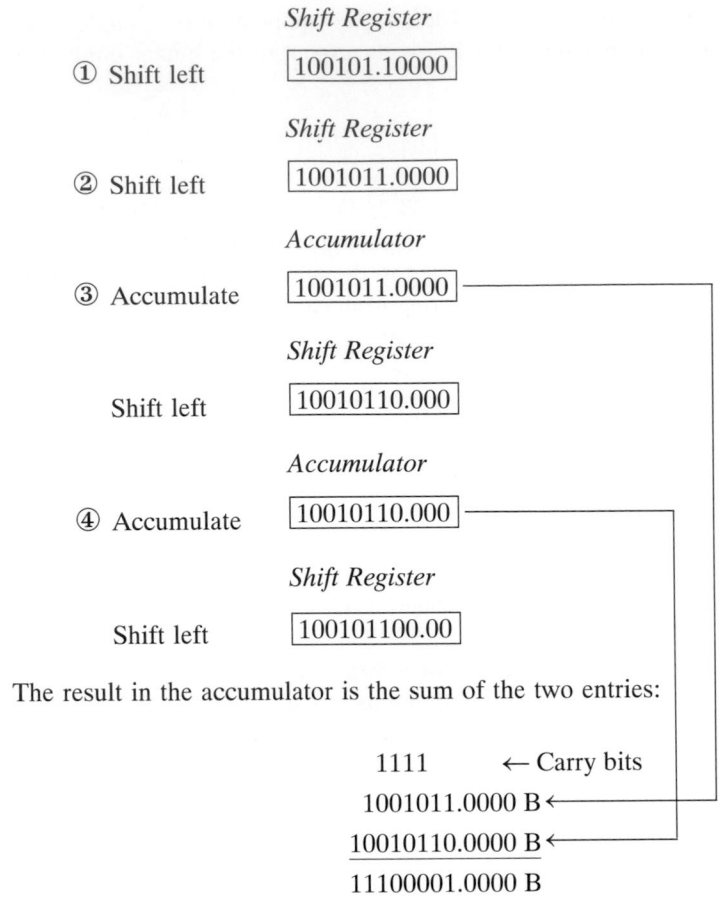

① Shift left

Shift Register

100101.10000

② Shift left

Shift Register

1001011.0000

③ Accumulate

Accumulator

1001011.0000

Shift left

Shift Register

10010110.000

④ Accumulate

Accumulator

10010110.000

Shift left

Shift Register

100101100.00

The result in the accumulator is the sum of the two entries:

```
    1111        ← Carry bits
 1001011.0000 B ←
10010110.0000 B ←
11100001.0000 B
```

This binary result is 225, the correct answer. This technique works well for integers and is similar to the method we use to multiply by a decimal number.

Multiplication by a Fraction

Multiplication by a fraction means that the result will be a smaller number than we start with. Let us multiply 26.75 by .2 D. We place 26.75 converted to binary in the shift register and use .2 D converted to binary as the control word. In this case we are expecting a smaller result, so we will be shifting right.

Let us first convert each number to binary.

$$
\begin{array}{r}
26.75 \\
\times\ .2 \\
\hline
5.35
\end{array}
\qquad
\begin{array}{r}
11010.11\ \text{B} \\
\times\ .0011001100110\overline{10011}\ \text{B}
\end{array}
$$

.2 D is a repeating binary sequence. It goes on forever. We will work down to five 1s (five accumulator entries) in the control word and see how close we are to the answer. So, we will consider that .2 D is .00110011001 B. This is rounded off, but it will illustrate the method. Greater accuracy can be obtained only by using more places.

The rules for using the control word are as follows: A "0" means shift right, and a "1" means shift right and then make an entry into the accumulator. Thus the control word is a set of instructions that runs like this:

$$. 0\ 0\ 1\ 1\ 0\ 0\ 1\ 1\ 0\ 0\ 1\ \ldots\ B$$

① Shift right ←⎯⎯⎯⎯⎯⎯⎯⎯⎯⎯⎯⎯⎯⎯⎯
② Shift right ←⎯⎯⎯⎯⎯⎯⎯⎯⎯⎯⎯⎯
③ Shift right and accumulate ←⎯⎯⎯⎯⎯
④ Shift right and accumulate ←⎯⎯⎯⎯
⑤ Shift right
⑥ Shift right
⑦ Shift right and accumulate
⑧ Shift right and accumulate
⑨ Shift right
⑩ Shift right
⑪ Shift right and accumulate ←⎯⎯⎯⎯⎯⎯⎯⎯⎯⎯⎯⎯⎯⎯⎯⎯⎯⎯

Notice that there will be five entries to the accumulator, one for each 1 in the control word.

Shift Register (contains 26.75)

$$\boxed{11010.11}$$

Remember, the bits shift right and the decimal point moves the other way.

The accumulator contains each entry and the totaled result. It is up to you to check the entire operation!

Accumulator

1		Carry Bits
1 1 1 1 1 1 1 1 1		
11.01011000000 B		Step ③
1.10101100000 B		Step ④
.00110101100 B		Step ⑦
.00011010110 B		Step ⑧
.0000001101011 B		Step ⑪
101.0101011110011 B		

The answer is checked as follows:

$$
\begin{aligned}
& 5. \\
+ \ & .25 \\
& .0625 \\
& .015625 \\
& .0078125 \\
& .00390625 \\
& .001953125 \\
& .00024414 \\
\underline{\ & .00012207} \\
& 5.342163085
\end{aligned}
$$

The result is close to but does not exceed 5.35. More accuracy is obtained by further use of the control word to additional significant places.

This is not only a good exercise but also great practice in adding binary numbers. However, you will agree, the sooner we make a digital circuit do this work, the better! And remember that we are adapting the circuitry of a machine operating in binary to a decimal world. The results are not always going to be easy to get, and they may not always be exact.

Multiplication of Decimal Numbers

The multiplication of 8.5×7.25 requires the use of both methods outlined previously. First, 8.5×7 is performed using the integer rules. Then, $8.5 \times .25$ is performed using the fractional rules. The results are added together. Simple numbers are chosen here to illustrate the point. A more complex example follows later.

· ·

EXAMPLE 2.16

Multiply 8.5×7.25.

$$
\begin{array}{r}
8.5 \\
\times\ 7.25 \\
\hline
61.625
\end{array}
\qquad
\begin{array}{r}
1000.1\ \text{B} \leftarrow \text{Shift register} \\
\times\ 0111.01\ \text{B} \leftarrow \text{Control word(s)}
\end{array}
$$

SOLUTION

First,

$$
\begin{array}{r}
1000.1\ \text{B} \\
\times\ 111.\ \text{B} \quad \leftarrow 0 = \text{Shift left}
\end{array}
$$

$\qquad\qquad\qquad\qquad\qquad 1 = \text{Accumulate and shift left}$

Shift Register

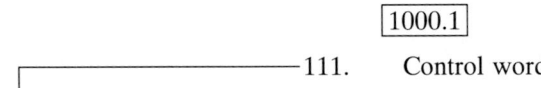

$\boxed{1000.1}$

——————————————————— 111.   Control word

① Accumulate and shift left
② Accumulate and shift left
③ Accumulate and shift left

Now we do the fractional part:

$$
\begin{array}{r}
1000.1\ \ \text{B} \\
\times\ \ \ \ \ .01\ \text{B}
\end{array}
$$

④ Shift right
⑤ Shift right and accumulate

Accumulator

1000.1000 B	Step ①
10001.0000 B	Step ②
100010.0000 B	Step ③
10.0010 B	Step ⑤
111101.1010 B	

Checking our result:

$$
\begin{array}{r}
0.5 \\
0.125 \\
1 \\
4 \\
8 \\
16 \\
+\ 32 \\
\hline
61.625 \quad \text{Check!}
\end{array}
$$

And now, a more complex example.

EXAMPLE 2.17

Multiply 63.7 × 13.8.

SOLUTION

We will show only the accumulator entries. It is left to you to work the problem.

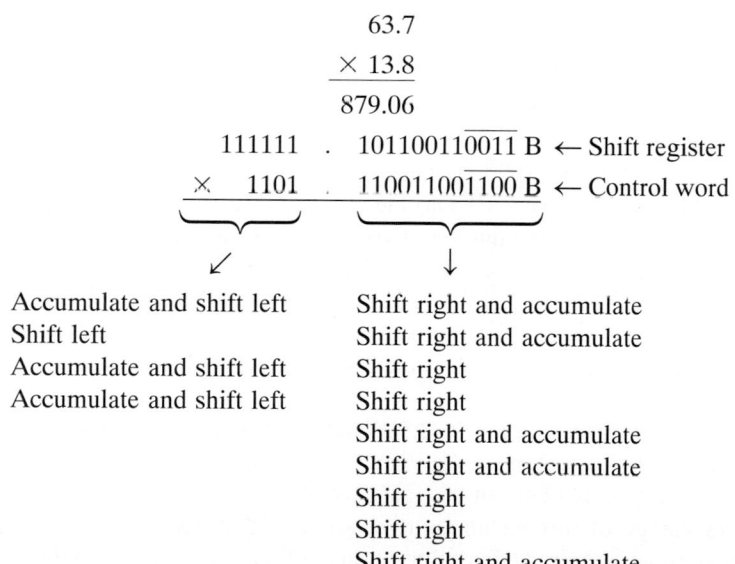

$$
\begin{array}{r}
63.7 \\
\times\ 13.8 \\
\hline
879.06
\end{array}
$$

111111 . 101100110011 B ← Shift register

× 1101 . 110011001100 B ← Control word

Accumulate and shift left   Shift right and accumulate
Shift left   Shift right and accumulate
Accumulate and shift left   Shift right
Accumulate and shift left   Shift right
                              Shift right and accumulate
                              Shift right and accumulate
                              Shift right
                              Shift right
                              Shift right and accumulate

11	1111.1011	0011	0011
1111	1110.1100	1100	1100
11111	1101.1001	1001	1001
1	1111.1101	1001	1001
	1111.1110	1100	1100
	1.1111	1101	1001
	.1111	1110	1100
	.0001	1111	1101
110110	1110.1111	1011	1111

(Result is 878.984130859)

In this problem, the carry bits are not shown for purposes of clarity. However, you can check the accumulator entries against your own. The result, which is for the entries shown, is close to the actual answer of 879.06 and does not exceed it. A closer result may be obtained by using more significant bits.

2.7 BINARY DIVISION

Binary division is accomplished in a variety of ways. The simplest method involves repetitive subtractions, similar to the repetitive additions used to multiply two numbers. We count the number of times we can subtract one value from the other. This is typically a slow and inaccurate process.

In a microprocessor, the division operation may actually be performed by using a multiplication. Think of dividing 45 by 5. This is the same as multiplying 45 by 0.2. The trick is to determine the inverse of 5 (0.2) and then use it in a multiply operation. Finding the inverse of a number can be accomplished by choosing an initial value (a guess) for the inverse, and then seeing if the guess multiplied by the original is equal to one. The guess is modified accordingly before continuing with the "divide" operation.

In this section we will examine a third method, which involves subtraction and *restoration* of the dividend. Consider the two unsigned numbers 210 and 8. Their respective binary values are 11010010 B and 1000 B. Figure 2.5 shows how these two binary numbers are divided.

Whenever the divisor (1000 B) can be subtracted from the current set of dividend bits (initially 1101 B), we place a 1 into the quotient. This is illustrated in Figure 2.5(b), 2.5(c), and 2.5(e). After subtracting the divisor, the new set of dividend bits is found by bringing down the next bit from the original dividend. When the current dividend is smaller than the divisor (as in Figure 2.5(d) and 2.5(f), where we have dividends of 100 B and 10 B respectively), we place a zero into the quotient and *restore* the dividend (the subtraction operation must be undone). This method is similar to what we do when performing long division.

The final result shown in Figure 2.5(f) indicates an answer of 11010 B, which corresponds to 26 D, the correct integer answer. Note that the remainder (10) corresponds to 0.25, since this is two (10) left out of eight (1000).

The advantage of this technique over simple repetitive subtraction is that, typically, fewer steps are required (five subtractions in Figure 2.5 versus 26 the simple way). The disadvantage, as we will see in later chapters, is the complexity of the hardware

$\dfrac{210}{8}$ ⟶ 1000 $\overline{)11010010}$ ⟵ Quotient

Divisor ⟶ │ └─ Dividend

(*a*) Initial values

```
            1      (1101 > 1000)
1000 | 11010010
   −  1000
      ────
       101
```
(*b*) Initial quotient bit

```
           11      (1010 > 1000)
1000 | 11010010
   −  1000
      ────
      1010
   −  1000
      ────
        10
```
(*c*) Second quotient bit

```
          110      (100 < 1000)
1000 | 11010010
   −  1000
      ────
      1010
   −  1000
      ────
       100
   −  1000
      ────
      0000  ⟵── Cannot subtract divisor
```
(*d*) Third quotient bit

```
         1101     (1001 > 1000)
1000 | 11010010
   −  1000
      ────
      1010
   −  1000
      ────
      1001
   −  1000
      ────
         1
```
(*e*) Fourth quotient bit

```
                    ┌─ Final quotient
        11010       (10 < 1000)
1000 | 11010010
   −  1000
      ────
      1010
   −  1000
      ────
      1001
   −  1000
      ────
        10  ⟵── This is the remainder
   −  1000
      ────
      0000  ⟵── Cannot subtract
```
(*f*) Fifth (final) quotient bit

FIGURE 2.5   Binary Division of Unsigned Integers

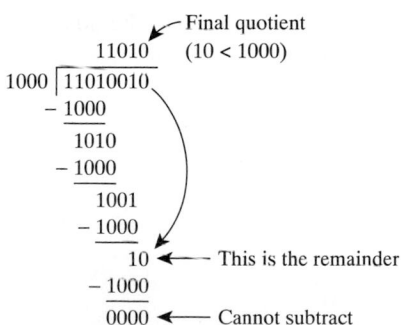

required to implement this technique. This brings up an important point: we often must choose between two different approaches to a solution. If speed is important, the restoration method is acceptable. If minimal hardware is a requirement, repetitive subtraction will suffice. This is why it is important to be exposed to many different ways of solving a problem.

2.8 BINARY CODED DECIMAL (BCD) ARITHMETIC

We have seen how computers can use binary numbers to perform math operations. All these methods use numbers in pure binary and frequently are subject to errors due to the imprecision of representing decimal numbers with a limited number of binary bits. For example, .2 D converts to the repeating binary sequence .0011001100110011 B. No computer can afford an infinite number of places to represent such a number. This causes certain rounding errors in large computers that must be "covered up." Additional programming or circuitry must be used to obtain answers that are apparently correct to the user.

Many microcomputers use languages that avoid these problems by making use of a 4-bit binary code called **binary coded decimal (BCD).** Each digit to be used in an operation is encoded using 4 bits in an 8421 code or 8421 weighting scheme. In this way, decimal numbers are stored in a computer using 0s and 1s but without the necessity of converting to pure binary. All the inaccuracies due to rounding disappear when using BCD arithmetic.

For example,

$$48 \text{ D} = 01001000 \text{ in BCD}$$

The 4 becomes 0100 and the 8 becomes 1000. So any decimal number is easily represented using BCD. Each digit requires 4 bits, making the size (number of bits) of a BCD number easily predictable. For instance, an 8-digit decimal number requires precisely 32 bits (8×4). Such estimations are not so easy when using pure binary.

Doing arithmetic with BCD numbers requires a few tricks, but they are easily learned and implemented. Let us see how this is done.

Addition Using BCD Numbers

In performing addition of BCD numbers, we must remember that the addition will actually be addition in hexadecimal. Example 2.18 illustrates this point.

••

EXAMPLE 2.18

Case I: Add 46 D + 23 D.

SOLUTION

$$
\begin{array}{rl}
46 \text{ D} & \quad \text{In BCD:} \quad 01000110 \\
+ \ 23 \text{ D} & \quad \phantom{\text{In BCD:}} \quad \underline{00100011} \\
\hline
69 \text{ D} & \quad \phantom{\text{In BCD:}} \quad \underline{01101001} \\
\end{array}
$$

$\underbrace{}_{6} \ \underbrace{}_{9}$ The correct answer!

••

However, when a carry occurs, things are different.

••

EXAMPLE 2.19

Case II: Add 46 D + 29 D.

SOLUTION

$$
\begin{array}{rrr}
46\ D & 01000110 & 46\ H \\
+\ 29\ D & +\ 00101001 & +\ 29\ H \\
\hline
75\ D & \underline{01101111} & 6F\ H \\
& \quad\underbrace{\quad}_{6}\ \underbrace{\quad}_{F} &
\end{array}
$$

In this case, 6F is *not* an acceptable answer. The sum of 6 and 9 result in a hex F. Since we are trying to represent decimal numbers, only the codes for 0–9 are acceptable. When an illegal code (A–F) appears, we must perform a correction.

The correction is to add 06 to the result:

$$
\begin{array}{rr}
01101111 & 6F\ H \\
+\ 00000110 & +\ 06\ H \\
\hline
\underline{01110101} & 75\ H \\
\underbrace{\quad}_{7}\ \underbrace{\quad}_{5} &
\end{array}
$$

••

Let us try several more examples to determine other necessary corrections.

••

EXAMPLE 2.20

Case III: Add 57 + 92.

SOLUTION

$$
\begin{array}{rrr}
57 & 01010111 & 57\ H \\
+\ 92 & +\ 10010010 & +\ 92\ H \\
\hline
149 & \underline{11101001} & E9\ H \\
& \quad\underbrace{\quad}_{E}\ \underbrace{\quad}_{9} &
\end{array}
$$

The E appears and is an illegal code. To obtain the correct result we add 60 H to the result.

$$
\begin{array}{rr}
11101001 & E9\ H \\
+\ 01100000 & +\ 60\ H \\
\hline
\underline{101001001} & 149\ H \\
\underbrace{\ }_{1}\ \underbrace{\ }_{4}\ \underbrace{\ }_{9} &
\end{array}
$$

The correction causes a carry into the third hex position or the ninth binary bit.

EXAMPLE 2.21

Case IV: Add 75 D + 49 D.

SOLUTION

75 D	01110101	75 H
+ 49 D	+ 01001001	+ 49 H
124 D	10111110	BE H

In this case, both parts of the result are over the code for 9 and are illegal. The correction is to add 66 H to the result. We see that the first "digit" (E) carries into the second "digit" (B). We also see the second "digit" *carry* into the third digit.

```
1111111 ← carries
10111110              BE H
+  01100110          +  66 H
  100100100            124 H
  ‿‿‿ ‿‿‿ ‿‿‿
   1   2   4
```

The correction then involves observing the result and deciding whether to add the binary representation 06 H, 60 H, or 66 H to obtain the correct result.

We are also able to handle decimals with this method without the problem of rounding errors.

EXAMPLE 2.22

Add 26.21 D + 29.46 D.

SOLUTION

26.21 D	00100110.00100001
29.46 D	+ 00101001.01000110
55.67 D	01001111.01100111

6 added as a correction ⟶ + 0110.

```
            01010101.01100111
            ‿‿‿ ‿‿‿ . ‿‿‿ ‿‿‿
             5   5  .  6   7
```

BCD Subtraction

Subtracting using BCD numbers is a bit trickier. A variation of complementing must be used, but in decimal. The *9's complement* of a decimal number is found by subtracting from the largest number in the base using the same number of places. The complement of 475 is found by subtracting from 999. The complement of 42 is found by subtracting from 99. A 6-digit number would be subtracted from 999999. The *10's complement* is found by adding 1 to the 9's complement.

To perform the subtraction 46 D − 29 D, we do the following:

$$46 \text{ D}$$
$$\underline{- \ 29 \text{ D}}$$
$$17 \text{ D}$$

First, find the 10's complement of 29.

$$99$$
$$\underline{- \ 29}$$
Add 1 $$\qquad 70$$
10's complement $$\qquad \underline{+ \ \ 1}$$
$$71$$

Now add 46 to 71.

$$46$$
$$\underline{+ \ 71}$$
$$117$$
Carry indicates a positive result ⟋   ⟍ Answer

To save the step of adding 1 at the end, we can subtract from 9A H instead of 99 H.

....

EXAMPLE **2.23**

Subtract 43 D from 72 D using binary.

72 D	01110010 B	72 H
− 43 D	− 01000011 B	− 43 H
29 D	?	?

SOLUTION

First subtract 43 H from 9A H.

$$9\text{A H}$$
$$\underline{- \ 43 \text{ H}}$$
$$57 \text{ H} \quad (\text{10's complement})$$

Now add 57 H to 72 H.

$$72 \text{ H}$$
$$\underline{+ \ \ 57 \text{ H}}$$
$$129 \text{ H}$$
⟍ Answer   (72 − 43 = 29)

....

2.9 TROUBLESHOOTING TECHNIQUES

There are many times when the techniques we learn in one part of our digital study are applied in another area. For example, the simple rules of binary addition are actually used to design a hardware circuit called a ***half-adder*** that implements them. The rules, in their original form, are shown in Figure 2.6(a). Note that the only carry generated occurs when two 1s are added.

Figure 2.6(b) shows the rules rearranged into a ***truth table.*** A truth table is a structure that organizes all of the input and output combinations of a digital circuit. On the left side are the two bits being added. They are called *A* and *B*. So, *A* and *B* both take on the values of 0 and 1 as shown. The right side of the truth table lists the outputs (Sum and Carry) and their states.

We often see patterns we recognize in the truth table. In Chapter 3 we will see that these patterns are often the output of a basic logic gate which can then be used in the design of a digital circuit. For instance, the Sum output is actually the output of an ***exclusive-OR (XOR) gate,*** and the Carry output is the output of an ***AND gate.*** Figure 2.6(c) shows how both gates are used to make the half-adder.

In general, the more we know about the basics, about binary numbers and how to work with them, the better off we will be when designing or troubleshooting digital circuitry.

FIGURE 2.6   Designing the Half Adder

$$
\begin{array}{cccc}
0 & 0 & 1 & 1 \\
+0 & +1 & +0 & +1 \\
\hline
0 & 1 & 1 & 11
\end{array}
$$

Carry to next stage

(*a*) Original rules of addition for two bits

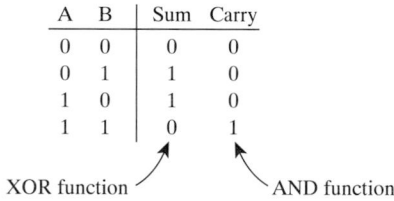

XOR function   AND function

(*b*) Rules rewritten as a truth table

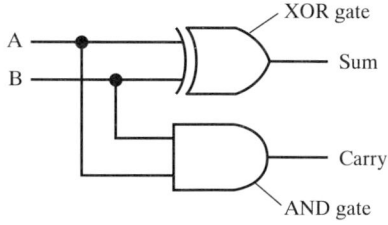

(*c*) Actual digital circuit (half adder)

SUMMARY

In this chapter we learned the simple rules for binary addition. We also learned how to multiply in binary by the use of a control word and an accumulator. Binary division was accomplished using a subtraction and restoration method. We covered BCD addition and subtraction and saw how BCD requires that some adjustments be made after an arithmetic operation. Finally, we saw how the basic operation of addition is actually performed in hardware.

STUDY QUESTIONS

1. Add the following numbers both in binary and in hexadecimal.
   (a) 37 + 29
   (b) 75 + 94
   (c) 26.125 + 18.675
   (d) 57.4 + 27.9

2. Find the signed range of numbers for each group of bits.
   (a) 3 bits
   (b) 6 bits
   (c) 10 bits
   (d) 16 bits

3. What are the signed and unsigned values of 10001010 B?

4. Convert the following numbers to binary and subtract using 1's complement arithmetic.
   (a) 88 − 26
   (b) 94 − 138
   (c) 18.7 − 14.8

5. Subtract the following numbers using 2's complement arithmetic.
   (a) 68 − 49
   (b) 57 − 83
   (c) 26.41 − 18.27

6. Subtract the following numbers using signed 2's complement arithmetic.
   (a) 83 − 47
   (b) 242 − 126
   (c) 148 − 206

7. Multiply the following numbers using the shift register and accumulator technique.
   (a) 206.4 × 36
   (b) 141.3 × .6
   (c) 27.5 × 14.3

8. Divide 150 by 10 using the subtract and restore method.

9. Repeat question 8 for 150 divided by 11.

10. Add the following numbers using BCD arithmetic. (Show the operations in binary.)
    (a) 26 + 49
    (b) 16.3 + 18.9
    (c) 93.2 + 46.8

11. Subtract the following numbers using BCD arithmetic. (Show the operations in binary.)
    (a) 27 − 13
    (b) 286 − 431
    (c) 14.3 − 17.9

12. Add the following numbers in binary.

 (a) 111010.1101 B
   100101.1001 B

 (b) 11010.1101   B
   1100.1011   B
    11.1001   B

13. Find the difference between the following pairs of numbers using signed 2's complement arithmetic.

 (a) 69 D
  − 47 D

 (b) 28 H
  − 79 H

14. Find the binary product of the following pairs of numbers using a shift register and accumulator. Show all accumulator entries.

 (a) 11101.101 B
  × 110.01 B

 (b) 10111.111 B
  × 1110.101 B

15. Divide each set of numbers. Then multiply your results by the original divisors to check your answers.

 (a) 10101101 B
  ÷ 11011 B

 (b) 101001 B
  ÷ 1001 B

16. Show that dividing a number by 10 B (using the subtract and restore method) gives the same answer as shifting the number right one bit position.

17. Add the following numbers using BCD arithmetic.

 (a) 47.12
  + 29.41

 (b) 69.99
  + 47.21

18. Subtract the following numbers using BCD arithmetic.

 (a) 27.4
  − 32.9

 (b) 19.61
  − 12.08

19. Multiply 546.85 × 453.2 using the shift register and accumulator technique discussed in this chapter.

20. Explain why the 10's complement of zero (in BCD) is also zero.

21. Show that the 2's complement of any number can be found by subtracting the number from the next highest power of 2. For example, the 2's complement of 5 can be found by subtracting 5 from 8, or the 2's complement of 1000 can be found by subtracting 1000 from 1024.

22. The 2's complement of a number may be found by using a *serial* method that follows these steps:

 1. Beginning with the LSB, transfer any 0s that exist in the original number to the result.
 2. Transfer the first 1 encountered.
 3. Invert all remaining bits up to and including the MSB, and transfer them to the result.

 Several applications of this method are illustrated in Figure 2.7.

FIGURE 2.7   **Serial 2's Complement Examples**

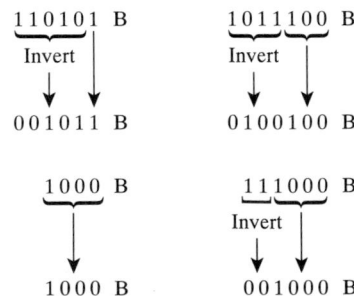

Use this technique to find the 2's complement of each number. Then use the original method to verify your answers.
(a) 10000 B
(b) 10001 B
(c) 10100 B
(d) 10101 B

23. Show that the value -6 can be represented by 1010 B, 11010 B, and 11111010 B.

24. Compare the largest unsigned, largest signed, and largest BCD numbers possible with 4, 8, and 16 bits.

25. Explain the meaning of the subtraction rules shown in Figure 2.8, then use them to subtract 10100 B and 1001 B.

FIGURE 2.8 **Rules for Binary Subtraction**

$$
\begin{array}{cccc}
0 & 0 & 1 & 1 \\
-0 & -1 & -0 & -1 \\
\hline
0 & 11 & 1 & 0 \\
\end{array}
$$

‎ ↑
└── 1 to borrow from next column

3

LOGIC FAMILIES

INSTRUCTIONAL OBJECTIVES

When finished with this chapter you should be able to:

1. Describe the differences between the three main logic families: TTL, CMOS, and ECL.
2. Explain the basic internal architecture of each logic family.
3. Identify the pin numbers on an IC.
4. Discuss why one logic family may be chosen over another.

SELF-EVALUATION QUESTIONS

Keep the following questions in mind and try to answer them when you have completed the chapter.

1. What is the range of voltage for a TTL zero? A TTL one?
2. What causes propagation delay?
3. Why is TTL called *saturated logic*?
4. To what does the term *noise margin* refer?

3.1 INTRODUCTION

Before we begin our examination of logic gates, the basic building blocks of digital circuits, let us take the time to familiarize ourselves with the different types of logic families available (TTL, CMOS, etc.), and many of the basic things we should know about the physical devices we will be using. This includes the acceptable voltage ranges for logic zeros and ones and the cause of gate propagation delay. We will see how the basic logic elements are constructed, and also learn how a digital circuit can be simulated with software.

3.2 LOGIC FAMILIES

Digital logic gates and their associated circuitry have been in a constant state of change and improvement. Thus it is appropriate to present a brief history and comparison of the series of families of logic that evolved. In the 1960s a form of logic that included all the typical logic gates (AND, OR, NAND, etc.) was developed. It used diodes and resistors and was referred to as diode-resistor logic (DRL). This family of logic was soon replaced by resistor-transistor logic (RTL), which was replaced by a family that is very popular today called transistor-transistor logic or *TTL*. Two other basic types that are popular and important now are emitter-coupled logic (*ECL*) and *CMOS* (complementary symmetry metal oxide semiconductor). Each has advantages and disadvantages.

TTL logic offers good speed at a reasonable price. ECL offers high speed (> 500 MHz) at a hefty price, while CMOS, which represents field effect transistor (*FET*) technology, offers low power consumption. Highly complex circuitry using tens of thousands of field effect transistors can perform digital miracles at relatively low power in the CMOS family. CMOS keeps undergoing improvements, and PMOS, NMOS, HMOS, and VMOS are also now available.

Regardless of the family of logic, you can find the logic gates you will require in any of the available families. Table 3.1 compares time and electrical characteristics of five logic families. Let us take a closer look at several of them.

TABLE 3.1   Comparison of Major Characteristics of Various Logic Families

LOGIC	BINARY ELEMENT CLOCK FREQUENCY, TYPICAL (MHz)	DELAY TIME (nsec)	POWER/ GATE (mW)	OPERATING VOLTAGE (V)
DRL	40	30	8	$5.0 \pm 10\%$
RTL	8	24	2.5	$3.6 \pm 10\%$
TTL	50	10	15	$5.0 \pm 10\%$
CMOS	25	25	Low (varies with f)	$+3$ to $+15 \pm 10\%$
ECL	400	4	40	$-5.2 \pm 20\%$

3.3 TTL

A very popular series of TTL (transistor-transistor logic) logic is the 54/74 series. The 54 series is intended for military use and operates over a wide temperature range ($-55°C$ to $+125°C$) to accommodate a variety of environments. The 74 series is a consumer-oriented version of the TTL family and is restricted to a narrower range of operating temperatures ($0-70°C$). The 7404 is a hex inverter, as is the 5405, with one difference being its operating temperature. For many applications this range is adequate, since television receivers and other consumer items operate in environments in which we all feel comfortable. It is strongly recommended that you have a TTL data book for a reference.

The familiar acronym **DIP** stands for dual in-line package, and a suffix (J, N, W, T) is affixed to the number to indicate the type and style of package: J for ceramic DIP, N for plastic DIP, W for ceramic flat pack, and T for flat pack. The 54/74 series comes in a variety of available subfamilies, designated by letter(s) after the 54 or 74 (e.g., 7400, 74S00, 74LS00, 74H00). Table 3.2 summarizes the capability of these improvements to the 54/74 series.

Figure 3.1 shows a typical dual in-line package. Two numbers appear on the top: SN74LS00 is the part number, and 9647 is the manufacturer's date code. The date code indicates that this particular integrated circuit was manufactured in the 47th week of 1996. This information is useful if it ever becomes necessary to identify a bad batch.

TABLE 3.2   Summary of Capabilities of the 5400/7400 Series

FAMILY	TYPICAL GATE PROPAGATION DELAY (nsec)	TYPICAL POWER DISSIPATION PER GATE (mW)	COMMENTS
74xxx	10	10	
74Lxxx	33	1	Low power
74Hxxx	6	22	High level
74Sxxx	3	19	Schottky
74LSxxx	10	2	Low-power Schottky
74ASxxx	1.5	22	Advanced Schottky
74ALSxxx	5	1	Advanced LS

FIGURE 3.1   Typical DIP

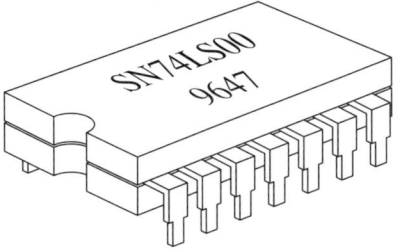

Working with TTL Gates

It is common to breadboard new digital circuits in a laboratory to see if they perform as planned. There are many things to know that help make the breadboarding (and design) experience trouble free.

The first thing you must know how to do is count the pins on the IC. Figure 3.2 shows several examples of how pins are numbered. In each case, the pin at the lower left is pin 1. This is readily found by looking at the shape of the package. One side may have a notch taken out of it, or there may be a small, round pit or dot near one corner. Usually, if you can read the information printed on the IC, pin 1 is in the lower-left corner.

Once we find the correct pin, the next thing to do is connect power and ground to the ICs in the circuit. Typically, TTL ICs have their ground pin in the lower-right location (pin 7 on a 14-pin DIP, pin 8 on a 16-pin DIP, etc.) and their supply voltage V_{cc} pin at the upper left (the highest numbered pin in the DIP). There are exceptions. The 7490 decade counter has V_{cc} on pin 5 and ground on pin 10. It is best to check the pinout of an IC before making connections. Sometimes a TTL device can withstand power connected to the wrong pins; other times it is very bad to accidentally miswire a pin. Part of the reason for this is the *totem-pole* transistors used in the output stage of a TTL gate.

FIGURE 3.2   TTL Package Pin Numbering

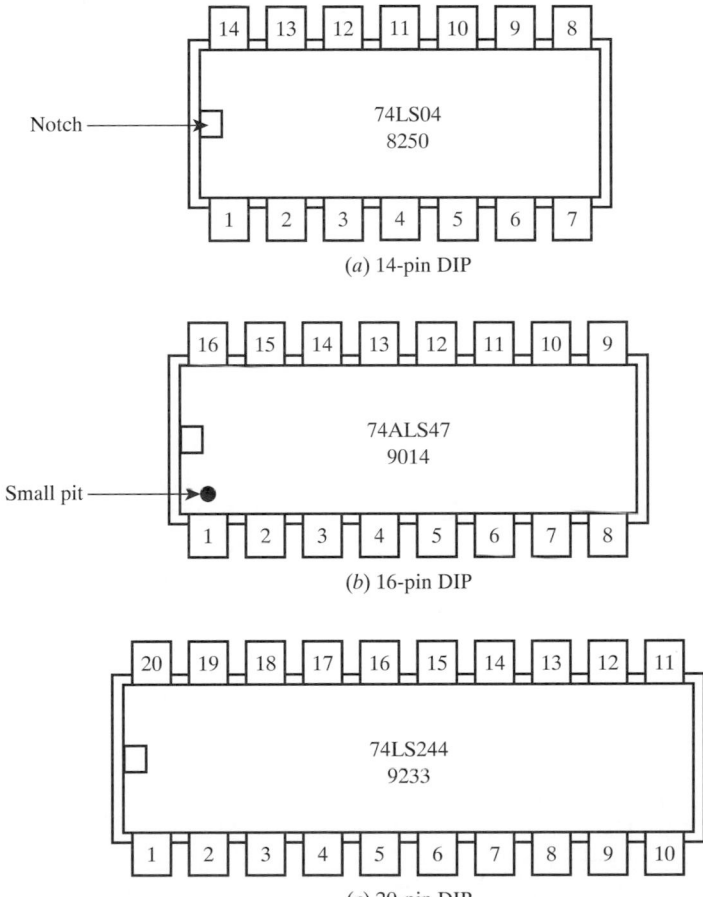

(*a*) 14-pin DIP

(*b*) 16-pin DIP

(*c*) 20-pin DIP

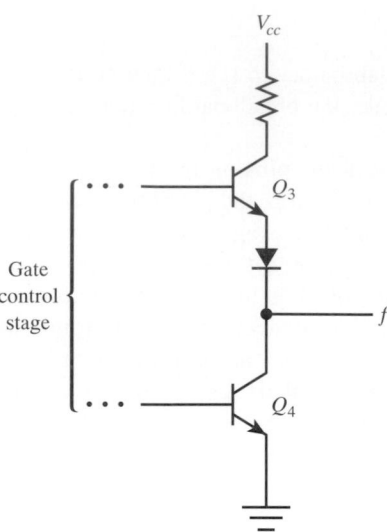

FIGURE 3.3   Totem-Pole TTL Output

The term comes from the fact that the transistors look like they are stacked on top of each other. Figure 3.3 shows an example of a totem-pole output. Two transistors, Q_3 and Q_4, control the state of the digital output. If Q_3 is on, the output is high (a logic one). If Q_4 is on, the output is low (a logic zero). Now, suppose that the IC is powered correctly, but the output of the gate is accidentally grounded. Will either transistor be damaged? Probably not. If Q_3 is on, the ground at the output will simply draw all available output current, which is limited by an internal resistor on the collector of Q_3. The chip may run hot, but it will not be damaged. If Q_4 is on when the output is accidentally grounded, nothing at all happens, since Q_4 has nothing in its collector path but ground.

If, on the other hand, the output is accidentally taken directly to V_{cc}, there will be a problem if the gate is currently outputting a zero (meaning Q_4 is on). The power supply will be connected directly across Q_4 and will most likely damage the transistor.

In general, it is best to leave unused outputs alone. No harm will come to them if they are left open.

Care must also be taken when working with a TTL input. Suppose that a logic one level is required on an input all the time. Putting five volts on the input will do the job. Figure 3.4 shows two ways to make a logic one on an input. In Figure 3.4(a), the input is connected directly to the +5 V power supply. Unfortunately, this is not a good idea. There are times when an IC fails when an input actually becomes shorted to ground

FIGURE 3.4   Making a Logic One on an Input

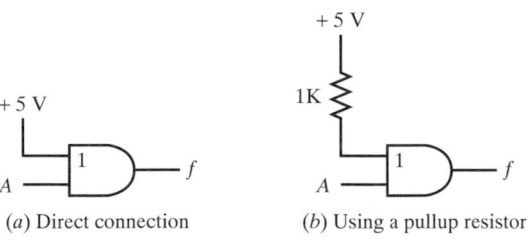

(*a*) Direct connection   (*b*) Using a pullup resistor *inside* the IC (maybe the silicon melted during a thunderstorm or there was a manufacturing defect). If this should happen to an input connected directly to +5 V, there could be sparks and smoke. Figure 3.4(b) shows a better way to provide the logic one level. Here a ***pullup resistor*** is used to pull the input up to a logic one level. The idea is that the input draws so little current, the voltage drop across the resistor will be insignificant, and the pin will have a good logic one level applied. If the input should self-destruct and go to ground internally, the power supply will only have to output a few milliamperes to the pullup resistor. This will prevent further damage and also allow us to find the problem during troubleshooting, since we will find zero volts at the bad pin instead of +5 V.

To put a logic zero on a pin we simply ground it.

Gate Loading

Anything connected to the output of a gate is considered a load, even the inputs of other gates (which require input currents in the appropriate directions to make 0s and 1s). When just a few gates are thrown together, there is no need to be concerned with the loading effects on the outputs. But in a design with a large number of common connections, it is a good idea to verify that the ***fan out*** of the gate is not exceeded. The fan out of a gate refers to the number of other inputs that it can drive. Figure 3.5 shows a NAND gate driving a set of N inputs. When the NAND gate outputs a logic zero, it becomes a current sink, and pulls a small amount of current out of each input to make zeros on them. The sum of the individual currents is the current that the driving gate must sink. This current has an upper limit specified by the manufacturer. For example, the low-level output current I_{OL} of the 74ALS04 hex inverter is 8 mA. For the gate inputs being driven low, let us assume the low level input current I_{IL} is −0.1 mA (the minus sign has to do with direction, not magnitude). So, dividing 8 mA by 0.1 mA gives 80, which suggests that up to 80 inputs may be driven from the output of one inverter. In actuality, it is a good idea to derate the current rating for reliability reasons, possibly up to 50 percent of maximum. This would reduce our count to 40, still a large number of inputs that can be driven.

But what about driving all those gates with a logic one? The high-level output current of the 74ALS04 is −0.4 mA (again, the minus sign is based on the direction). Dividing by a high-level input current of 20 μA, we get a fan out of 20. If this value is also derated, we are down to a fan out of 10 for the inverter. The derated fan out of 40 for making 0s is set aside, since the logic one fan out is smaller.

FIGURE 3.5 Gate Loading

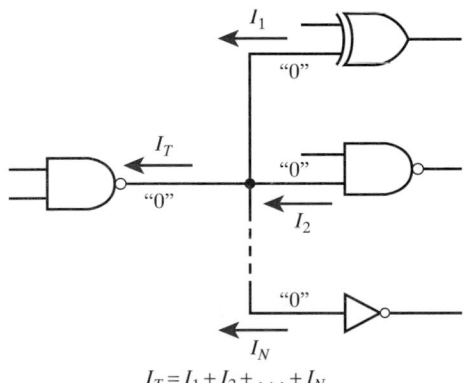

$$I_T = I_1 + I_2 + \ldots + I_N$$

Gate Delay

Every TTL gate has a small amount of built-in delay, called ***propagation delay,*** that causes the output signal to change slightly later in time than the input. Figure 3.6 shows the delay between changes in an input waveform and the resulting output waveform. The gate delay is in the ns range, and is not a factor in low-speed digital circuitry. High-speed circuits, with fast clocks, may need to be carefully designed around the gate delay. For example, if the gate delay is 5 ns, the input to the gate cannot be changed at a rate faster than 200 MHz. When many gates are used in combination, their individual delays add up and further limit the speed at which the input signals can change.

Gate delay does, however, have some positive benefits. Some digital circuits depend on the delays for normal operation, such as switch debouncers, shift registers, and oscillators.

Where does gate delay come from? Figure 3.7 provides an answer by showing the internal schematic of a TTL inverter. The gate delay in the inverter is a result of the *switching time* of the four transistors. When the input A is low, transistors Q_1 and Q_3 are on (actually, they are saturated). The high-level output is provided by Q_3. Q_2 and Q_4 are off. When the input changes to a logic one level Q_1 shuts off, which allows Q_2 to turn on (by yanking some current through the base-collector junction of Q_1). With Q_2 on, its emitter current turns Q_4 on, which makes the output low and causes Q_3 to turn off. All of these state changes in the transistors require a small amount of time to move charges around in the various junctions. This is what causes the gate delay. Table 3.3 lists some typical gate delays for several TTL devices. The table shows times for low-level to high-level delays (T_{PLH}) and high-level to low-level delays (T_{PHL}). For design purposes you may want to use the maximum times for each gate. This should help eliminate any timing problems.

FIGURE 3.6   Propagation Delay (*td*) Between Input and Output

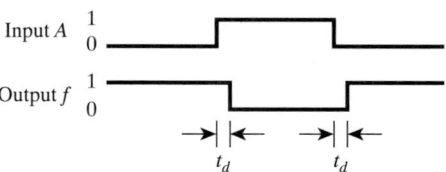

FIGURE 3.7   Schematic of an Inverter

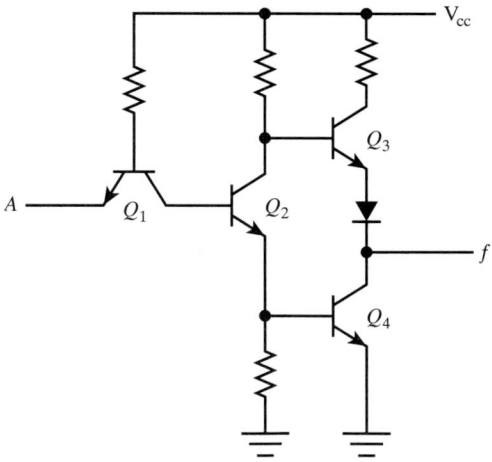

TABLE 3.3 Gate Propagation Delays

DEVICE	$T_{PLH(ns)}$	$T_{PHL(ns)}$
74ALS00/04	3/11	2/8
74ALS02	3/12	3/10
74ALS08	4/14	3/10
74ALS32	3/14	3/12

Note: All values are min/max.

Logic Levels

In a digital circuit, the binary values 0 and 1 must be represented using a pair of voltage levels. Each logic family has a different standard, and great care must be used when attempting to mix logic families. Usually one family is selected for design based on its desirable characteristics. Table 3.4 shows the voltages used by several families.

Logic families that use positive voltages are called ***positive level*** families. Logic families that use the most positive value to represent a binary "1" are called ***positive logic***. ***Negative level*** families operate on negative voltages. Logic families in which the most positive voltage is a "0" are called ***negative logic***. TTL is both *positive level* and *positive logic*. ECL is *negative level* and *positive logic*.

The TTL family uses a +5-V power supply ($+5 \pm 10\%$). Even though a "0" is represented with 0 V and a "1" is represented with 5 V, a certain range is acceptable. A logic zero on the input of a TTL gate can be a level from 0 to 0.8 V. A logic "1" on the input can be from 2 to 5 V.

The range of allowable TTL levels (see Figure 3.8) is defined differently for the output of a TTL gate. The allowable range for a logic zero at the output is 0 to 0.4 V. The range for a one at the output is 2.4 to 5 V.

The difference between the input maximum "zero" and the output maximum "zero" is called the ***noise margin*** of the family. It is also the difference between the output minimum "one" and the input minimum "one." For TTL, the noise margin is 400 mV. This allows the gate to function correctly with up to 400 mV of noise (peak voltage) existing at an input. This noise may be induced in a wire connected from a TTL output to a TTL input by some external influence. Figure 3.9 illustrates this.

Table 3.5 summarizes various ways of representing and referring to the logic levels for a "0" or "1."

Within a digital circuit, then, logic levels will be found that represent a binary 0 or a binary 1. In TTL circuitry these levels will generally be 0 or +5 V or within the specified limits from 0 to 0.8 and 2 to 5 V.

TABLE 3.4 Voltages Used by Several Logic Families

FAMILY	VOLTAGE (V)	
	LOGIC 0	LOGIC 1
TTL	0 V	+5 V
ECL	−1.7 V	−.9 V*
CMOS	0 V	3–14 V

*ECL uses −5.2 V supply voltage.

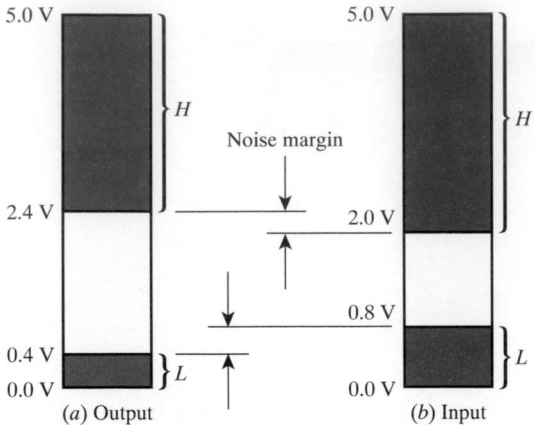

FIGURE 3.8 **Range of Allowable TTL Levels**

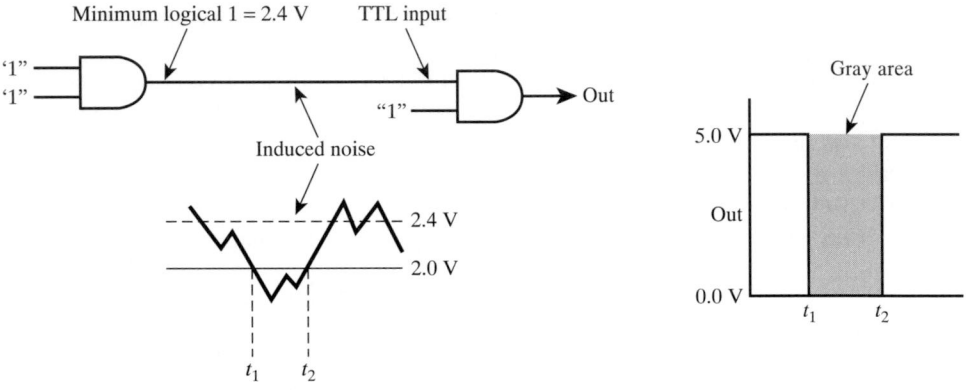

FIGURE 3.9 **Noise Induced in Wire Connected from TTL Output to TTL Input**

TABLE 3.5 **Logic Levels for "0" and "1"**

"0"	"1"	
Low	High	
Down	Up	
Off	On	
False	True	
Open	Closed	
0 V	+5 V	← TTL
0 V	3 V to 14 V	← CMOS
−1.7 V	−.9 V	← ECL
0 V	3.6 V	← RTL (obsolete)

TABLE 3.6   Number of Gates/Chip

CATEGORY	GATES
SSI	1 to 10
MSI	10 to 99
LSI	100 to 999
VLSI	1000+

Gate Density

Gate density refers to the number of gates contained within a single integrated circuit. Table 3.6 lists the common terms relating to circuit densities. Small-scale integration (**SSI**) is used for basic logic gates. Medium-scale integration (**MSI**) is used for counters, adders, comparators, and shift registers. Large-scale integration (**LSI**) devices are digital clock chips, arithmetic-logic units, UARTs, and static RAMs.

Very-large-scale integration (**VLSI**) is commonly used for fabricating microprocessors and other high speed, highly complex digital circuitry such as digital signal processors, real-time data compression, and high-density dynamic RAMs. Today, millions of transistors are being placed on the small silicon chip that rests inside the sturdy package. It is not uncommon to see miniature fans mounted directly on top of an IC to cool it. Processors such as the Pentium, which contains over three million transistors, will completely shut down if overheated. So, in addition to equating gate density with complexity, we must also equate it with power consumption.

3.4 CMOS

CMOS stands for complementary metal-oxide semiconductor, and refers to the type of transistors used to construct logic elements. These transistors are metal oxide semiconductor (**MOS**) FETs (field effect transistors), and operate differently from the bipolar-junction (NPN) transistors used in TTL circuitry. Figure 3.10 shows a comparison of the BJT and the MOS FET. The MOS FET is a voltage-controlled device, meaning that the voltage on the gate-source terminals controls the drain-source current. A silicon-dioxide (SiO_2) layer forms an insulating barrier on the gate input. This produces a gate resistance in the 10 to 100 Megohm range. Thus, the gate does not draw any current. Instead, the voltage between the gate and the source terminals controls the amount of charge carriers in the *channel,* the N-type semiconductor the drain and source connect to. By varying V_{GS}, the drain current (I_D) is controlled. This control requires very little power, one of the nice characteristics of the CMOS logic family. Also, because MOS FETs are voltage controlled devices, the power supply voltage for a CMOS device is less restricted than the 5 V TTL limit. Recall from Table 3.1 that CMOS operates anywhere from 3 V up to 15 V. This makes CMOS logic ideal for use in an automotive application, where the vehicle's 12 V battery voltage can be used to power the CMOS circuitry.

The "complementary" portion of CMOS refers to the use of both N-channel and P-channel MOS FETs to construct a gate. Using complementary devices allows the same set of voltages to turn one device on and the other off. This is illustrated in the CMOS

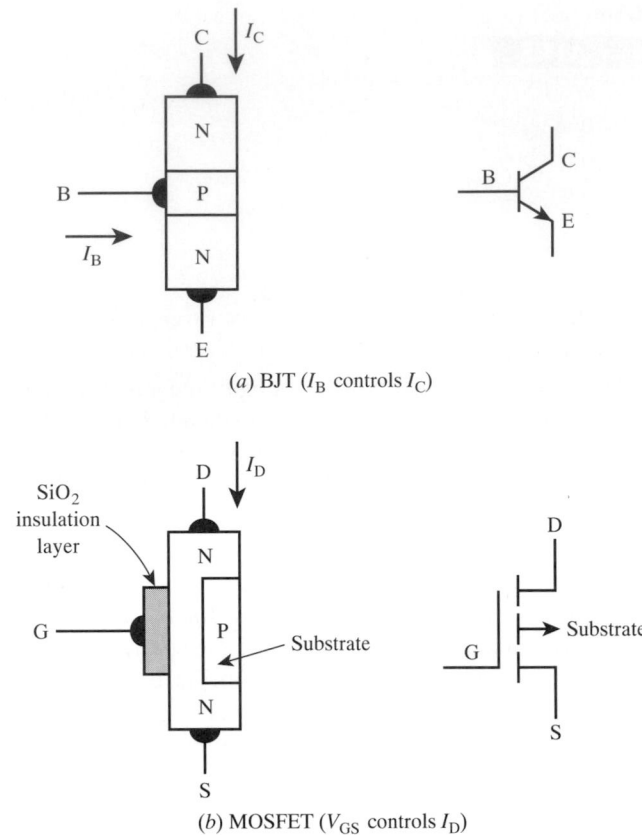

(a) BJT (I_B controls I_C)

(b) MOSFET (V_{GS} controls I_D)

FIGURE 3.10   Comparing a BJT with a MOS FET

FIGURE 3.11   CMOS Inverter

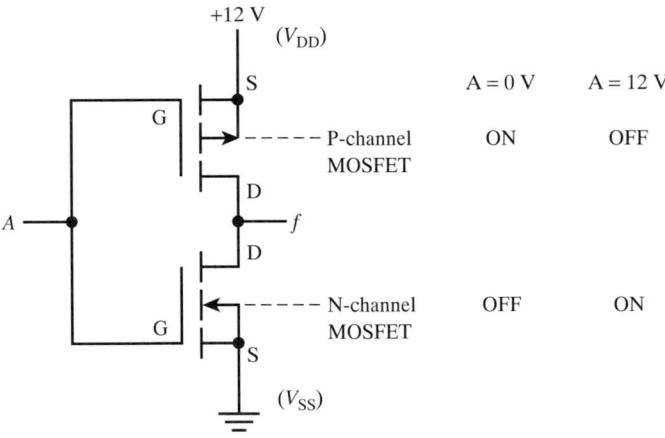

inverter shown in Figure 3.11. When the input A is low, the P-channel MOS FET is turned on (V_{GS} is -12 V) and the N-channel MOS FET is off (V_{GS} is 0 V). The P-channel MOS FET has a low channel resistance when it is on, so the output is pulled up to 12 V.

When the input A goes high, the P-channel MOS FET is turned off (V_{GS} is 0 V), and the N-channel MOS FET turned on (V_{GS} is 12 V). This pulls the output low.

Other logic functions are made by using additional pairs of P-channel and N-channel MOS FETs.

Families

There are several CMOS logic families, as indicated in Table 3.7. The original series was the 4000 series which was manufactured by RCA. This family is slow compared to TTL and is not pin compatible. The 74C series is pin compatible with many TTL devices, offering a low-power replacement, but still not as fast as TTL. The speed is improved in the 74HC/HCT and 74AC/ACT advanced CMOS families. 74HCT devices are direct substitutes for their corresponding 74LS parts. 74AC/ACT components are not pin compatible with TTL because their pins have been rearranged to reduce noise.

Connecting CMOS and TTL Devices

For CMOS devices that are not electrically compatible with TTL, an interfacing circuit must be used to connect a TTL output to a CMOS input, and vice versa. Figure 3.12 shows how this is done. In Figure 3.12(a), a TTL gate drives a CMOS input. Typically the TTL gate needs help making a logic one on the CMOS input. Using a pullup resistor makes it easier to get the required voltage level. If the CMOS device is operated at a higher power supply voltage than 5 V, a level shifter is needed.

When a CMOS output drives a TTL input, as in Figure 3.12(b), a buffer is required to improve the output current capability of the CMOS gate. There are both CMOS and

TABLE 3.7   **CMOS Logic Families**

FAMILY	POWER SUPPLY	SPEED	TTL PIN COMPATIBLE
4000	3–15 V	Slow	No
74 C	2–6 V	Slow	Yes
74 HC/HCT	2–6 V	Fast	Yes
74 AC/ACT	2–6 V	Fast	No

FIGURE 3.12   **Interfacing CMOS and TTL**

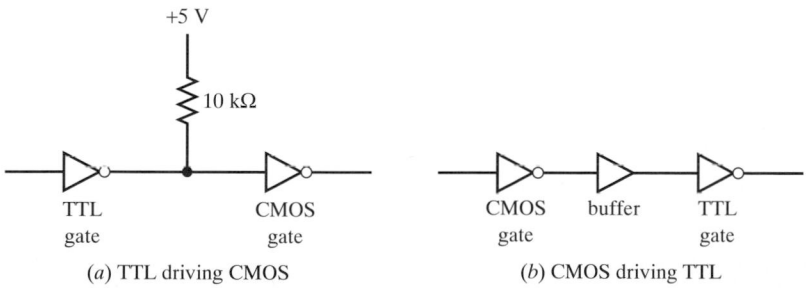

(a) TTL driving CMOS   (b) CMOS driving TTL

TTL buffers suitable for this purpose. In some cases, it may not be necessary to use the buffer at all, but that depends on the specific configuration and careful examination of all the currents involved.

3.5 ECL

The first difference we note about the emitter-coupled logic (ECL) family is its negative power supply voltage. As shown in Figure 3.13(a), the negative supply biases the differential amplifier pair. Unlike a TTL transistor, which is driven into saturation, the ECL transistor merely adjusts its collector current up or down slightly, always trying to balance with its matching pair somewhere in their active regions. This can be done faster than the off-saturated-off changes in a TTL transistor, giving ECL a nice speed advantage over TTL. Power is another matter. TTL transistors turn on and off, and therefore do not use power constantly. ECL transistors are always on and thus always use power. This disadvantage is more than offset by ECL's speed advantage.

Figure 3.13(b) shows the phase relationship between the input A and the two outputs Vo_1 and Vo_2. The complementary symmetry between the outputs is due to the operation of the matched transistor pair's constant current configuration. Vo_1 is out of phase with input A (the inverting function), and Vo_2 is in phase. Before the outputs can be used they must be level-shifted back to the appropriate ECL values. This is a result of the biasing rules for the matched pair.

The two main ECL families are the 10K and 100K series, with other specialized types made by several different manufacturers.

3.6 ASICs

Application-specific integrated circuits (*ASICs*) are custom integrated circuits designed to be a complete solution to a particular application. For example, an ASIC may control all of the bus activity between several shared microprocessors, perform image/data

FIGURE 3.13   Basic ECL Circuit

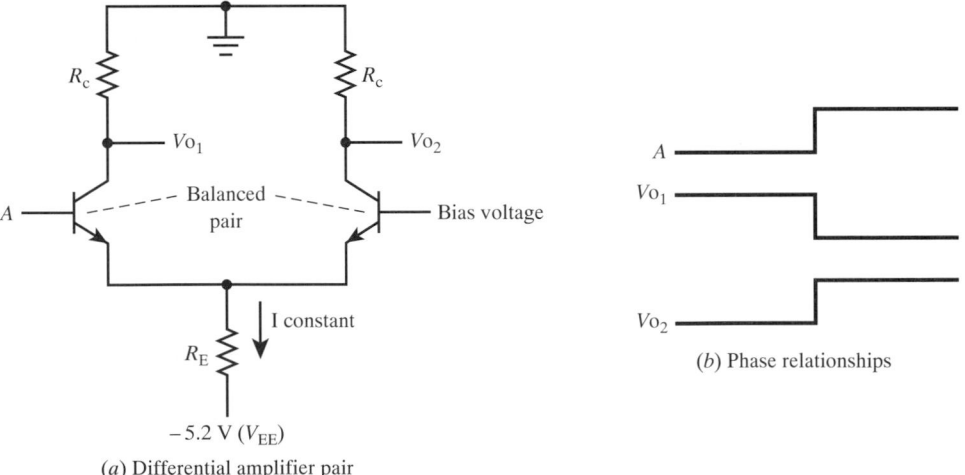

(*a*) Differential amplifier pair

(*b*) Phase relationships

compression/expansion, control an automated fuel-pump with credit-card features, play chess, or provide 8-channel 10 Megabit switching for an Ethernet hub. An ASIC may have hundreds of inputs or outputs, or only a few. No matter what the application, an ASIC may provide a cost-effective solution. Typically, the design of the circuit to be contained within the ASIC is specified in a hardware language, such as Verilog or VHDL (see Chapter 9 for more details). Plenty of companies will take the design and produce a working chip, or even design the entire chip for you. It would be a good exercise to search the Internet for ASIC-related topics, and see what people are building with them.

3.7 TROUBLESHOOTING TECHNIQUES

Let us discuss a few more items of importance concerning breadboarding a digital circuit. First, as a reminder, it is always a good idea to use pullup resistors to force a logic one level. Leaving a TTL input open and hoping that the gate will pull it high internally is not a good practice.

Second, there are times when a properly wired breadboard circuit does not work, even after all integrated circuits have been changed and moved (in case there are internal breadboard faults in a certain location). Sometimes this is due to a lack of *bypass* capacitors across the power and ground pins of the integrated circuit. When the transistors inside a TTL device change state, a small amount of noise and current fluctuation occurs at the power supply connections to the IC. Connecting a 0.01 to 0.1 μF capacitor across the power pins reduces the switching effects and may make the difference between a working circuit and a nonworking circuit.

Third, before you plug an IC into its socket or breadboard location, examine its pins. Are they all straight? Are they all there? Is pin 1 in the right place? After you insert the IC, check for bent pins again. It is easy to curl a pin underneath the IC during insertion and not even know it.

Another way to set up and test a new circuit is to do it using software, with a circuit simulation package such as Electronics Workbench or Pspice. Figure 3.14 shows a screen shot of the Electronics Workbench program simulating a simple digital circuit (with keyboard-controlled switches providing the 0s and 1s at the inputs and a logic indicator showing the state of the output). Electronics Workbench contains many of the common logic devices in its parts libraries, has several useful virtual instruments (logic analyzer, oscilloscope, and pattern generator), and is easy to learn and use. Try loading the INTRO file into Electronics Workbench. It is the circuit contained in Figure 3.14. Experiment with the switches, play with the wiring, add/remove logic gates, and familiarize yourself with the software. There is even a provision that allows a fault to be placed into the circuit, hidden from the user, to provide some troubleshooting practice. Truly, Electronics Workbench is a valuable addition to your study of digital electronics.

SUMMARY

In this chapter we examined several features of the three most common logic families: TTL, CMOS, and ECL. Basic information such as pin numbering, power connections, input/output voltage levels, and the internal operation of each type of logic family was provided. We saw that one-chip ASIC technology also plays an important role in digital circuitry, along with circuit simulation software such as Electronics Workbench.

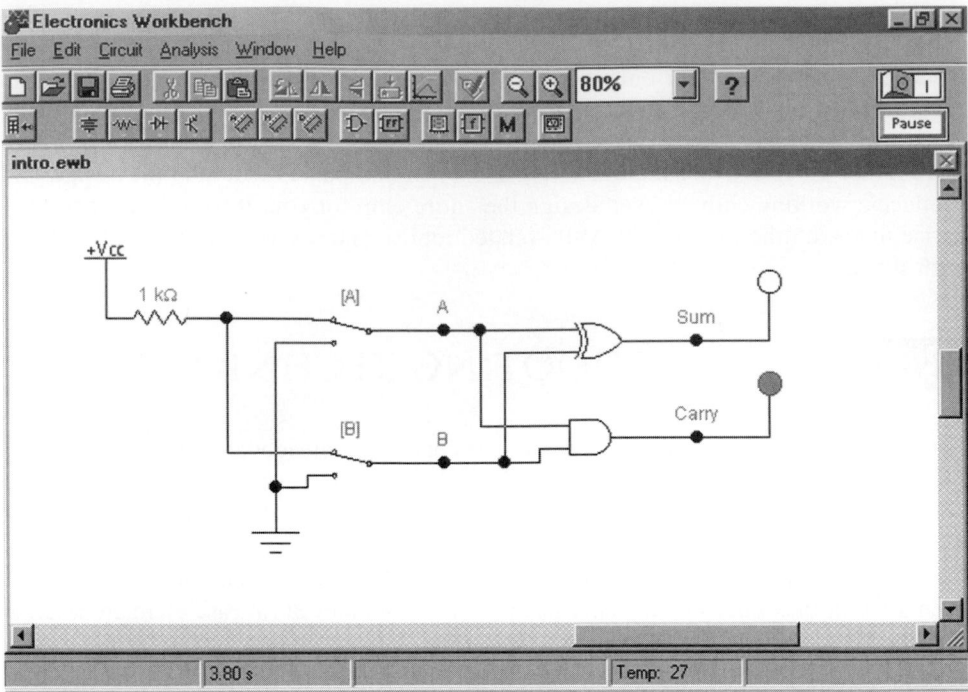

FIGURE 3.14   Electronics Workbench Screen Shot

STUDY QUESTIONS

1. What are the three main logic families?

2. What are two differences between TTL and CMOS logic gates?

3. Why is ECL faster than TTL?

4. What are two differences between 5400 TTL logic and 7400 TTL logic?

5. What does DIP stand for?

6. Where is pin 1 located on an integrated circuit?

7. What does a date code of 9711 mean?

8. Where are the power and ground pins typically located on a TTL package?

9. Is it acceptable to ground a TTL output?

10. Is it acceptable to take a TTL output directly to +5 V?

11. Why is the TTL output circuitry called totem-pole?

12. What is the proper way to place a logic one on a TTL input?

13. What is fan out?

14. An output capable of sinking 4 mA drives a set of 0.2 mA inputs. What is the fan out?

15. Is the fan out for a logic zero level the same as the fan out of a logic one?

16. What is propagation delay? What causes it?

17. What is a gate's noise margin?

18. What is the range of acceptable voltages for a logic zero and a logic one:
    (a) On an input?
    (b) On an output?

19. Under which type of classification (SSI, LSI, etc.) would each logic function fall?
    (a) Quad XOR gate
    (b) DMA controller for an audio card
    (c) Octal D flip-flop
    (d) Microcontroller with A/D converter, timers, RAM, and EPROM

20. Why is power consumption so important?

21. What does *CMOS* stand for?

22. How is a MOS FET controlled?

23. Why does no current flow into the gate of a MOS FET?

24. Are TTL and CMOS packages pin-for-pin compatible?

25. What is the general building block of an ECL circuit?

26. Why does ECL use so much power?

27. What does ASIC stand for? Give an example of where you might use an ASIC.

When finished with this chapter you should be able to:

1. Understand the function and operation of the simplest gates: inverter, buffer, AND, OR, NAND, NOR, XOR.
2. Apply Boolean algebra rules and truth tables.
3. Use truth tables to determine whether a gate is good or bad.

4. Substitute one type of gate for another.
5. Use DeMorgan's theorem.
6. Understand open-collector and tri-state output gates.
7. Understand pulse rise time, fall time, and pulse width.
8. Understand the binary adder circuitry.

Keep the following questions in mind and try to answer them when you have completed the chapter.

1. Why is a noise margin important?
2. What is an open-collector output, and what is it used for?
3. What is propagation delay?

4. What is meant by the expression "tri-state logic"?
5. Why is a bounceless switch circuit necessary?
6. What role does the truth table play?
7. What is the difference between a half adder, a full adder, and a parallel adder?

4.1 INTRODUCTION

This chapter will familiarize you with a number of the basic building blocks of digital electronics. These building blocks are logic gates and include the following types:

1. Inverter.
2. Buffer.
3. AND gate.
4. OR gate.
5. Exclusive OR gate.
6. NAND gate.
7. NOR gate

All digital devices, no matter how complicated, are constructed from these basic logic gates.

We will examine the operation of *combinational* logic circuitry as well, where several basic gates are connected in different ways to perform some operation.

4.2 TRUTH TABLES

The operation of any logic gate is described by its *truth table*. A truth table contains the various binary input combinations and their associated outputs. Figure 4.1(a) shows a one-input truth table. The input column shows one input, *A,* and one output, *f.* The first line of the truth table shows that when *A* is low, *f* is high. The second line shows the reverse.

Figure 4.1(b) shows a two-input truth table. Notice that with two inputs we now have four combinations of input patterns. Input *A* is considered the MSB of the input pair. In general, the maximum number of patterns to consider is 2^N, where N is the number of digital inputs.

Figure 4.1(c) illustrates the use of the ***don't care*** input value. Here, an *x* is used to denote an input whose state does not matter. The *x* could be a one, or a zero, or the impossible combination of both at the same time. The third line of the truth table indicates that when the $\overline{E}$ input is high (inactive), the output will be low regardless of the state of the *D* input. The overbar on the $\overline{E}$ input indicates that it is an *active low* input, an input that causes something to occur when it is low. When high, no action is taken. Accordingly, an *active high* input (no overbar) causes its action when it is high, and is considered inactive when low.

You will find that the truth table adds another layer of understanding to a complicated circuit, or serves as a starting point in a new design.

FIGURE 4.1 **Truth Tables**

A	f
0	1
1	0

A B	f
0 0	0
0 1	1
1 0	1
1 1	0

$\overline{E}$ D	f
0 0	1
0 1	0
1 x	0

(*a*) One-input truth table   (*b*) Two-input truth table   (*c*) Truth table containing a don't care (*x*) input

4.3 THE BUFFER

The *buffer* is a very simple logic gate. At first glance it appears to perform no function at all. It produces the same level that is applied to it. Figure 4.2 shows this gate, its Boolean expression, and its truth table. Such a gate, however, is used to buffer a signal. It minimizes the effect of the noise on a line by acting as a digital amplifier. If a signal has become lower in amplitude because it has had to drive a number of other TTL inputs, a buffer brings it back up. For example, a TTL "1" can be as low as 2 V and still be considered a "1." It is, however, marginal. Passing through a buffer gives the signal new life. A "1" is again a 5-V level.

FIGURE 4.2   7407 Hex Buffer

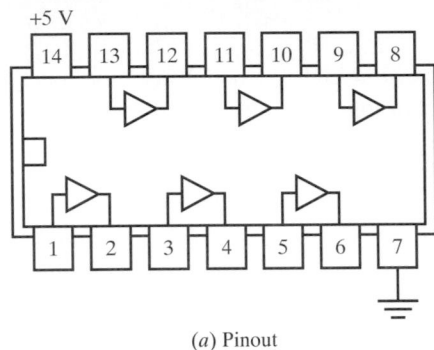

A	f
0	0
1	1

$f = A$

(*b*) Truth table and Boolean expression

(*a*) Pinout

FIGURE 4.3   The Logic Inverter (*a*) Inverter Symbol (*b*) Resistor–Transistor Logic Gate (*c*) 7404 Hex Inverter

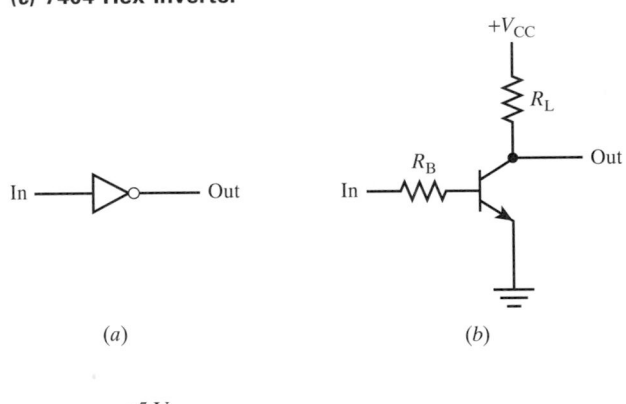

(*a*)   (*b*)

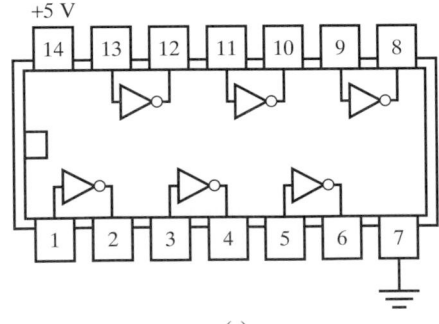

(*c*)

4.4 THE INVERTER

Figure 4.3 shows the logic symbol for an *inverter* and the corresponding RTL schematic. The inverter's sole function is to change a zero to a one, or a one to a zero. The RTL gate acts as a simple transistor switch. When the input is pulled low (to ground for a zero) the transistor turns off and the output floats up to a 1 level. On the other hand, when a "1" is applied to the input, the transistor turns on and the output is pulled low. This type of circuit always uses more current than necessary to turn the transistor on and is referred to as a *saturated* form of logic. **Saturated logic** is slower than nonsaturated logic because of the added time needed to remove excess electrical charge from the base of the transistor. Figure 4.4 shows modern TTL, CMOS, and ECL inverters.

FIGURE 4.4 (*a*) **TTL Inverter** (*b*) **CMOS Inverter** (*c*) **ECL Inverter**

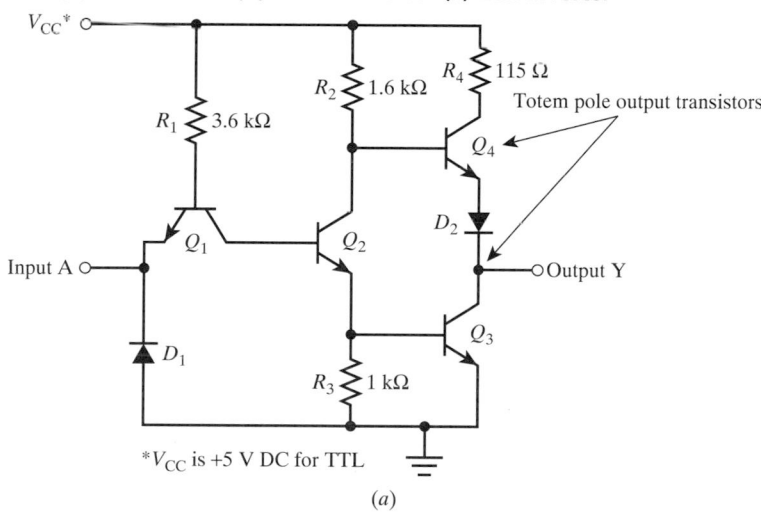

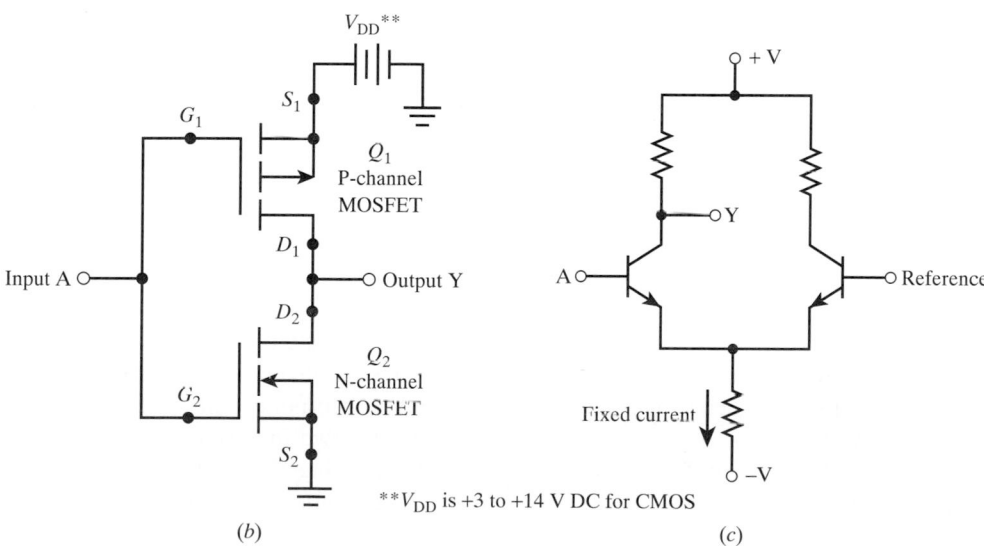

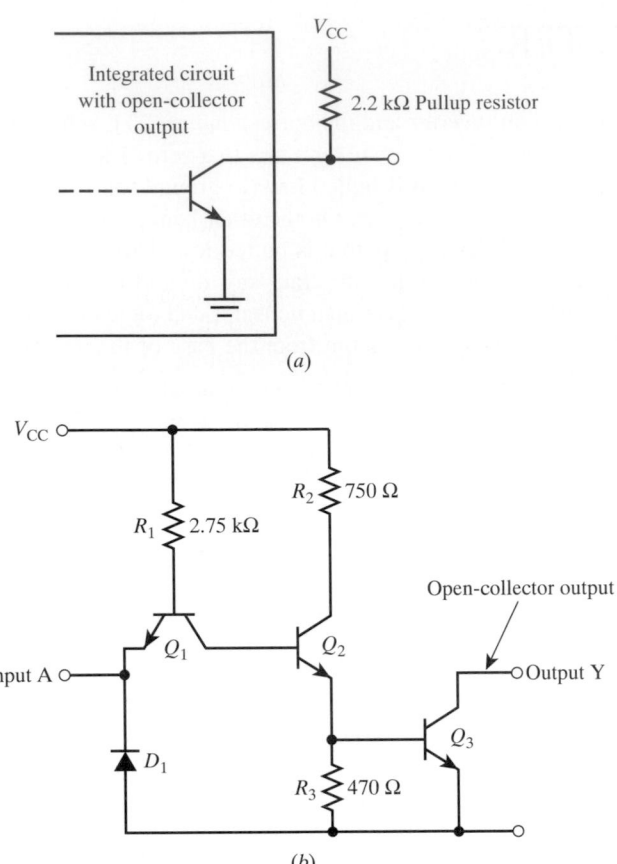

FIGURE 4.5 (*a*) Open-Collector Output (*b*) 7405 Inverter (Open-Collector) Stage

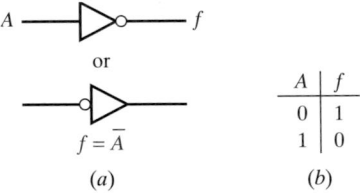

A	f
0	1
1	0

(*a*) (*b*)

FIGURE 4.6 (*a*) Logic Inverter and (*b*) Truth Table

The TTL family features a number of integrated circuits with ***open-collector*** (OC) outputs. In such circuitry (OC outputs), the outputs usually require an external load resistor to pull the output high. Such a resistor is called a *pullup* resistor. Figure 4.5 shows a 7405 inverter stage with an OC output.

Figure 4.6 shows the inverter, its Boolean expression, and the associated truth table.

The circle on the output of the symbol indicates the inverting or complement operation. The same function is performed if the symbol incorporates the circle on the input side. Only one circle is proper, since two circles (or two inversions) would get you back where you started. The logic expression accompanying Figure 4.6(a) describes in algebraic terms the behavior of the gate. This type of algebra, called ***Boolean algebra,*** is named for the work done by George Boole in the mid-19th century and is a form of symbolic logic used in the design of digital circuitry. The line over the $\overline{A}$ is the inverting or complement function and means NOT A. The expression $f = \overline{A}$ is read, "f will be a '1' when A is *not* a '1.'"

The Schmitt Trigger

Often it is necessary to square up a signal for use by a digital circuit. For instance, a 60-Hz AC sine wave is often used to synchronize digital circuitry and video display circuitry. The 16-msec period of the waveform is very slow in comparison to a microcomputer's abilities. A 60-Hz pulse waveform is needed to accurately time events and operate logic circuitry. A Schmitt trigger switches at one specific voltage, not a range of voltages. Most TTL gates expect a logic zero to be from 0 to 0.8 V and also expect a logic one to be from 2 to 5 V. The region from 0.8 to 2 V is a forbidden zone or gray area. A signal in this area will not produce predictable results on a standard TTL gate. The Schmitt trigger, however, switches on at a specific voltage (the lower threshold voltage) and switches off at another voltage (the upper threshold voltage). Figure 4.7 shows a sine wave input to a Schmitt inverter.

The Schmitt trigger is not concerned with voltages in the gray area, since it has already switched. The Schmitt trigger will switch back when the input signal passes the upper threshold voltage. Figure 4.8 shows a simple circuit to square up a sine wave input for use by logic circuitry.

FIGURE 4.7 (*a*) Schmitt Trigger and (*b*) Sine Wave Input

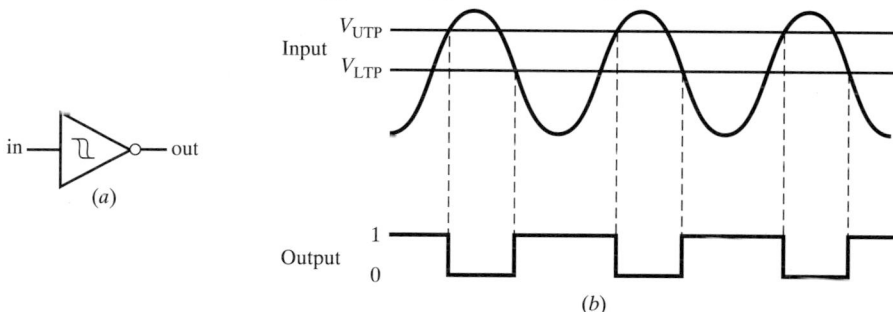

FIGURE 4.8 Schmitt Trigger Used to Square Up a Sine Wave

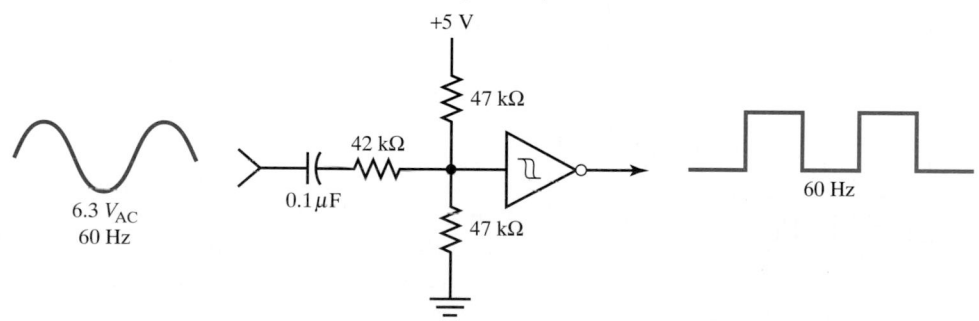

4.5 THE AND GATE

The function of an AND gate is to produce a "1" at the output only when all inputs are at this level. If even one input is low, the output will be low. Figure 4.9 shows the symbol for a two-input AND gate and associated Boolean expression and truth table. AND gates come with many more inputs if necessary. The truth table lists the four possible input combinations that can be applied to the gate. Both inputs could be low (0), or one or the other could be high (1), or both could be high. The truth table lists every possible input combination. There are *no more* with only two inputs. The number of entries on a truth table is always 2^N, where N is the number of inputs on the gate.

The Boolean equation reads, "f is a '1' when A is a '1' AND B is a '1.'" The Boolean equation for the AND operation can be written in more than one way: $f = A \times B$ or $f = A \cdot B$ or $f = AB$. In any case, a multiply is taken as the word AND to specify that particular logic function. Figure 4.10 shows a 7408 AND gate package.

Figure 4.11 shows a three-input AND gate and a truth table that is $2^N = 2^3 = 8$ lines long.

FIGURE 4.9 (*a*) Two-Input AND Gate (*b*) Truth Table (*c*) Boolean Expression

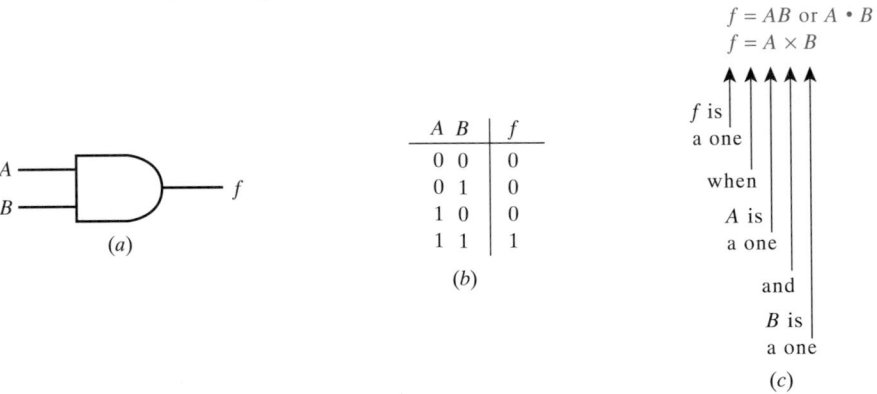

A	B	f
0	0	0
0	1	0
1	0	0
1	1	1

(*b*)

$f = AB$ or $A \cdot B$
$f = A \times B$

f is a one
when
A is a one
and
B is a one

(*c*)

FIGURE 4.10 7408 Quad AND Gate

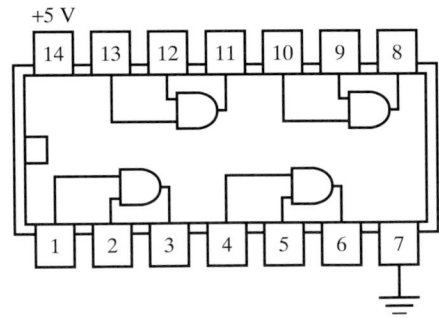

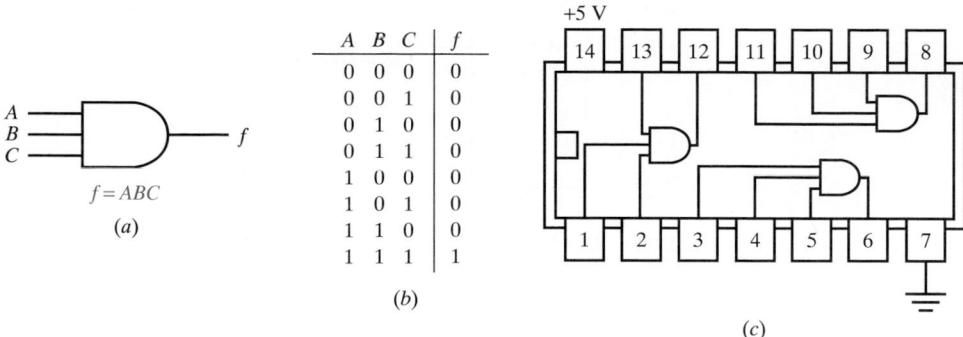

A	B	C	f
0	0	0	0
0	0	1	0
0	1	0	0
0	1	1	0
1	0	0	0
1	0	1	0
1	1	0	0
1	1	1	1

$f = ABC$

(a)

(b)

(c)

FIGURE 4.11   (*a*) **Three-Input AND Gate** (*b*) **Truth Table** (*c*) **7411 Triple Three-Input AND Gate**

4.6   THE OR GATE

The ***OR gate*** will produce an output if any input is high. A plus sign ($+$) is used to represent the OR operation in Boolean algebra. Figure 4.12 shows a two-input OR gate and truth table. The equation reads, "f is a '1' when A is a '1' OR 'B' is a '1' or both A and B are '1.'"

One application of the OR gate is shown in Figure 4.13. Here the OR gate is used to chip enable a memory device during a read operation. The OR gate output goes low only when both IO/$\overline{\text{M}}$ and $\overline{\text{RD}}$ are low. Control signals such as IO/$\overline{\text{M}}$ and $\overline{\text{RD}}$ are commonly found in microprocessor-based systems.

Figure 4.14 shows a three-input OR gate and its eight-line truth table. The same logic function can be made by cascading two 2-input OR gates together, as indicated in Figure 4.14(c).

FIGURE 4.12   7432 Quad Two-Input OR Gate

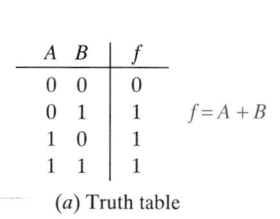

A	B	f
0	0	0
0	1	1
1	0	1
1	1	1

$f = A + B$

(*a*) Truth table

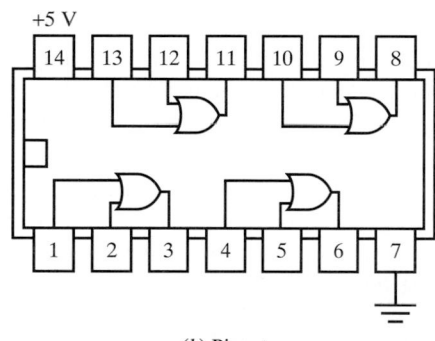

(*b*) Pinout

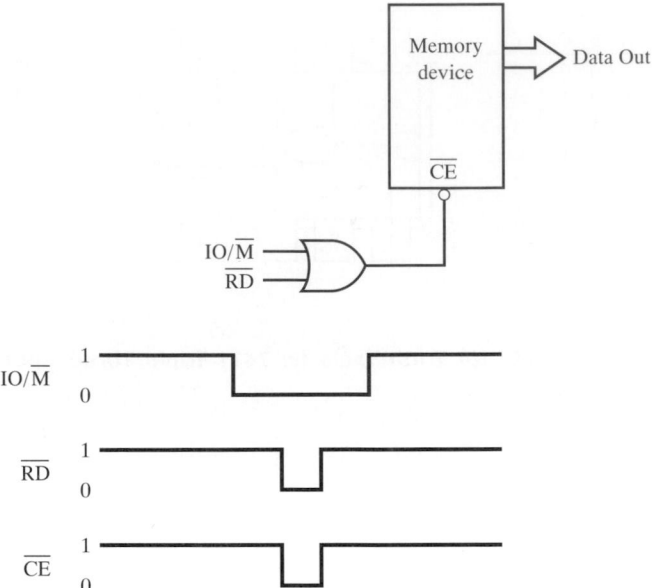

FIGURE 4.13 Enabling a Memory Device

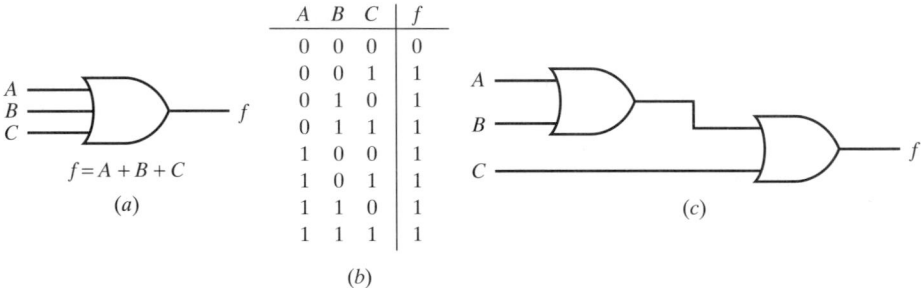

$f = A + B + C$

(a)

(b)

(c)

FIGURE 4.14 (a) Three-Input OR Gate (b) Truth Table (c) Alternate Circuitry

4.7 NAND AND NOR GATES

The *NAND gate* is an inverted AND gate. The inversion is at the output of the AND. Figure 4.15 shows two- and three-input NAND gates and their truth tables. A long line indicates that the inversion is at the output on the Boolean expression. The *NOR gate* is an inverted OR gate. The inversion takes place at the output of the OR gate. Figure 4.16 shows the NOR gate in two forms. Care must be used in writing the Boolean expressions when showing the inversions. Short lines may *never* be replaced by a single long line.

    Caution: Two shorts do not make a long! When writing Boolean expressions, great care must be taken to prevent lines (inversions) from running together. An inverted inversion cancels itself ($\overline{\overline{A}} = A$), but $\overline{A}\,\overline{B} \neq \overline{AB}$! This can be shown by comparing the truth tables for $\overline{A}\,\overline{B}$ and $\overline{AB}$ as in Figure 4.17. Since the truth tables are different, we conclude that $\overline{A}\,\overline{B} \neq \overline{AB}$. So, use care in drawing the lines!

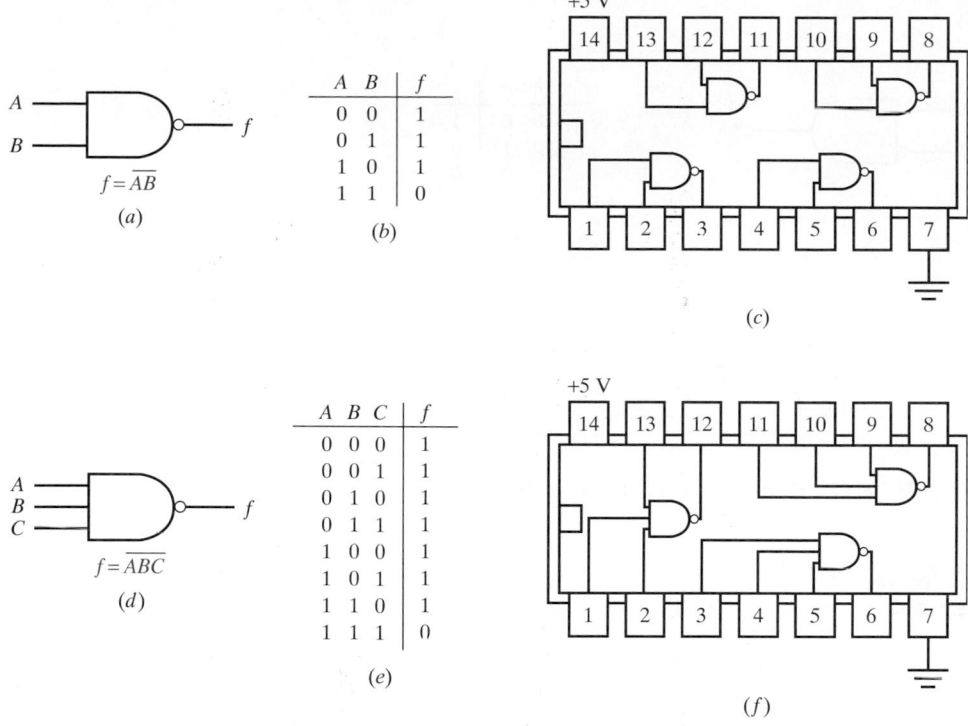

FIGURE 4.15 (*a*) Two-Input NAND Gate and (*b*) Truth Table; (*c*) 7400 Quad Two-Input NAND Gate; (*d*) Three-Input NAND Gate and (*e*) Truth Table; and (*f*) 7410 Triple Three-Input NAND Gate

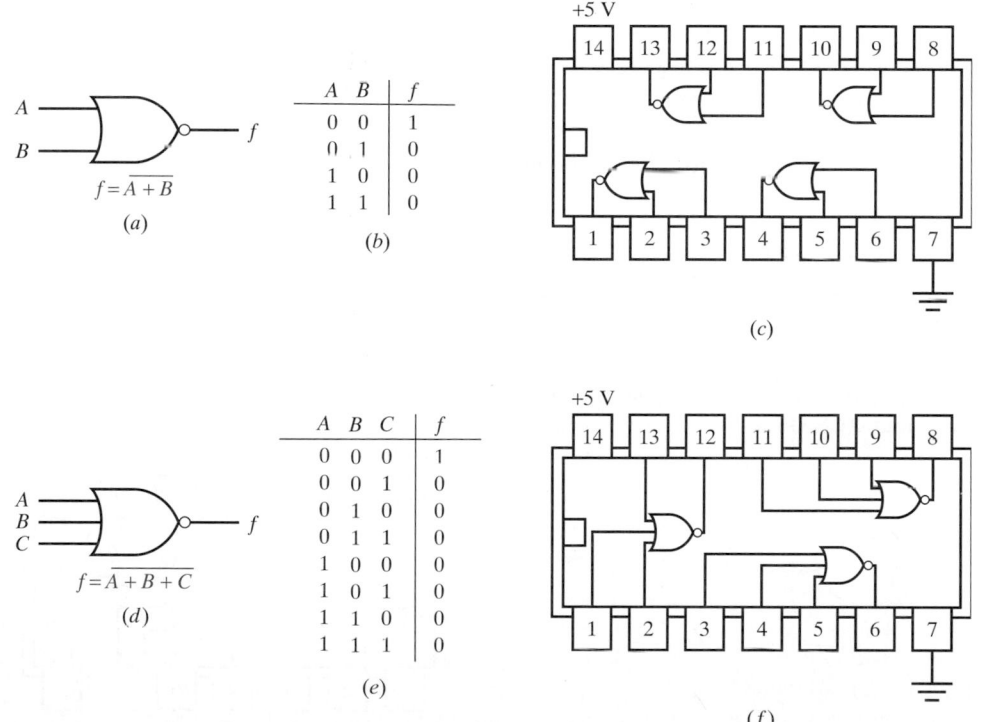

FIGURE 4.16 (*a*) Two-Input NOR Gate and (*b*) Truth Table; (*c*) 7402 Quad Two-Input NOR Gate; (*d*) Three-Input NOR Gate and (*e*) Truth Table; (*f*) 7427 Triple Three-Input NOR Gate

81

$f = \overline{A}\,\overline{B}$

(a)

A	B	f
0	0	1
0	1	0
1	0	0
1	1	0

(b)

$f = \overline{AB}$

(c)

A	B	f
0	0	1
0	1	1
1	0	1
1	1	0

(d)

FIGURE 4.17   Comparison of $\overline{A}\,\overline{B}$ and $\overline{AB}$

4.8   EXCLUSIVE OR AND EXCLUSIVE NOR GATES

The exclusive OR gate is a special OR gate. The output will be high if either input is high, but *not* both. Figure 4.18 shows such a gate and its logic symbol. The new symbol $\oplus$ represents the exclusive OR function. This equation can be written in two ways, $f = A \oplus B$ and $f = A\overline{B} + \overline{A}B$. The second expression is read: "f is a '1' when A is a '1' and B is a '0' OR when A is a '0' and B is a '1.'" You may also say that the output is a one when the inputs are *different*.

The Boolean equation for the exclusive OR gate may also be written like this: $f = A\overline{B} + \overline{A}B$. Figure 4.19 shows how this Boolean equation is implemented. You will agree that using one exclusive OR gate is easier than using five logic gates (two inverters, two AND gates, and one OR gate). Our familiarity with the equation may, however, allow us to simplify a more complex equation by recognizing the *form* of the exclusive OR equation.

Figure 4.20(a) provides the truth table of the exclusive NOR gate. It is easy to see that the exclusive NOR gate outputs a one when its inputs *are the same*, just as the exclusive OR gate did when its inputs were different. Figure 4.20(b) provides the pinout of the exclusive NOR gate.

Note: When using Electronics Workbench, right-clicking on a gate brings up a context-sensitive menu. Choosing Help from the menu displays the truth table for the gate.

FIGURE 4.18   (a) Two-Input Exclusive OR Gate (XOR) and (b) Truth Table; (c) 7486 Quad Two-Input XOR Gate

$f = A \oplus B$
$\quad = A\overline{B} + \overline{A}B$

(a)

A	B	f
0	0	0
0	1	1
1	0	1
1	1	0

(b)

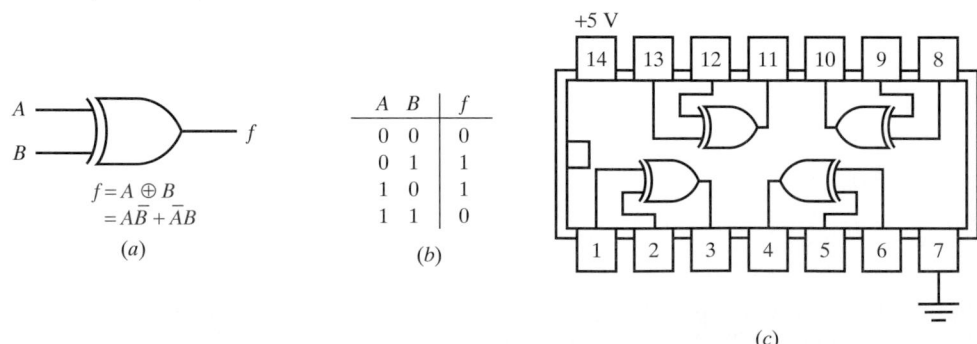

(c)

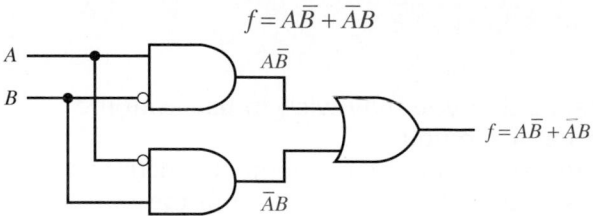

FIGURE 4.19 Modeling the XOR Function

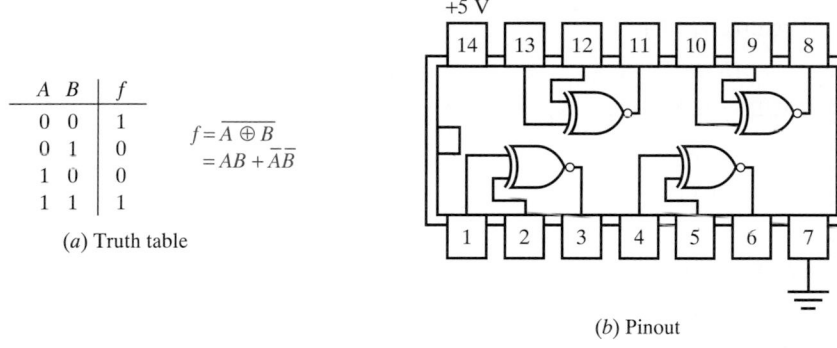

A	B	f
0	0	1
0	1	0
1	0	0
1	1	1

$$f = \overline{A \oplus B}$$
$$= AB + \overline{A}\overline{B}$$

(*a*) Truth table

(*b*) Pinout

FIGURE 4.20   74266 Quad Two-Input XNOR Gate

C/C++ HELPER

The TRUTHTAB program on the companion CD-ROM is designed as an online reference for the truth tables of all the basic logic gates. The user is provided with a menu of choices that allows individual gates, or all of them, to be accessed. A sample execution follows.

```
C> TRUTHTAB
Select a gate:
1: Inverter
2: AND
3: OR
4: NAND
5: NOR
6: XOR
7: XNOR
8: All
Choice? 2
AND
0  0 ┊ 0
0  1 ┊ 0
```

(continued)

```
1   0  ⋮  0
1   1  ⋮  1
```

TRUTHTAB is a handy reference that you may wish to use until the truth tables become second nature to you.

The TRUTHTAB program uses a set of C++ member functions to implement each of the basic gates. Here is a portion of the TRUTHTAB program, showing all of the function definitions.

```cpp
int GATES::NOT(int A)
{
    return(!A);
}
int GATES::AND(int A, int B)
{
    return(A && B);
}
int GATES::OR(int A, int B)
{
    return(A || B);
}
int GATES::NAND(int A, int B)
{
    return( NOT( AND(A,B) ) );
}
int GATES::NOR(int A, int B)
{
    return( NOT( OR(A,B) ) );
}
int GATES::XOR(int A, int B)
{
    return( AND(A,NOT(B)) || AND(NOT(A),B) );
}
int GATES::XNOR(int A, int B)
{
    return( NOT( XOR(A,B) ) );
}
```

The member functions are called with the appropriate values (0/1) in the **A** and **B** parameters, and return a zero or one according to their own Boolean expressions. The ! operator in C++ is used for inversion, && is the logical AND operation, and || is the logical OR. Notice how the XOR function is written. The C++ statement corresponds to $A\overline{B} + \overline{A}B$.

4.9 BOOLEAN ALGEBRA

We have seen the Boolean equations for all of the basic logic gates. Table 4.1 summarizes the equations. Our familiarity with the Boolean equations for the basic gates will help us work with more complicated Boolean expressions.

As with ordinary mathematical equations, Boolean equations may be manipulated by a set of legal operations, and have properties that we must be aware of. Table 4.2 lists some Boolean identities that are useful when simplifying a Boolean expression. To see how the identity works, plug in values of zero and one for A and see what you get. For example, in the identity $A + 0 = A$, if we let A equal zero we get $0 + 0 = 0$, and when A is a one we get $1 + 0 = 1$. The output is the same as the A input, hence $A + 0 = A$. This identity also indicates that ORing an input signal with zero results in the same signal at the output.

We will make more use of these identities in Chapter 5.

Gate Substitutions and DeMorgan's Theorem

Frequently in designing logic circuits, a particular type of gate is needed and we are reluctant to use another integrated circuit. If an inverter is needed and a spare NOR gate is available, then an answer is at hand. This section shows how one gate may substitute for another without changing the logic. Figure 4.21 illustrates several ways to build an inverter. Each circuit will replace an inverter as shown. Tying inputs together, which frequently increases the load on a gate, is not a good idea. The better approach is shown. In the case of the NAND gate, unused inputs are tied high through a 2.2-KΩ resistor. This is done to prevent a faulty input from disabling the gate.

TABLE 4.1 Basic Logic Gate Summary

Buffer	$A \longrightarrow\!\!\!\triangleright\!\!\!\longrightarrow f$	$f = A$
Inverter	$A \longrightarrow\!\!\!\triangleright\!\!\circ\!\!\longrightarrow f$	$f = \overline{A}$
AND	$A \atop B$ $\longrightarrow f$	$f = AB$
NAND	$A \atop B$ $\longrightarrow f$	$f = \overline{AB}$
OR	$A \atop B$ $\longrightarrow f$	$f = A + B$
NOR	$A \atop B$ $\longrightarrow f$	$f = \overline{A + B}$
XOR*	$A \atop B$ $\longrightarrow f$	$f = A \oplus B = A\overline{B} + \overline{A}B$
XNOR*	$A \atop B$ $\longrightarrow f$	$f = \overline{A \oplus B} = \overline{A}\,\overline{B} + AB$

*XOR and XNOR are abbreviations for exclusive OR and exclusive NOR.

TABLE 4.2   Boolean Expressions and Identities

EXPRESSIONS	
TERMS	**MEANING**
A	A is a 1
$\overline{A}$	A is a 0
$=$	when
$\times$, $\cdot$, Multiply	AND
$+$, Sum	OR

IDENTITIES
$A \cdot 0 = 0$
$A \cdot 1 = A$
$A \cdot A = A$
$A \cdot \overline{A} = 0$
$A + 0 = A$
$A + 1 = 1$
$A + A = A$
$A + \overline{A} = 1$
$A \oplus 0 = A$
$A \oplus 1 = \overline{A}$
$A \oplus B = A\overline{B} + \overline{A}B$
$\overline{A \oplus B} = \overline{A}\,\overline{B} + AB$
$\overline{\overline{A}} = A$

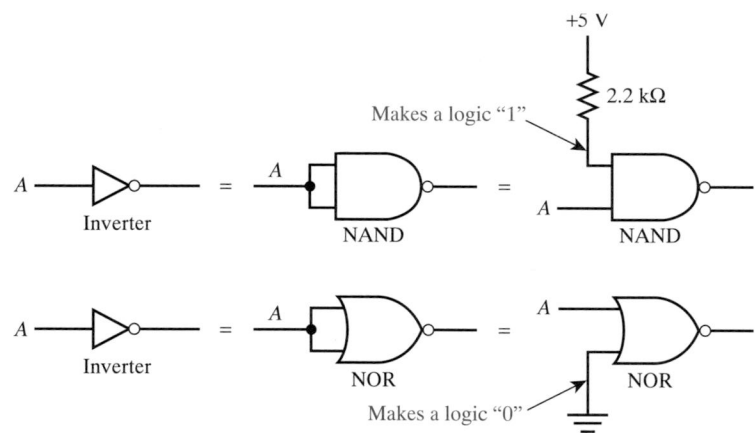

FIGURE 4.21   Inverters

It is a relatively simple matter to construct OR gates from NOR gates by adding an inverter. The same may be done to construct AND gates from NAND gates. These substitutions may be seen in Figure 4.22.

Another 19th century mathematician, Augustus DeMorgan, a friend of George Boole, is known for two theorems that find extensive use in digital design:

1. $\overline{A}\,\overline{B} = \overline{A + B}$
2. $\overline{AB} = \overline{A} + \overline{B}$

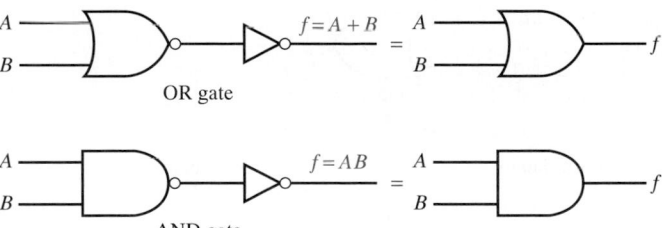

FIGURE 4.22   Substitutions of OR and AND Gates

Theorem 1 shows that the AND gate may be replaced by a NOR gate, provided the inputs are complemented. Theorem 2 shows that a NAND gate may replace an OR gate if the inputs are complemented. Figure 4.23 shows these gate equivalents.

Table 4.3 lists some interesting conversions using ***DeMorgan's theorem.*** Study it to become more familiar with these gate substitutions. In the next chapter we will be using all the logic gates and the gate substitution ideas in more detail. These form the basis for all digital circuit design.

Figure 4.24 shows the relations between various logic gates using DeMorgan's theorem. For instance, in Figure 4.24(a) we can see that a NAND gate can convert to an OR gate if the inputs are inverted. We also see that an AND gate can convert to a NOR gate if the inputs are inverted. Clearly, the difference between an AND gate and a NAND gate is that the outputs are inverted. Figure 4.24(b) shows the same relationships in equation form.

You may be interested to see the original form of DeMorgan's theorem as he developed it in the early 1800s. It reads as follows:

> The negative or contradictory of an alternative proposition is a conjunction in which the conjuncts are the contradictions of the corresponding alternates,

FIGURE 4.23   DeMorgan's Theorems Illustrating Gate Equivalence

DeMorgan 1:

$$f = \overline{A}\,\overline{B} \qquad = \qquad f = \overline{A + B}$$

DeMorgan 2:

$$f = \overline{A B} \qquad = \qquad f = \overline{A} + \overline{B}$$

TABLE 4.3   Gate Conversions Using DeMorgan's Theorem

$\overline{A + B}$	$=$	$\overline{A}\,\overline{B}$
$\overline{\overline{A} + B}$	$=$	$A\,\overline{B}$
$A + B$	$=$	$\overline{\overline{A} \cdot \overline{B}}$
$A + \overline{B}$	$=$	$\overline{\overline{A} \cdot B}$
$\overline{AB + BC}$	$=$	$\overline{AB} \cdot \overline{BC}$
$\overline{AB + BC}$	$=$	$\overline{A} + \overline{B} + \overline{B} + \overline{C} = \overline{A} + \overline{B} + \overline{C}$

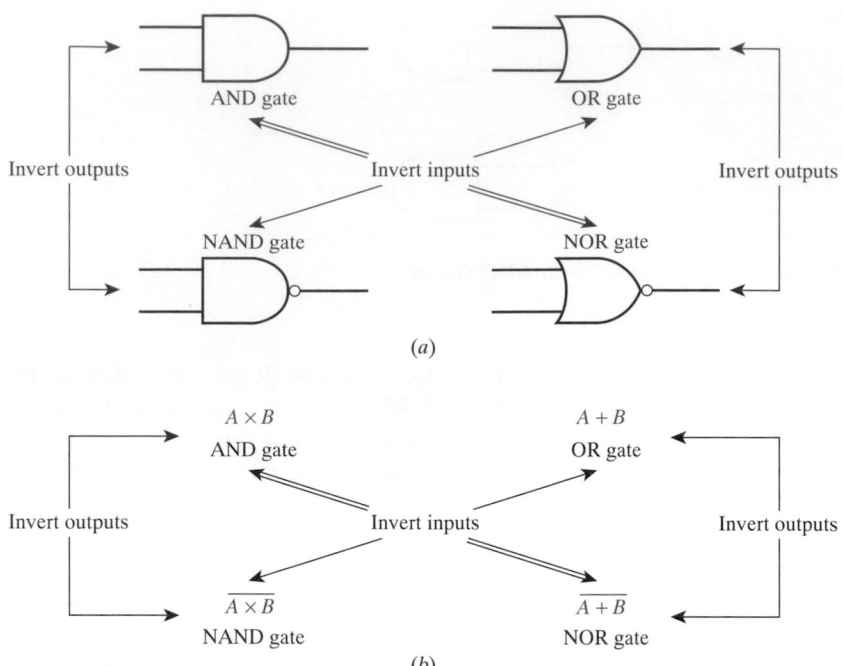

FIGURE 4.24 **Gate Relationships**

and that the negative of a conjunctive is an alternative proposition in which the alternates are the contradictories of the corresponding conjuncts.

Then again, you may prefer $\overline{A} \cdot \overline{B} = \overline{A + B}$ and also $\overline{A} + \overline{B} = \overline{A \cdot B}$. We include this material for its historical significance.

4.10 PRACTICAL APPLICATIONS

A number of functional circuits can be easily constructed to show the usefulness of logic gates. Let us examine a few simple applications.

Digital Oscillator

The digital oscillator produces a free-running pulse train. Figure 4.25 shows a capacitor-controlled ring oscillator and a crystal oscillator.

LED Driver

Another useful circuit is an LED (light-emitting diode) driver. Either a buffer or an inverter may be used to turn on the LED. This circuit is very useful as an indicator of the state of a logic gate's output. Figure 4.26 shows these circuits. The circuit is called a driver circuit because it is able to provide the drive current required by the LED. The resistor limits the current to protect both the LED and the logic gate. Such driver circuits are used in quantity to operate indicator lights.

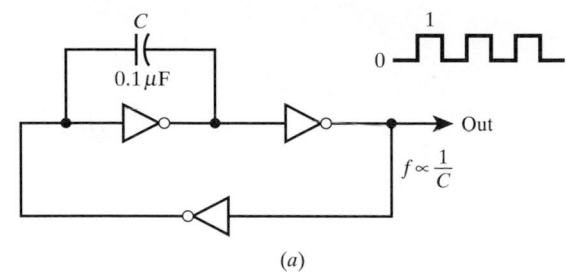

(a)

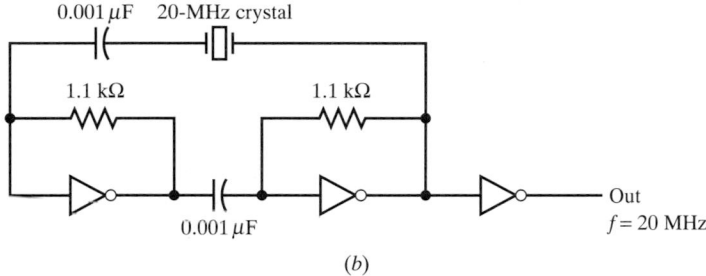

(b)

FIGURE 4.25 Digital Oscillators: (a) Ring Oscillator Using 7404 Inverters (b) Crystal Oscillator Using 74LS04 Inverter

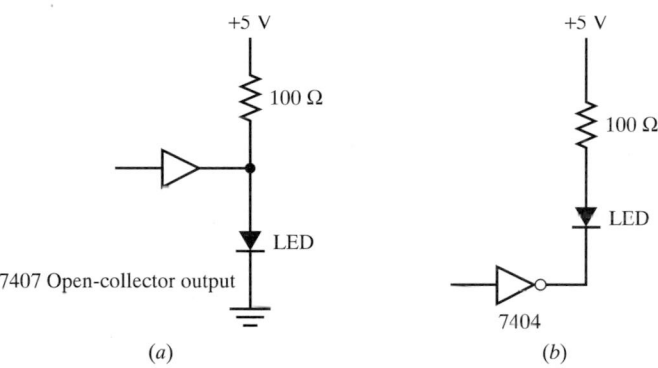

(a)   (b)

FIGURE 4.26 LED Drivers: (a) Buffer Type and (b) Inverter Type

One of the main reasons for producing open-collector outputs is to be able to turn high-current indicators on and off. While TTL operates on a +5 V DC, it is possible to operate a higher voltage indicator. The 7406 is an open-collector gate that can switch up to 30 V DC. Figure 4.27 shows the circuit. A logic "1" on the input drives the output low, providing a ground for the lamp to turn it on.

Wired OR Gate

An open-collector output also can be used when a wire is common to many circuits. The wire turns a single indicator lamp on. At any point on the line, an open collector may pull the line down, thus lighting the lamp. For example, a warning indicator might be set off

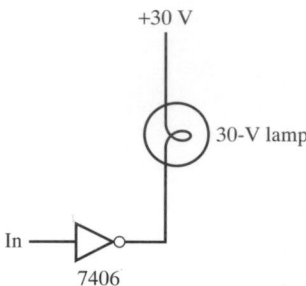

FIGURE 4.27 **High-Voltage Lamp Driver**

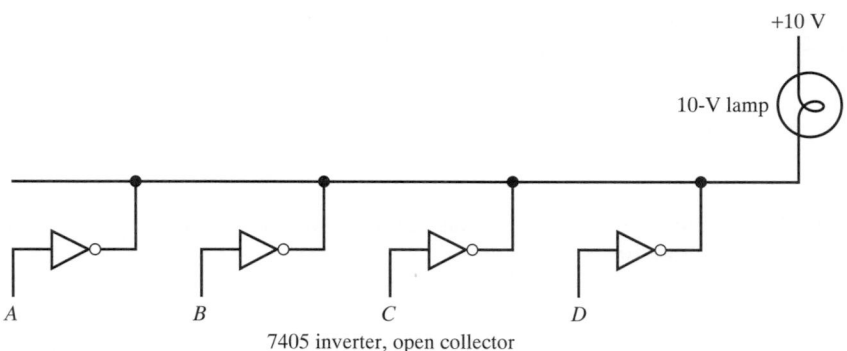

FIGURE 4.28 **Wired OR Function**

from several different sources. Figure 4.28 illustrates this important principle. Any open-collector gate can turn on the lamp. This function is called a ***wired OR*** function. A logic "1" on *A* or *B* or *C* or *D* will turn on the light.

In a later chapter we will make use of a very special type of logic gate related to this discussion of the wired OR. This special gate is referred to as a ***tri-state logic*** gate. This special-purpose gate has the ability to do one of three things to a digital signal line:

1. Apply a "0" to the line.
2. Apply a "1" to the line.
3. Disconnect from the line (high impedance state).

Tri-state gates have a control line to connect or disconnect. If disconnected, the gate is said to be tri-stated. A tri-state inverter is shown in Figure 4.29.

FIGURE 4.29 **(a) Tri-State Logic Gate (b) Equivalent Three-State Output**

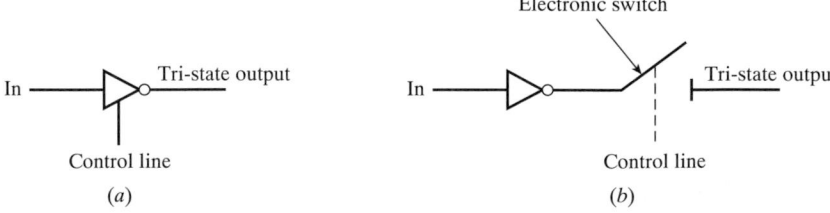

The Bounceless Switch

Logic circuitry demands that a pulse have a very clean edge. In most cases, this edge must rise in less than 10 nsec. Unfortunately, every mechanical switch is slow and subject to **contact bounce.** Figure 4.30 shows the effect of a common mechanical switch on closing and the desired waveform. Figure 4.30(c) shows the standard measurement for rise time. As Figure 4.31 shows, the rise time for a pulse is the time to travel from 10% of maximum to 90% of maximum or from 0.5 to 4.5 V in a TTL circuit. Similarly, the fall time (t_f) is the time to go from 90% to 10% of the maximum. The pulse width (t_{pw}) is the time between 50% points.

For a TTL gate the rise time should be less than 10 nsec, and it must be clean. Figure 4.32 shows two ways to completely clean up the waveform from a mechanical switch. The circuits both use a single-pole, double-throw switch (SPDT). One advantage of these circuits is that each has two separate outputs that are mutual complements. This circuit is also referred to as a DC latch because of its locking action.

A variation of the bounceless switch circuit can produce a latching circuit to set off an alarm. A separate switch is required to perform the reset operation. Figure 4.33 shows the alarm circuit complete with LED indicator.

FIGURE 4.30 Contact Bounce: (*a*) Dirty Switch (*b*) Resulting Waveform (*c*) Desired Waveform

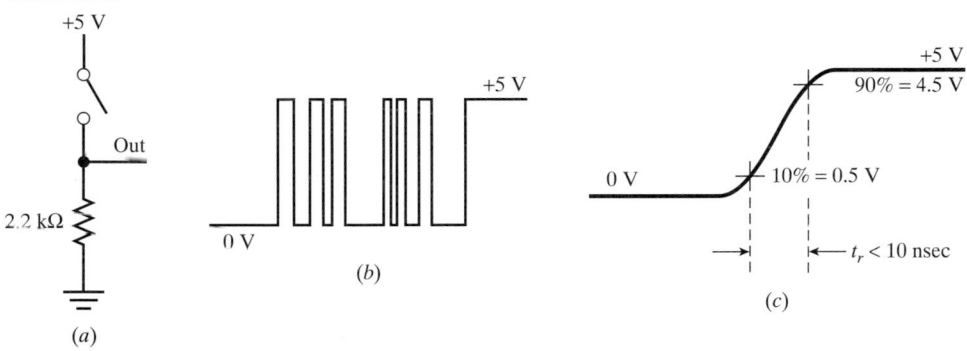

FIGURE 4.31 Rise Time t_r and Fall Time t_f of a Pulse

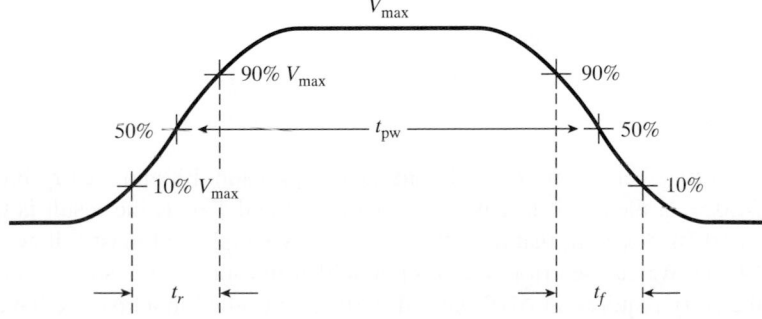

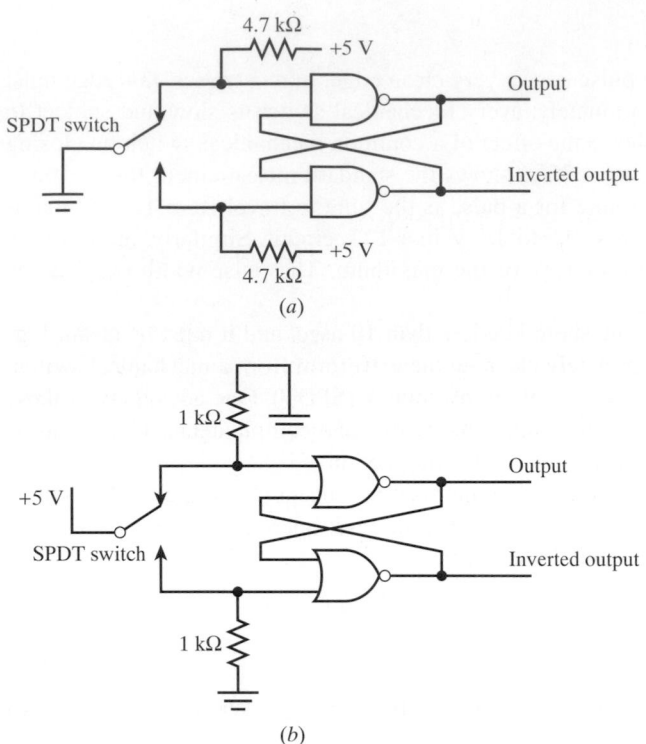

FIGURE 4.32 **Bounceless Electronic Switches:** (*a*) **7400 NAND Switch** (*b*) **7402 NOR Switch**

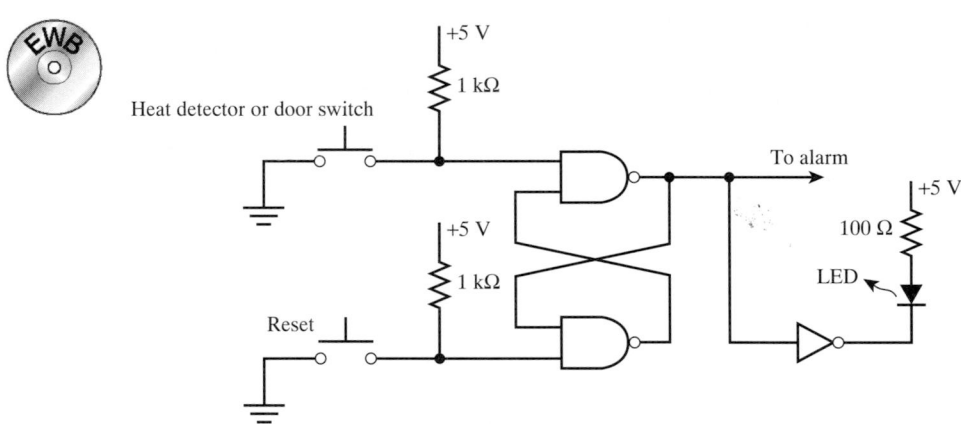

FIGURE 4.33 **Alarm Circuit with Indicator**

The Binary Adder

To add two binary bits together, we must be able to have both a sum bit and a carry bit. For instance, to add $0 + 1$, the result is 1 with 0 to carry. To add $1 + 1$, the result is 0 with 1 to carry (i.e., 10 B). A circuit that does this is relatively simple and must follow a truth table (Figure 4.34). We can see from the truth table that the sum requires an exclusive OR gate and the carry requires an AND gate. If these are installed in a box, we have

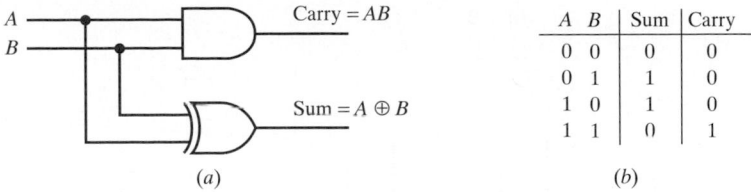

FIGURE 4.34   Binary Adder Circuit: (*a*) Half Adder and (*b*) Truth Table

A B	Sum	Carry
0 0	0	0
0 1	1	0
1 0	1	0
1 1	0	1

a packaged binary adder called a *half adder circuit*. A **full adder** includes a carry input as well and can be made from a pair of half adders (HA) as shown in Figure 4.35. Figure 4.35(c) shows a full adder circuit in Electronics Workbench that you can examine. Press the A, B, and C keys to see the outputs of the full adder change state. The full adder (FA) can be used with other full adders to build a parallel adder. Such an adder can add large binary numbers together quickly. To illustrate its usefulness, in Figure 4.36 we add two numbers: 9 and 14.

We are now ready to design some digital circuitry using the basic logic gates. Boolean algebra and truth tables will be important aspects of our designs.

FIGURE 4.35   Full Adder Circuit: (*a*) Two Half Adders and (*b*) Full Adder Truth Table (*c*) Electronics Workbench ADDER Circuit

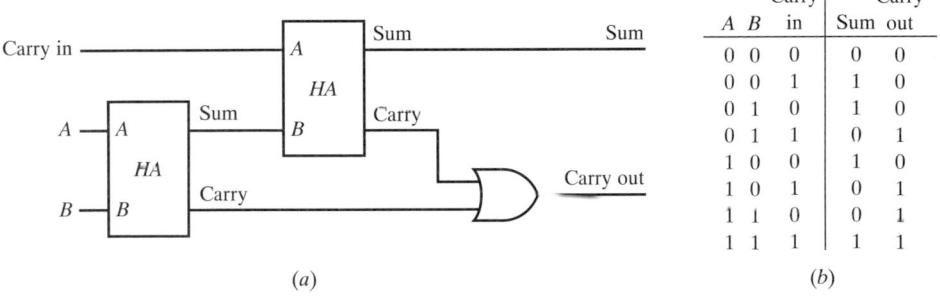

A	B	Carry in	Sum	Carry out
0	0	0	0	0
0	0	1	1	0
0	1	0	1	0
0	1	1	0	1
1	0	0	1	0
1	0	1	0	1
1	1	0	0	1
1	1	1	1	1

(*a*)   (*b*)

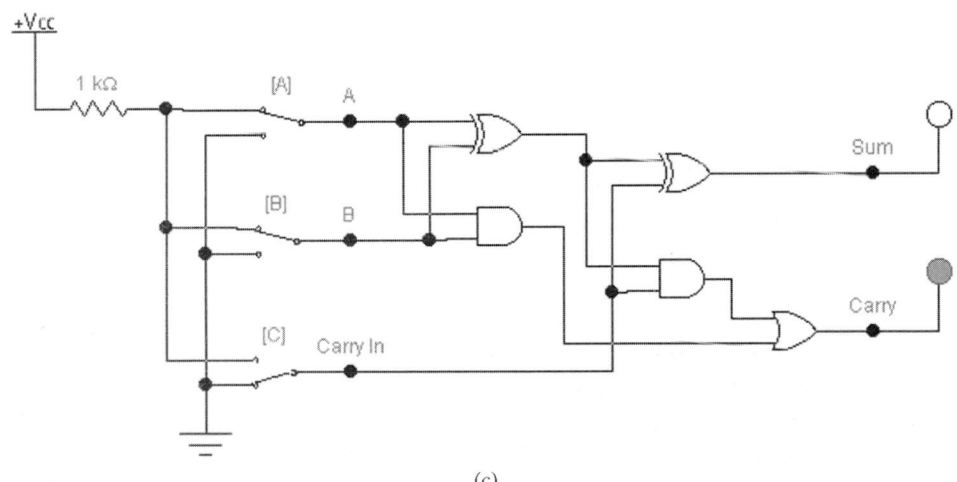

(*c*)

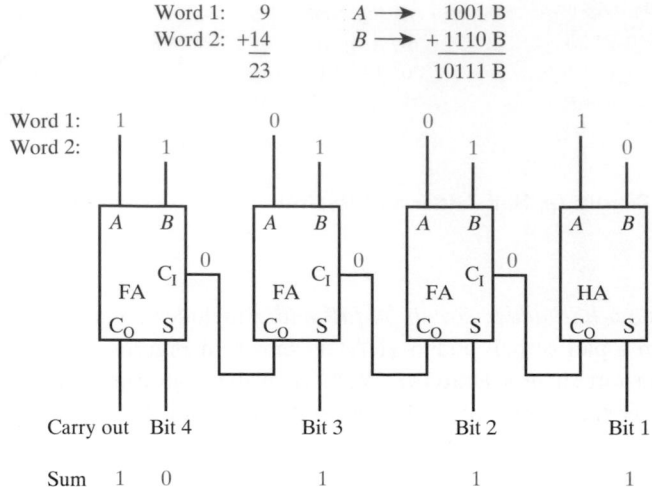

```
Word 1:    9        A ─→    1001 B
Word 2: +14          B ─→  + 1110 B
        ──                ────────
        23                 10111 B
```

Word 1:   1   0   0   1
Word 2:   1   1   1   0

FIGURE 4.36   **Four-Bit Parallel Adder**

4.11 TROUBLESHOOTING TECHNIQUES

In this section we will see how the output of a logic gate is controlled by a small delay called a propagation delay and how it affects the operation of digital circuits. We will also see how to determine whether a gate is good or bad.

Propagation Delay

One very important aspect of digital logic gates that is often overlooked (but becomes very important in high-speed applications) is the effect of gate propagation delay time. Figure 4.37 demonstrates that the output of a 74LS00 NAND gate does not go low the

FIGURE 4.37   **Propagation Delay**

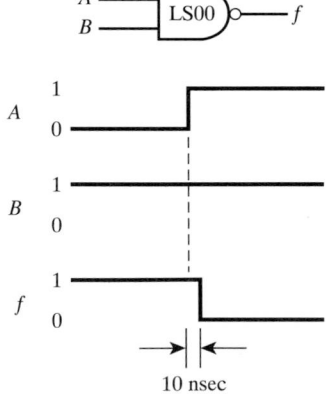

instant the input goes high. This actually occurs 10 ns later. This 10-ns period is the propagation delay time of the 74LS00 and is a function of the time needed to turn transistors on and off inside the gate. Propagation delay can cause disastrous results if a circuit is not designed correctly. For example, the designer of the circuit in Figure 4.38 ignored the propagation delays (use 10 ns per gate to figure it out for yourself) and got an extra pulse, or glitch, of 20 ns. It is important to keep track of the gate delays when designing a digital circuit.

Troubleshooting Logic Gates

Learning the behavior of the various logic gates is important and can lead to the identification of faulty gates in digital circuitry. For instance, if a NAND gate has two inputs that are both low and an output that is also low, is this a possible faulty gate? The answer is yes! A NAND gate should have a "1" at its output when both inputs are low.

FIGURE 4.38 Errors Due to Propagation Delay: (*a*) Simple Digital Circuit (*b*) Desired Timing (*c*) Actual Timing Due to Gate Delays (10 ns Each)

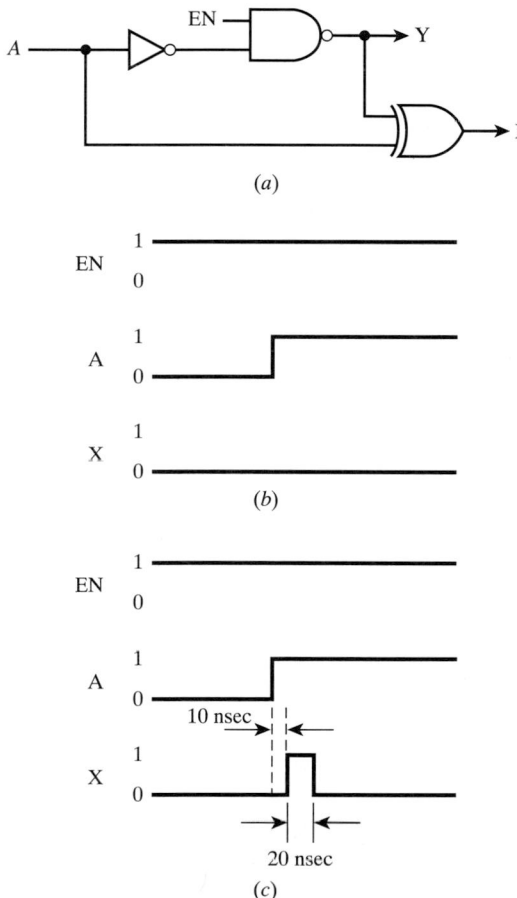

EXAMPLE **4.1**

Identify each of the logic gates in Figure 4.39 as either good or possibly bad.

1.

2.
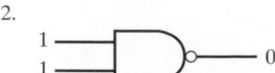

3.
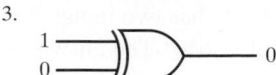

4.

5.

FIGURE 4.39   For Example 4.1

SOLUTION

1. Bad gate—output should be low.
2. Good gate.
3. Bad gate—output should be high.
4. Bad gate—no inversion.
5. Bad gate—output should be low.

EXAMPLE **4.2**

In the circuit shown in Figure 4.40, are there any bad gates for the levels indicated?

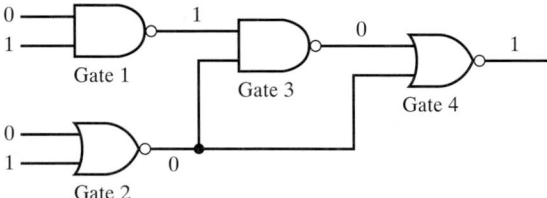

FIGURE 4.40   For Example 4.2

SOLUTION

Yes, gate 3 is faulty.

Logic gates usually do not have DC or static signals applied to them. More often the signals are constantly changing, and we refer to the inputs as dynamic. Such a signal is a series of pulses. If a series of pulses is applied to an inverting type of gate, the pulses become inverted as they pass through the gate. This is normal because each "0" is complemented to a "1" and each "1" is complemented to a "0." Consider a two-input AND gate. If the A input is high, pulses on the B input pass through the AND gate. However, if A is low, no pulses pass, since the AND operation allows the output to be high only when all inputs are high. We often look for pulses at an output to determine if a gate is good or bad, and not just a single high or low state.

SUMMARY

The logic gates are the building blocks of the digital electronics industry, and an understanding of each gate and its various uses is essential to later work with microprocessors. Figure 4.41, which summarizes the logic gates and shows them in their different forms when DeMorgan's theorem is

FIGURE 4.41   Summary of Logic Gates

Logic Symbols		Truth Tables		
		Input A	Input B	Output f
OR $\qquad$ $f = \overline{\overline{A} \cdot \overline{B}}$ $\qquad$ $f = A + B$		0	0	0
		0	1	1
		1	0	1
		1	1	1
AND $\qquad$ $f = A \cdot B$ $\qquad$ $f = \overline{\overline{A} + \overline{B}}$		0	0	0
		0	1	0
		1	0	0
		1	1	1
NAND $\qquad$ $f = \overline{A \cdot B}$ $\qquad$ $f = \overline{A} + \overline{B}$		0	0	1
		0	1	1
		1	0	1
		1	1	0
NOR $\qquad$ $f = \overline{A} \cdot \overline{B}$ $\qquad$ $f = \overline{A + B}$		0	0	1
		0	1	0
		1	0	0
		1	1	0
Exclusive OR $\qquad$ $f = A \oplus B = A\overline{B} + \overline{A}B$		0	0	0
		0	1	1
		1	0	1
		1	1	0

applied, constitutes a review of information presented in this chapter. Study these logic gates well, as they are used over and over again in this text. The basic TTL building blocks are the inverter, the AND gate, and the OR gate. Specialized gates such as the NAND, NOR, and XOR can be made from combinations of these gates. With an understanding of truth tables and simple Boolean algebra, it is possible to design digital logic circuits to perform important tasks or functions such as oscillation, latching, switch debouncing, and adding. Finally, it is important to keep track of propagation delay when designing a digital circuit.

STUDY QUESTIONS

General

1. Reword each of the following Boolean expressions to read in English describing the function. For example, $f = AB$ becomes "f is a '1' when A is a '1' and B is a '1.'"
   (a) $f = A\,\overline{B}\,C$
   (b) $f = A + \overline{B} + C$
   (c) $f = A\,B\,\overline{C} + A\,\overline{B}\,C$
   (d) $f = \overline{AB} + \overline{A}\,C$

2. Draw truth tables for a four-input NOR gate and for a four-input NAND gate.

3. Show by comparing truth tables that $\overline{\overline{A}}\,\overline{\overline{B}} \neq A\,B$.

4. Identify each of the logic gates shown in Figure 4.42 as either good or bad, given the logic levels indicated.

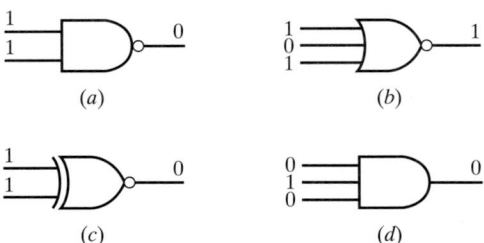

FIGURE 4.42 For Question 4.4

5. Use DeMorgan's theorem to change each equation to NAND logic only.
   (a) $f = \overline{A}\,\overline{B} + \overline{C}\,D$
   (b) $f = A\,\overline{B}\,C + \overline{A}\,B\,C$
   (c) $f = A\,B + A\,\overline{B} + \overline{A}\,B\,C$

6. Use DeMorgan's theorem to change each equation to NOR logic only.
   (a) $f = \overline{A\,B} + \overline{B}\,C$
   (b) $f = A\,C\,D + \overline{A}\,\overline{B}\,C$
   (c) $f = (A + \overline{B})(B + \overline{C})$

7. Determine the Boolean equation for the circuit shown in Figure 4.43.

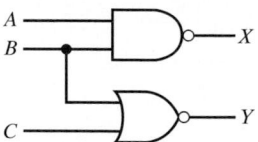

FIGURE 4.43   For Question 4.7

8. Determine the Boolean equation for the circuit shown in Figure 4.44. Can the circuit be simplified?

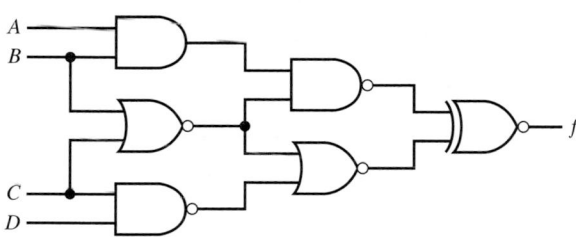

FIGURE 4.44   For Question 4.8

9. Draw the timing diagram for the circuit shown in Figure 4.45. Assume that all gates have a delay of 10 ns.

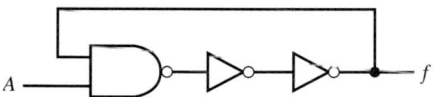

FIGURE 4.45   For Question 4.9

10. In the circuit shown in Figure 4.46, it is desired that the *R* output go high 40 ns after the inputs go to 1001 (*ABCD*). Will this happen? Assume that all gates have a delay of 15 ns.

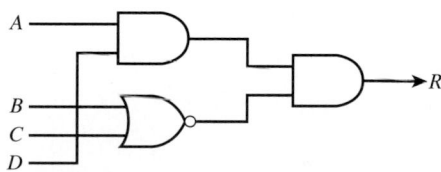

FIGURE 4.46   For Question 4.10

11. The circuitry of Figure 4.47 has five inputs and four outputs. Given the inputs on the truth table, follow the signals through the circuitry to determine the values of f_1 through f_4. If the output is PULSES or $\overline{\text{PULSES}}$, indicate using the symbols P and $\overline{P}$. Otherwise, just label the output 0 or 1.

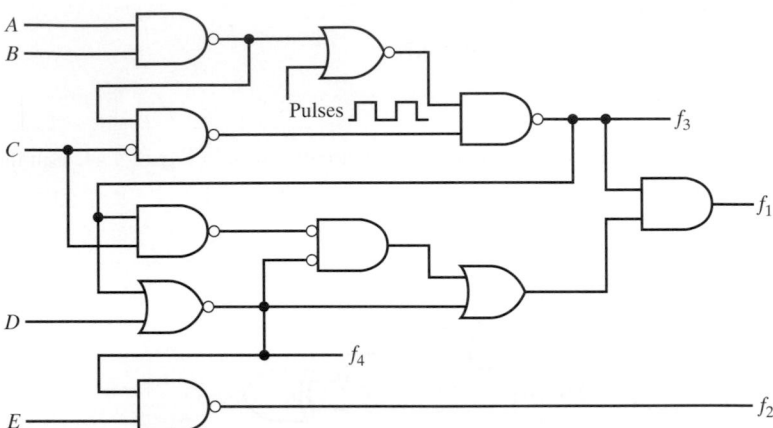

A	B	C	D	E	f_1	f_2	f_3	f_4
0	0	0	0	0				
1	1	1	1	1				
1	0	1	0	1				
0	1	0	1	0				
1	1	0	0	0				
0	0	1	0	1				
0	1	1	0	0				
1	1	1	0	1				
0	1	0	1	1				
0	0	0	1	1				

FIGURE 4.47 For Question 4.11. For each truth table entry, fill in f_1 through f_4.

12. Given the waveforms A and B shown in Figure 4.48, sketch the resulting waveforms from the following 2-input gates: AND, OR, NAND, NOR, XOR, and from a NAND gate with one input inverted.

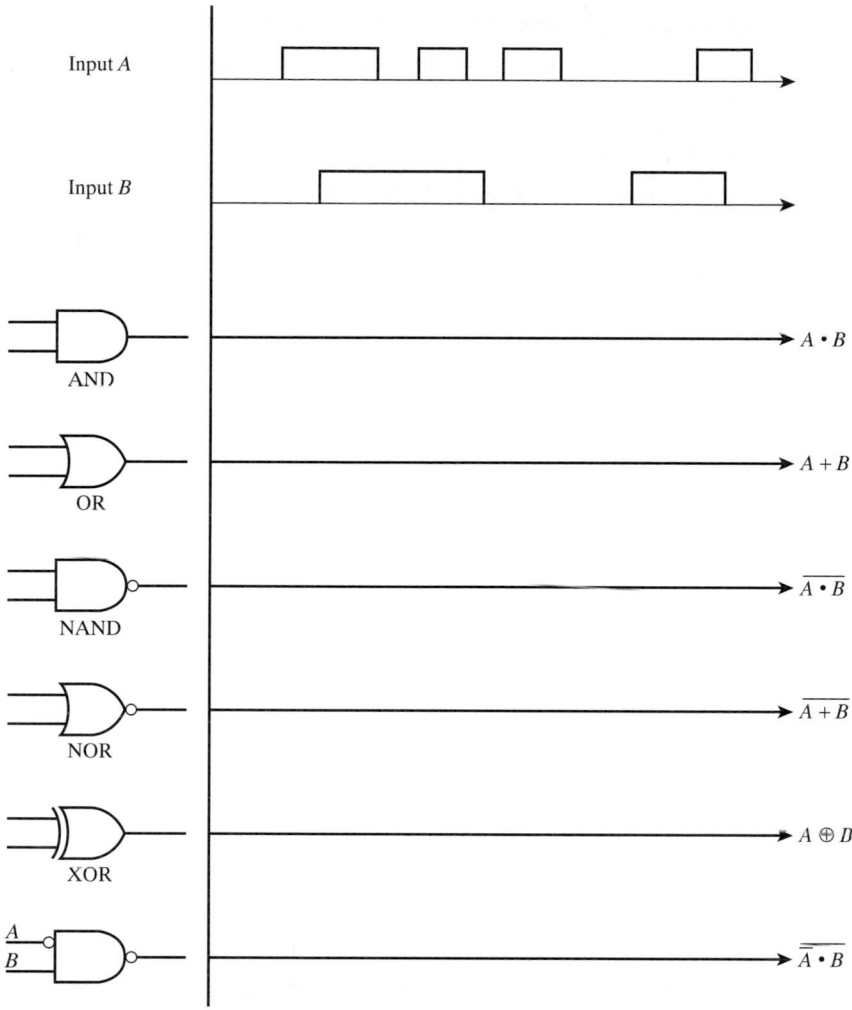

FIGURE 4.48 **For Question 4.12**

13. What is the truth table of the circuit in Figure 4.49? Can this circuit be simplified?

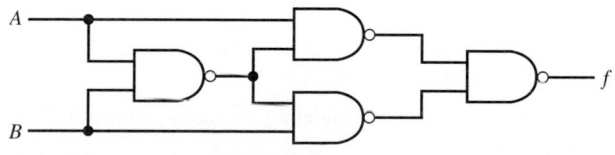

FIGURE 4.49 **For Question 4.13**

14. Show, using its truth table, how a NAND gate can be used as an inverter.

15. Repeat question 14 for a NOR gate and an exclusive OR gate.

16. How many lines does a five-input truth table have? What about an eight-input table?

17. What is an open-collector gate? What is needed to get it to work?

18. What is a pullup resistor? What logic level is associated with it?

19. Draw the schematic for each of these Boolean equations:
    (a) $f = AB + A\overline{B}$
    (b) $f = AB\overline{C} + AC$
    (c) $f = \overline{A} + AB + AC$

20. Evaluate each equation in question 19 for these values of A, B, and C:
    (a) $A = 0, B = 1, C = 0$
    (b) $A = 1, B = 0, C = 1$

21. Determine the truth table for the half *subtractor*. Design the required circuitry.

22. An automobile has digital sensors that report the following states:

Key:	Door:	Lights:
0–Not in ignition	0–Open	0–Off
1–In ignition	1–Closed	1–On

Furthermore, a buzzer is controlled by a digital signal as well (0–Off, 1–Buzzing).
    Show the digital circuitry necessary to implement these conditions for sounding the buzzer:
    (a) The key is in the ignition and the door is open.
    (b) The key is not in the ignition and the lights are on.

23. Exclusive OR gates can be used to generate a *parity* bit for a group of 1s and 0s. For example, the number 1000 has odd parity because there are an odd number of 1s. 1010 has even parity because there are an even (2) number of 1s. Figure 4.50 shows how parity is generated for a four-bit number. What is the output of the circuit for both example numbers? How is even parity represented at the output?

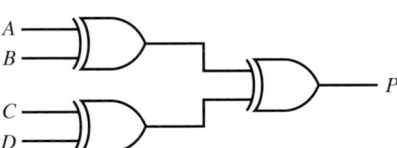

FIGURE 4.50   For Question 4.23

24. Show how parity is generated for an eight-bit number.

25. A technician needs an OR gate to complete a design, but there are no spare OR gates available, and no additional ICs may be added to the design. Spare inverters and NAND gates are available. What can be done?

Electronics Workbench

26. Use Electronics Workbench to simulate the circuit shown in Figure 4.19. Verify that the circuit performs the exclusive OR operation.

27. Check your answer to question 4.11 by simulating the circuit shown in Figure 4.47 using Electronics Workbench.

COMBINATIONAL LOGIC CIRCUIT DESIGN

INSTRUCTIONAL OBJECTIVES

When finished with this chapter, you should be able to:

1. Understand the function of a digital black box (DBB) and logic circuit design.

2. Use truth tables, symbolic logic, Boolean algebra, and Boolean reduction to reduce the gate cost and package count of a DBB.

3. Use Karnaugh mapping as a second method of reducing DBB logic (most useful with four or more variables).

4. Use DeMorgan's theorem for converting DBB logic to all-NAND or all-NOR logic.

5. Understand the operation of seven-segment LED displays.

6. Understand the purpose of programmable array logic.

7. Use the Quine-McCluskey method to reduce a logic function.

8. Determine the cost of building a logic circuit and sketch the circuitry from logic equations.

SELF-EVALUATION QUESTIONS

Keep the following questions in mind and try to answer them when you have completed the chapter.

1. What is meant by "sum of products" and by "product of sums"?

2. What are the rules for reducing a Boolean equation?

3. How are the ones (or zeros) from a truth table output column used in a Karnaugh map to reduce a Boolean equation?

4. Why is it important not to overuse terms in reading from a Karnaugh map?

5. Why is it important to try to reduce a Boolean equation?

6. Why is Boolean reduction very difficult on problems involving more than four variables? How does the Quine-McCluskey method help?

7. Do a sum-of-products result and a product-of-sums result produce the same logic function?

8. What is the purpose of gate array logic? What is available for circuits using fewer than 100 logic gates?

5.1 INTRODUCTION

This chapter explores several methods of digital design to produce circuitry that will behave in certain desirable ways. We use *combinations* of basic gates to obtain more complex behavior. In each case, the design begins with the examination of a truth table outlining the desired logic function. The truth table yields a Boolean equation that is simplified and reduced to a logic drawing, which produces the logic function. We then practice the techniques learned by designing and observing logic circuitry. The subject of gate array logic introduces a fast way of building logic circuitry that is compact and easy to use.

5.2 THREE-INPUT DIGITAL BLACK BOXES

The construction of a DBB begins with the problem of designing the logic circuitry to make a particular logic function work. When the design is complete, we will know what circuitry to put into the box. The basic box consists of three inputs and one output. We begin by defining when the output is to be a "1" ($f = 1$). The following expression defines our first design problem: $f = m(0, 1, 6, 7)$.

This expression is read "f is to be a '1' for lines number 0, 1, 6, and 7" (on the truth table). This is a neat way of defining the project. The m numbers* are the binary count each line on the truth table represents. Figure 5.1 shows the truth table and resulting logic function.

Each line on the table represents a binary count. For m_0 (the count of line 0), $A = 0, B = 0$, and $C = 0$. These occur at the same time, and we write this as an AND function: $\overline{A}\,\overline{B}\,\overline{C}$. This is one of four times that f is to be a "1." Each of the four occurrences causes f to be a "1" and the four occurrences are ORed together. The logic function is then: $f = \overline{A}\,\overline{B}\,\overline{C} + \overline{A}\,\overline{B}\,C + A\,B\,\overline{C} + A\,B\,C$. Figure 5.2 shows the circuitry that is required.

The circuit is constructed to have three inputs, *A, B,* and *C,* and a single output. It will fit neatly into the box, but what a lot of wiring! And there are a lot of logic gates. We will soon see that there is a much better way to build the same box. The improved method is called ***Boolean reduction.***

---

*The use of the letter m (as in m numbers) comes from an obsolete form of truth table notation. Terms of the form $A\,B\,C, A\,\overline{B}\,C$, and $\overline{A}\,\overline{B}\,C$ are called minimum terms (or min terms). Terms of the form $A + B + C, A + \overline{B} + \overline{C}$, and $\overline{A} + \overline{B} + \overline{C}$ are called maximum terms (or max terms). For a min term to be a "1," each variable must be active. This is the case of the AND operation. For the max term to be a "1," only one variable needs to be active. This is the OR operation.

FIGURE 5.1 (*a*) **Three-Input Digital Black Box and** (*b*) **Truth Table**

m	A	B	C	f	
0	0	0	0	1	$\leftarrow \overline{A}\,\overline{B}\,\overline{C}$
1	0	0	1	1	$\leftarrow \overline{A}\,\overline{B}\,C$
2	0	1	0	0	
3	0	1	1	0	
4	1	0	0	0	
5	1	0	1	0	
6	1	1	0	1	$\leftarrow A\,B\,\overline{C}$
7	1	1	1	1	$\leftarrow A\,B\,C$

$f = \overline{A}\,\overline{B}\,\overline{C} + \overline{A}\,\overline{B}\,C + A\,B\,\overline{C} + A\,B\,C$

(*a*)   (*b*)

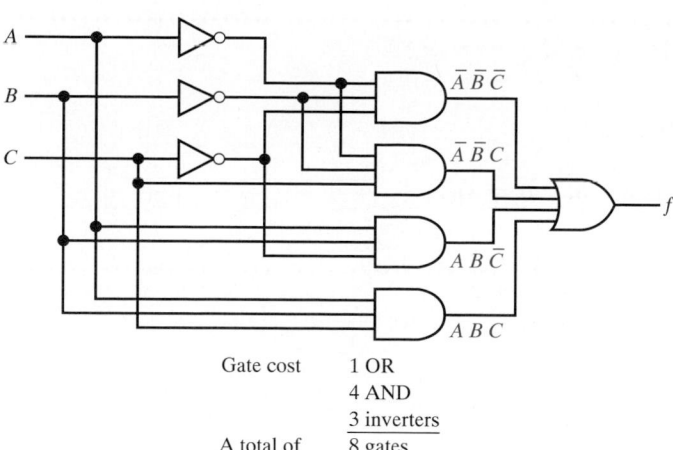

Gate cost 1 OR
           4 AND
           3 inverters
A total of 8 gates

FIGURE 5.2   Digital Black Box Circuitry

5.3   REDUCING LOGIC FUNCTIONS

If you have a function $WX + W\overline{X}$, it may be factored to $W(X + \overline{X})$ the same way we would factor a mathematical equation. Since $X + \overline{X} = 1$, the expression reduces to just W. Let's do that again (see Figure 5.3). You can see from the figure that as factoring and simplifying proceed, the number of required logic gates dwindles:

Step ① requires one OR, one inverter, and two AND gates for a total of four gates.

Step ② requires one OR, one inverter, and one AND gate for a total of three gates.

Step ③ requires *no* gates, just a piece of wire! We will make *very* good use of Boolean reduction.

This method of reduction tells us to factor out terms whenever possible. It also allows for the simplification of digital circuitry.

FIGURE 5.3   Boolean Reduction

Steps in Boolean reduction   Circuit

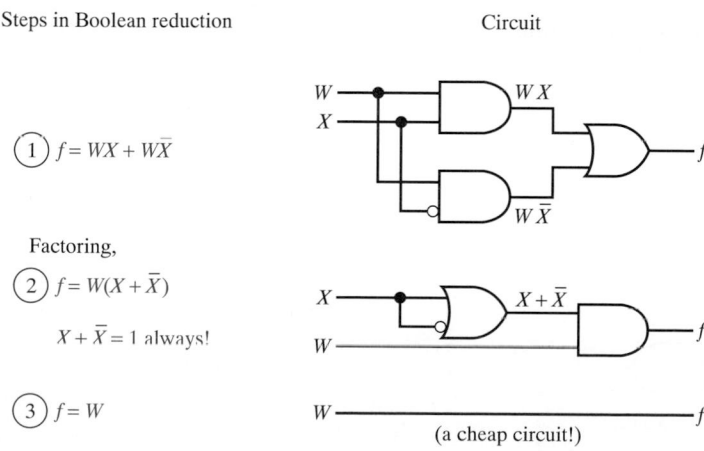

① $f = WX + W\overline{X}$

Factoring,

② $f = W(X + \overline{X})$

   $X + \overline{X} = 1$ always!

③ $f = W$

(a cheap circuit!)

EXAMPLE 5.1

Reduce $f = A B + A \overline{B}$.

SOLUTION

$f = A(B + \overline{B}) = A(1) = A$ (since $B + \overline{B} = 1$).

EXAMPLE 5.2

Reduce $f = A B C + A B \overline{C}$.

SOLUTION

$f = A B (C + \overline{C}) = A B (1) = A B$.

EXAMPLE 5.3

Reduce $f = A \overline{B} \overline{C} + \overline{A} \overline{B} \overline{C}$.

SOLUTION

$\overline{B} \overline{C}$ is common to both terms, so it is factored out: $f = (A + \overline{A}) \overline{B} \overline{C} = (1) \overline{B} \overline{C} = \overline{B} \overline{C}$.

Let us formalize this method of reduction by saying that we always compare two similar terms looking for *one* variable that is different. To compare two terms, all the variables (letters) must be present. For instance, in $f = A B \overline{C} + A B D$, there is no hope of comparing because the letters are different.

Note: $f = A B \overline{C} + A B D$ can be factored, but no terms drop out: $f = A B (\overline{C} + D)$. However, we *do* save one logic gate by factoring, a useful thing to remember (see Figure 5.4).

To continue with Boolean reduction, we can compare two terms that have *one* element different. The different element is eliminated. A few examples appear in Table 5.1.

FIGURE 5.4   Saving a Gate by Factoring

$f = AB\overline{C} + ABD$ looks like this:

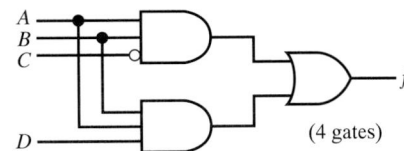

(4 gates)

$f = AB(\overline{C} + D)$ looks like this:

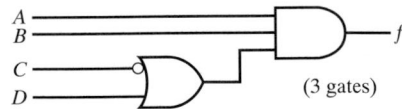

(3 gates)

TABLE 5.1 Some Simple Reductions

LOGIC FUNCTION	REDUCED FUNCTION
$f = A\overline{B}C + ABC$	$f = AC(B + \overline{B}) = AC$
$f = A\overline{B}CD + A\overline{B}\overline{C}D$	$f = A\overline{B}D$
$f = A\overline{B}C + \overline{A}\overline{B}C$	$f = \overline{B}C$
$f = \overline{A}\overline{B}C + \overline{A}B\overline{C}$	$f = \overline{A}\overline{B}$
$f = \overline{A}\overline{B}C + AB\overline{C}$	$f = \overline{A}\overline{B}C + AB\overline{C}$*
$f = \overline{A}BC\overline{D} + \overline{A}B\overline{C}\overline{D}$	$f = \overline{A}B\overline{D}$
$f = \overline{A}B\overline{C} + AB\overline{C}$	$f = B\overline{C}$
$f = B\overline{C}D + \overline{B}\overline{C}D$	$f = \overline{C}D$
$f = \overline{A}BD + \overline{A}\overline{B}D + ABC$	$f = \overline{A}D + ABC$
$f = A\overline{B}C + A\overline{B}\overline{C} + ABD$	$f = A\overline{B} + ABD$

*No reduction—can compare only terms that have one variable different.

We have discussed two methods of reducing the number of logic gates required:

① Boolean reduction to eliminate a different term from a pair.

② Factoring to drop one gate.

Remember to use both methods when designing a new circuit.

5.4 THE COST OF DIGITAL DESIGN

Our concern with the number of gates required to build a circuit is well justified. If 10,000 black boxes are to be built by a manufacturer, a reduction of one gate per black box will produce significant savings not only in logic gates but in wiring, sockets, and work hours.

We speak of two types of cost in the design of logic circuitry:

① *Gate cost*—the number of gates required.

② *Package cost*—the number of ICs required.

Let us reexamine the logic circuit of Figure 5.2, where $f = \overline{A}\,\overline{B}\,\overline{C} + \overline{A}\,\overline{B}\,C + AB\overline{C} + ABC$. The gate cost was eight gates. To build such a circuit from the 54/74 series, we might require the following: two 7411 triple three-input AND gates, one 7404 hex inverter, and one 7425 dual four-input NOR (we'll have to make an OR by inverting a NOR). The cost of that circuit is eight logic gates and four TTL packages. If we apply Boolean reduction to that circuit, we should find considerable savings in both types of cost.

Continuing our example, we will compare the terms a pair at a time, trying all combinations. To be sure that no comparison is missed, we compare the first with second through last, then the second with third through last, and so on. Here is the function again:

$$f = \overline{A}\,\overline{B}\,\overline{C} + \overline{A}\,\overline{B}\,C + AB\overline{C} + ABC$$
$$\underbrace{\qquad\qquad}_{\overline{A}\,\overline{B}}\qquad\underbrace{\qquad\qquad}_{AB}$$

The first two terms compare and reduce, and so do the last two. The new function is $f = \overline{A}\,\overline{B} + AB$.

This is *greatly* reduced from the original. We use a term twice only if there is a lone term that does not compare in any other way. We do *not* always use every comparison, lest the function get bigger. Remember, the object is to build a working circuit at the lowest possible cost. The new term requires one OR gate, two AND gates, and two inverters—only five gates and a lot less wire, since there are two input gates instead of three.

Figure 5.5 shows the new logic diagram. This new function reads as follows: "*f* is a '1' when $A = 0$ and $B = 0$ or when $A = 1$ and $B = 1$." Checking the truth table (Figure 5.6) reveals that this is just what we asked for! When $A = 0$ and $B = 0$, *f* is to be a "1"; when $A = 1$ and $B = 1$, *f* is to be a "1." When reading the 1s from the truth table we are using a method called the *sum of the products* (S/P). *Note:* the circuit can be further simplified by recognizing that $\overline{A}\,\overline{B} + A\,B$ is the alternate form of $\overline{A \oplus B}$, the exclusive NOR function. So, we actually require a *single* logic gate to do the same job as the original eight logic gates.

An alternate method of logic design is the *product of the sums* (P/S) method. In P/S we read the zeros from the table and want to find when the output *f* is at a "0." The problem is specified as follows: $\overline{f} = m(2, 3, 4, 5)$. The expression reads, "*f* is a '0' for counts 2, 3, 4, and 5." We reduce the function using Boolean reduction. Reading from the same truth table:

$$\overline{f} = \overline{A}\,B\,\overline{C} + \overline{A}\,B\,C + A\,\overline{B}\,\overline{C} + A\,\overline{B}\,C$$

$$\underbrace{\qquad}_{\overline{A}\,B} \qquad \underbrace{\qquad}_{A\,\overline{B}}$$

$\overline{f} = \overline{A}\,B + A\,\overline{B}$, and inverting both sides, $f = \overline{\overline{A}\,B + A\,\overline{B}}$. The product of sums result is *very* different from the sum of product result, but both are valid and satisfy the design requirement for our digital black box. The result is inverted to "*f* is a '1'" form by complementing each side of the equation. The result requires one NOR gate, two AND gates, and two inverters, as shown in Figure 5.7. Both methods are used in design, the simpler of the two being the winner.

FIGURE 5.5   DBB for $f = m$ (0, 1, 6, 7)

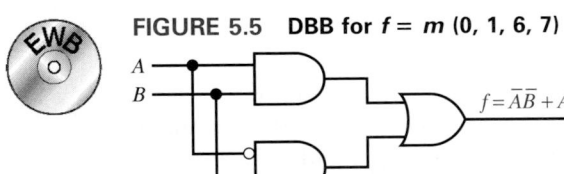

$f = \overline{A}\overline{B} + AB$

FIGURE 5.6   Truth Table

m	A	B	C	f	
0	0	0	0	1	$\leftarrow f = \overline{A}\,\overline{B}$
1	0	0	1	1	$\leftarrow$
2	0	1	0	0	
3	0	1	1	0	
4	1	0	0	0	$f = \overline{A}\,\overline{B} + A\,B$
5	1	0	1	0	
6	1	1	0	1	$\leftarrow f = A\,B$
7	1	1	1	1	$\leftarrow$

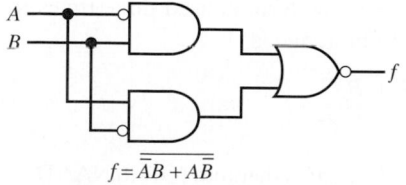

$f = \overline{\overline{A}B + A\overline{B}}$

FIGURE 5.7 Product of Sums Result

If we apply DeMorgan's theorem to the product-of-sums result such that we convert the output NOR gate to an AND gate, the equation becomes

$$\text{P/S} \qquad f = \overline{\overline{\overline{A}\,B} + \overline{A\,\overline{B}}} = \overline{\overline{A}\,B} \cdot \overline{A\,\overline{B}}$$

If we now change the NAND gates to OR gates using DeMorgan's theorem, we have

$$\text{P/S} \qquad f = \overline{\overline{A}\,B} \cdot \overline{A\,\overline{B}} = (A + \overline{B})\,(\overline{A} + B)$$

This form appears to be a product of sums and explains the name of the P/S method.

Let us try another example and a new twist by using DeMorgan's theorem on the result to use *all*-NAND or *all*-NOR logic.

••

EXAMPLE 5.4

Design a three-input DBB for $f = m(0, 2, 3, 6)$.

SOLUTION

The sum-of-products information is given and we derive the P/S information from what is left (see Figure 5.8). Develop the truth table (Figure 5.9) and read both ones and zeros.

A ———
B ——— DBB ——— f
C ———

$\overline{f} = m(1, 4, 5, 7)$

FIGURE 5.8 Three-Input DBB

m	A	B	C	f
0	0	0	0	1
1	0	0	1	0
2	0	1	0	1
3	0	1	1	1
4	1	0	0	0
5	1	0	1	0
6	1	1	0	1
7	1	1	1	0

S/P $\quad f = \overline{A}\,\overline{B}\,\overline{C} + \overline{A}\,B\,\overline{C} + \overline{A}\,B\,C + A\,B\,\overline{C}$

$\qquad\quad f = \overline{A}\,\overline{C} + \overline{A}\,B + B\,\overline{C}$

P/S $\quad \overline{f} = \overline{A}\,\overline{B}\,C + A\,\overline{B}\,\overline{C} + A\,\overline{B}\,C + A\,B\,C$

$\qquad\quad \overline{f} = \overline{B}\,C + A\,\overline{B} + A\,C$

FIGURE 5.9 Truth Table

The reductions are trickier this time. In both S/P and P/S one term is used three times to our advantage. This common term allows the others to be reduced.

$$\text{S/P} \quad f = \overline{A}\,\overline{C} + \overline{A}\,B + B\,\overline{C}$$
$$\text{P/S} \quad f = \overline{B}\,C + A\,\overline{B} + A\,C$$

Let us change the S/P result to all NAND logic using DeMorgan's theorem (OR to NAND— invert the inputs). The S/P equation contains a 3-input OR gate, which converts into a 3-input NAND gate with inverted inputs. Each input is a term from the original equation.

$$\text{S/P} \quad f = \overline{\overline{}\cdot\overline{}\cdot\overline{}}$$
$$f = \overline{\overline{A\,\overline{C}}\cdot\overline{\overline{A}\,B}\cdot\overline{B\,\overline{C}}} \qquad \text{all NAND gates}$$

Let us change the P/S result to all-NOR logic (AND to NOR—invert the inputs). The three AND terms are converted into three NOR terms with inverted inputs.

$$\text{P/S} \quad f = \overline{ + + }$$
$$f = \overline{\overline{(+)} + \overline{(+)} + \overline{(+)}}$$
$$f = \overline{\overline{(B + \overline{C})} + \overline{(\overline{A} + B)} + \overline{(\overline{A} + \overline{C})}}$$

Study the *solution* carefully to see how the conversions are made, a step at a time. Let's try another example.

EXAMPLE 5.5

Design a three-input DBB for $f = m(0, 2, 3)$.

SOLUTION

The other half of the problem must be: $\overline{f} = m(1, 4, 5, 6, 7)$.

$$\text{S/P} \quad f = \overset{0}{\overline{A}\,\overline{B}\,\overline{C}} + \overset{2}{\overline{A}\,B\,\overline{C}} + \overset{3}{\overline{A}\,B\,C}$$
$$f = \overline{A}\,\overline{C} + \overline{A}\,B$$

To NAND:

$$f = \overline{\overline{A}\,\overline{C} \cdot \overline{A}\,B}$$

$$\text{P/S} \quad \overline{f} = \overset{1}{\overline{A}\,\overline{B}\,C} + \overset{4}{A\,\overline{B}\,\overline{C}} + \overset{5}{A\,\overline{B}\,C} + \overset{6}{A\,B\,\overline{C}} + \overset{7}{A\,B\,C}$$
$$\overline{B}\,C \qquad A\,\overline{C} \quad A\,B$$
$$\overline{f} = \overline{B}\,C + A\,\overline{C} + A\,B$$
$$f = \overline{\overline{B}\,C + A\,\overline{C} + A\,B}$$

To NOR:

$$f = \overline{\overline{(B + \overline{C})} + \overline{(\overline{A} + C)} + \overline{(\overline{A} + \overline{B})}}$$

In the P/S reduction, two extra comparisons are ignored. These extra terms are not needed to reduce the function. Using them *all* is a mistake; the extra terms would defeat the purpose of reducing circuitry.

In making comparisons, we try to use each term once. This causes the term to be reduced in size. If a term must be used again, it is only done to help another term become smaller. In Example 5.5, the S/P result reduced as follows: The m_0 and m_2 terms are replaced with $\overline{A}\,\overline{C}$. To shorten the m_3 term, it is necessary to use the m_2 term again. This reduces the m_2 and m_3 terms to $\overline{A}\,B$. Although m_2 already had been "covered" once, it was used again to reduce m_3.

In the P/S result, the possibility of overcomparing exists. Since m_1 and m_5 reduce to $\overline{B}\,C$ and m_6 and m_7 reduce to $A\,B$, this leaves m_4 to be reduced if possible. Here it is used with m_6 to reduce to $A\,\overline{C}$. This completes the reduction, since each term has been used at least once. There are two other possible comparisons: m_4 with m_5 and m_5 with m_7. We call these "redundant" terms because using them is unnecessary (we've already reduced all five elements) and would bring the cost back up. Once an original term has been "covered," it is not necessary to cover it again.

The terms m_4 and m_5 (not used) *could* have been used *instead* of m_4 and m_6 (again, not both). We just need to "cover" m_4.

An alternate result then could be

$$\overline{f} = \overline{B}\,C + A\,\overline{B} + A\,B$$

In fact, this is a better result, since it is reduced even more. The last two terms combine, and the new function is

$$\overline{f} = \overline{B}\,C + A$$

This shows that Boolean reduction may lead to the choice of wrong pairs, producing a result that is not the shortest one possible. This example is a clue that a better method is desirable. The next section introduces the more desirable approach called Karnaugh mapping, but first let's try Boolean reduction once again.

· ·

EXAMPLE 5.6

Design a three-input DBB for $f = m(1, 2, 5)$.

SOLUTION

We deduce that $\overline{f} = m(0, 3, 4, 6, 7)$ for P/S. Figure 5.10 gives the truth table.

A	B	C	f
0	0	0	0
0	0	1	1
0	1	0	1
0	1	1	0
1	0	0	0
1	0	1	1
1	1	0	0
1	1	1	0

FIGURE 5.10   Truth Table

$$\text{S/P} \quad f = \overset{1}{\overline{A}\,\overline{B}\,C} + \overset{2}{\overline{A}\,B\,\overline{C}} + \overset{5}{A\,\overline{B}\,C}$$

$$\overline{B}\,C$$

$$f = \overline{B}\,C + \overline{A}\,B\,\overline{C}$$

To NAND:

$$f = \overline{\overline{\overline{B}\,C} \cdot \overline{\overline{A}\,B\,\overline{C}}}$$

$$\text{P/S} \quad \overline{f} = \overset{0}{\overline{A}\,\overline{B}\,\overline{C}} + \overset{3}{\overline{A}\,B\,C} + \overset{4}{A\,\overline{B}\,\overline{C}} + \overset{6}{A\,B\,\overline{C}} + \overset{7}{A\,B\,C}$$

$$\overline{B}\,\overline{C} \qquad B\,C \quad A\,\overline{C} \;\; \text{Redundant}$$

$$\overline{f} = \overline{B}\,\overline{C} + B\,C + A\,\overline{C}$$

So, $f = \overline{\overline{B}\,\overline{C} + B\,C + A\,\overline{C}}$.

To NOR:

$$f = \overline{\overline{(B + C)} + \overline{(\overline{B} + \overline{C})} + \overline{(\overline{A} + C)}}$$

The ability to select the proper terms in Boolean reduction improves with practice. The use of redundant terms will increase the cost of a logic circuit, and this is not what we wish to do. Figure 5.11 presents some more examples for you to study to better learn which terms should be selected and which ones should be left alone.

FIGURE 5.11   Boolean Algebra Reduction

1. $(\overline{A}\overline{B}C) + (\overline{A}\overline{B}\overline{C}) + (A\overline{B}C) + (A\overline{B}\overline{C})$

$f = 1 + 3$	Correct
$f = 2 + 4$	Correct
$f = 1 + 3 + 4$	Incorrect

2. $(ABC) + (A\overline{B}\overline{C}) + (A\overline{B}C) + (\overline{A}BC)$

$f = 2 + 3$	Correct
$f = 1 + 2 + 3$	Incorrect

3. $(\overline{A}BC) + (ABC) + (AB\overline{C}) + (A\overline{B}\overline{C})$

$f = 1 + 3$	Correct
$f = 1 + 2 + 3$	Incorrect

4. $(A\overline{B}C) + (\overline{A}BC) + (ABC) + (AB\overline{C})$

$f = 1 + 2 + 3$

5. $(ABC) + (\overline{A}\overline{B}C) + (AB\overline{C}) + (\overline{A}B\overline{C}) + (\overline{A}BC)$

$f = 1 + 3 + 4$	Correct
$f = 2 + 3 + 4$	Correct

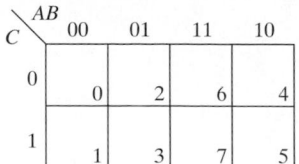

FIGURE 5.12   Karnaugh Map

5.5 KARNAUGH MAPPING

It is inevitable in black box design that another input will be required. With a fourth input the equations become more complex and we need help in finding terms that will be mutually comparable. The Karnaugh map is a visual method for spotting terms that are similar and yet offer a difference of one variable. Figure 5.12 shows a Karnaugh map for the three-input DBB. The three-input box has eight levels on the truth table ($2^N = 2^3 = 8$). There is a place for each m number on the Karnaugh map. Each of these eight boxes represents a possible input combination. The labeling of the Karnaugh map is important. The AB terms are across the top of the map (00, 01, 11, 10). Terms that are next to each other are different by only one bit. In this way, and because of the careful labeling, terms that are side by side differ by only one variable. Similarly, the C terms (0, 1) are down the side. Two boxes vertically adjacent also differ by one term. Let's try an example.

EXAMPLE 5.7

Simplify $f = m(0, 1, 4, 6)$.

SOLUTIONS

Boolean reduction:

$$f = \overline{A}\,\overline{B}\,\overline{C} + \overline{A}\,\overline{B}\,C + A\,\overline{B}\,\overline{C} + A\,B\,\overline{C}$$

$$\overline{A}\,\overline{B} \qquad\qquad A\,\overline{C}$$

$$f = \overline{A}\,\overline{B} + A\,\overline{C}$$

Karnaugh map method: See Figure 5.13.

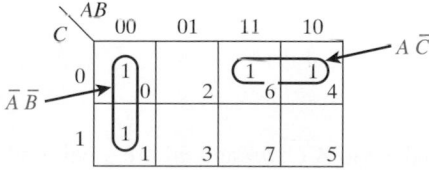

FIGURE 5.13   Example Solution

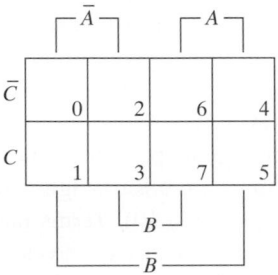

FIGURE 5.14   Three Ways to Label a Karnaugh Map

The Karnaugh map shows that m_0 and m_1, representing a vertical pair, are to be compared. Since $A = 0$ and $B = 0$ are the same, the C term is eliminated. A filled column points to $\overline{A}\,\overline{B}$ as the result. The map also shows that m_4 and m_6 differ by only one term. The B term is eliminated. Terms m_0 and m_4 also compare, but are not needed. This shows that the map can compare from one side to the other. For example, m_0 and m_4 compare, also m_1 and m_5. Since m_0 was used with m_1, it is "covered." It need not be used again. The same is true for m_4. Since m_4 was compared with m_6, it need not be reused. To do so would bring the cost of the circuit up again. We use the Karnaugh map to direct us to terms that can be compared:

1. We look for pairs, horizontal or vertical, but not diagonal. Diagonal terms have a difference of two terms and are of no use.

2. Groups of four are also important. A group of four can be a full row or a square.

Figure 5.14 shows three similar ways to label a Karnaugh map. Each way of showing the map has advantages. The object is to select one and become adept at map reading. Let's read some sample Karnaugh maps by doing a few examples.

•••

EXAMPLE 5.8

Design a three-input DBB in simplest form for $f = m(0, 2, 4, 5)$.

SOLUTION

See Figure 5.15. Terms m_0 and m_2 are a pair and reduce to $\overline{A}\,\overline{C}$; m_4 and m_5 are a pair and reduce to $A\,\overline{B}$; m_0 and m_4 are a pair, but are not needed, since m_0 and m_4 are already "covered." Thus $f = \overline{A}\,\overline{C} + A\,\overline{B}$ is the answer.

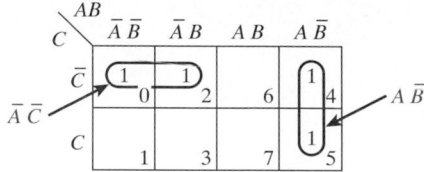

FIGURE 5.15 Karnaugh Map

••

EXAMPLE 5.9

Design a three-input DBB in simplest form for $f = m(1, 3, 4, 7)$.

SOLUTION

See Figure 5.16. Terms m_1 and m_3 are a pair and reduce to $\overline{A}\, C$; m_3 and m_7 are a pair and reduce to $B\, C$; m_4 does not compare with any other term. Thus the S/P answer is: $f = \overline{A}\, C + B\, C + A\, \overline{B}\, \overline{C}$. An alternate result is obtained from reading the zeros from the map: m_0 and m_2 compare to $\overline{A}\, \overline{C}$; m_2 and m_6 compare to $B\, \overline{C}$; m_5 is alone.

$$\text{P/S} \qquad \overline{f} = \overline{A}\,\overline{C} + B\,\overline{C} + A\,\overline{B}\,C$$
$$f = \overline{\overline{A}\,\overline{B} + B\,\overline{C} + A\,\overline{B}\,C}$$

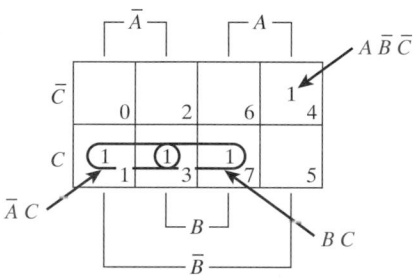

FIGURE 5.16 Karnaugh Map

••

EXAMPLE 5.10

Design a three-input DBB in simplest form for $f = m(1, 2, 3, 6, 7)$.

SOLUTION

See Figure 5.17. Terms m_1 and m_3 reduce to $\overline{A}\, C$; m_2 and m_6 and m_3 and m_7 are two pairs side by side and reduce to B. Thus $f = \overline{A}\, C + B$.

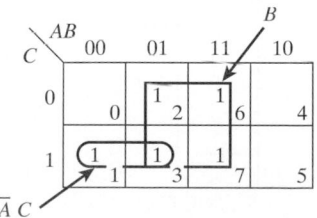

FIGURE 5.17 Karnaugh Map

••

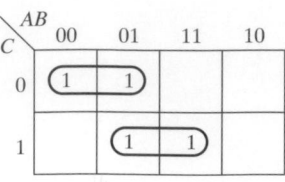

S/P $f = \overline{A}\,\overline{C} + B\,C$
P/S $\overline{f} = A\,\overline{C} + \overline{B}\,C$
$f = \overline{A\,\overline{C} + \overline{B}\,C}$

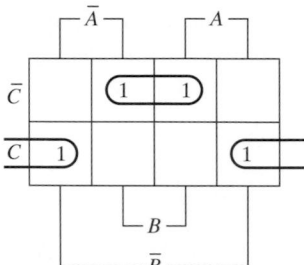

S/P $f = \overline{B}$
P/S $\overline{f} = B$
$f = \overline{B}$

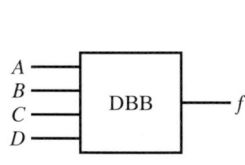

S/P $f = B\,\overline{C} + \overline{B}\,C = B \oplus C$
P/S $\overline{f} = \overline{B}\,\overline{C} + B\,C = \overline{B \oplus C}$
$f = \overline{\overline{B}\,\overline{C} + B\,C}$

FIGURE 5.18 **Sample Karnaugh Maps**

FIGURE 5.19 **Four-Variable Karnaugh Maps**

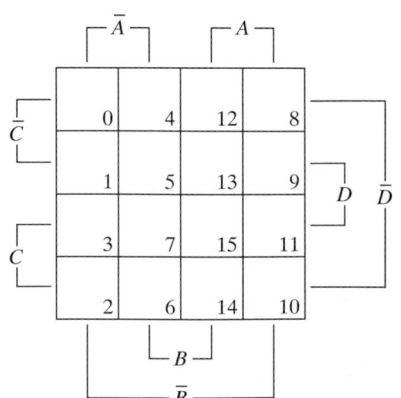

These three examples are done using Karnaugh maps labeled in different ways. By experimenting, you will develop your own preference. Figure 5.18 shows three maps and the appropriate solutions. Read each map. It becomes easy with practice!

A four-input digital black box has $2^4 = 16$ entries on a truth table. This means the Karnaugh map is twice as large. Figure 5.19 shows a four-input Karnaugh map in the different labeling schemes. The labeling is also important on the four-variable map. The m numbers are shown in each box, and adjacent boxes differ by one variable. To read the map, look for:

1. Vertical or horizontal pairs.
2. Rows or columns of four.
3. Squares of four.
4. Groups of eight.

The left and right columns of the map are next to each other (they wrap around). The top and bottom rows are next to each other. Diagonal terms do not compare. Here are some examples.

··

EXAMPLE 5.11

Design a four-input DBB in simplest form to produce $f = m(1, 5, 6, 7, 9, 13, 14, 15)$.

SOLUTION

Filling in the Karnaugh map, we obtain the map shown in Figure 5.20.

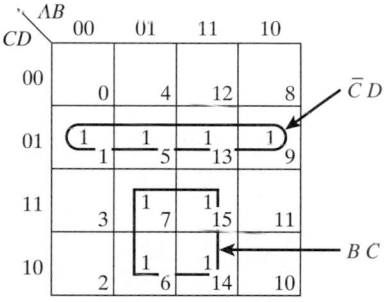

FIGURE 5.20 Karnaugh Map

Sum of Products: Terms m_1, m_5, m_{13}, and m_9 are a group of four $= \overline{C} D$; m_7, m_6, m_{15}, and m_{14} are a group of four $= B\,C$. Therefore, $f = B\,C + \overline{C}\,D$ (S/P). The result $\overline{C}\,D$ is obtained by observing that m terms 1, 5, 13, and 9 fill a row; $\overline{C}\,D$ is the common term. The common term for m's 6, 7, 14, and 15 is $B\,C$. This can be done in parts: $m_6 + m_7 = \overline{A}\,B\,C$, $m_{14} + m_{15} = A\,B\,C$. The two pairs reduce to $B\,C$.

Product of Sums: Reading zeros from the map: m_0, m_4, m_8, and m_{12} form a row of four $= \overline{C}\,\overline{D}$; m_2, m_3, m_{10}, and m_{11} form a square $= \overline{B}\,C$. Therefore $\overline{f} = \overline{C}\,\overline{D} + \overline{B}\,C$ and $f = \overline{\overline{C}\,\overline{D} + \overline{B}\,C}$.

····· C/C++ HELPER ·····

The K3 program on the companion CD-ROM solves three-input Karnaugh maps by exhaustively testing the map for groups of 4 and 2. There are six groups of four and twelve groups of two. The Boolean equation for the map is displayed as the result. A sample execution looks like this:

```
C> K3
Enter line 0 ---> 1
Enter line 1 ---> 0
Enter line 2 ---> 1
Enter line 3 ---> 0
Enter line 4 ---> 1
Enter line 5 ---> 1
Enter line 6 ---> 0
Enter line 7 ---> 0
F = A'C' + AB'
```

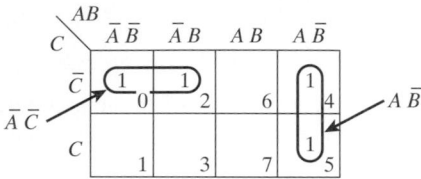

The equation corresponds to $F = \overline{A}\,\overline{C} + A\,\overline{B}$. The ' after a variable is used to represent inversion. This answer is the same as that for the map shown in Figure 5.15 (reproduced above).

The sample execution for the map in Figure 5.17 (reproduced below) is as follows:

```
C> K3
Enter line 0 ---> 0
Enter line 1 ---> 1
Enter line 2 ---> 1
Enter line 3 ---> 1
Enter line 4 ---> 0
Enter line 5 ---> 0
Enter line 6 ---> 1
Enter line 7 ---> 1
F = B + A'C
```

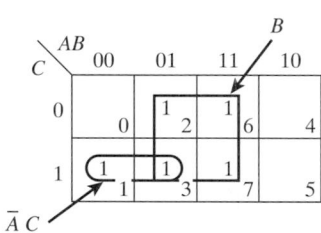

Once again, the answer is correct.

Try modifying K3 to solve four-input maps, or accept the input data in a different format (such as an entire row at a time).

EXAMPLE 5.12

Design a four-input DBB in simplest form to produce $f = m(0, 2, 5, 8, 9, 10, 13, 15)$.

SOLUTION

Using an alternate form of the map, we obtain the map in Figure 5.21.

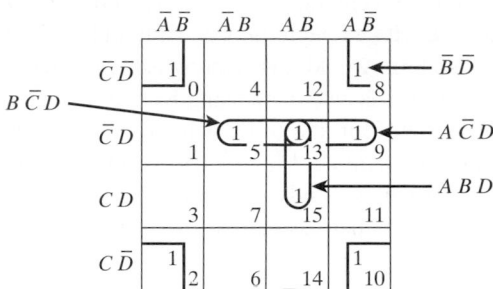

FIGURE 5.21   Karnaugh Map

Sum of Products: Square: $m_0, m_2, m_8, m_{10} = \overline{B}\,\overline{D}$. Pairs: $m_5, m_{13} = B\,\overline{C}\,D$, $m_{13}, m_{15} = A\,B\,D$, $m_{13}, m_9 = A\,\overline{C}\,D$. Thus $f = \overline{B}\,\overline{D} + B\,\overline{C}\,D + A\,B\,D + A\,\overline{C}\,D$.

Product of Sums: Square: $m_4, m_{12}, m_6, m_{14} = B\,\overline{D}$. Pairs: $m_1, m_3 = \overline{A}\,\overline{B}\,D$, $m_3, m_7 = \overline{A}\,C\,D$, $m_3, m_{11} = \overline{B}\,C\,D$. Thus $\overline{f} = B\,\overline{D} + \overline{A}\,\overline{B}\,D + \overline{A}\,C\,D + \overline{B}\,C\,D$ and $f = \overline{B\,\overline{D} + \overline{A}\,\overline{B}\,D + \overline{A}\,C\,D + \overline{B}\,C\,D}$.

EXAMPLE 5.13

Design a four-input DBB in simplest form to produce $f = m(1, 3, 9, 11, 12, 14)$.

SOLUTION

See Figure 5.22.

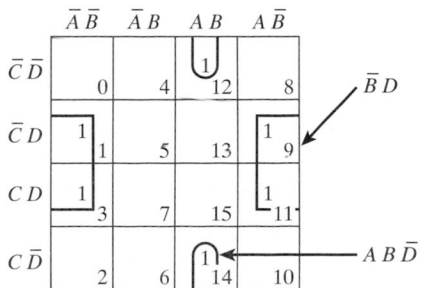

FIGURE 5.22   Karnaugh Map

Sum of Products: Square: $m_1, m_3, m_9, m_{11} = \overline{B}\,D$. Pair: $m_{12}, m_{14} = A\,B\,\overline{D}$. Thus $f = \overline{B}\,D + A\,B\,\overline{D}$.

Product of Sums: Square: $m_5, m_7, m_{13}, m_{15} = B\,D$. Column: $m_4, m_5, m_6, m_7 = \overline{A}\,B$. Square: $m_0, m_2, m_8, m_{10} = \overline{B}\,\overline{D}$. Thus $\overline{f} = B\,D + \overline{A}\,B + \overline{B}\,\overline{D}$ and $f = \overline{B\,D + \overline{A}\,B + \overline{B}\,\overline{D}}$.

5.6 LARGE KARNAUGH MAPS

We have seen that in the design of a four-input digital black box Karnaugh maps are very useful in determining which terms differ by only one element. This approach is a great improvement over Boolean reduction and the labor of figuring out which terms to compare. The map shows directly which terms compare by their positions relative to one another. The object is always to produce a simplified logic circuit. Overcomparing is always a danger. If any term is used more than once, it must always be with a lone term. Once a term has been used once, it need not be used again, unless a lone term would miss comparison otherwise.

Designing five- and six-input digital black boxes is also done using Karnaugh maps. To do this, the 4×4 map just studied is used two or four at a time. After six inputs it is best to find a computer to do the simplification. For example, Electronics Workbench can simplify Boolean expressions with up to eight variables. Study question 5.26 illustrates an example solution. Many inputs and complex logic equations can be solved using computer programs written for that purpose.

Consider the design of a five-variable DBB. Such a box would require a long truth table and would have $2^N = 2^5 = 32$ different input combinations. Using a pair of 4×4 Karnaugh maps solves this problem with one 4×4 for variable A and the other 4×4 map for variable $\bar{A}$. The two maps are envisioned as being one on top of the other to give depth.

....

EXAMPLE 5.14

Design a five-input DBB for $f = m(1, 5, 8, 9, 13, 22, 23, 24, 30, 31)$.

SOLUTION

First a look at the truth table, then the Karnaugh map, as shown in Figure 5.23.

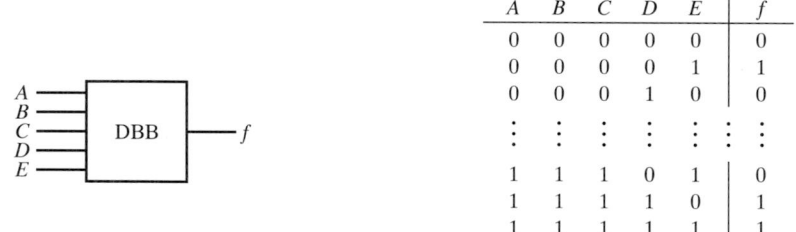

A	B	C	D	E	f
0	0	0	0	0	0
0	0	0	0	1	1
0	0	0	1	0	0
⋮	⋮	⋮	⋮	⋮	⋮
1	1	1	0	1	0
1	1	1	1	0	1
1	1	1	1	1	1

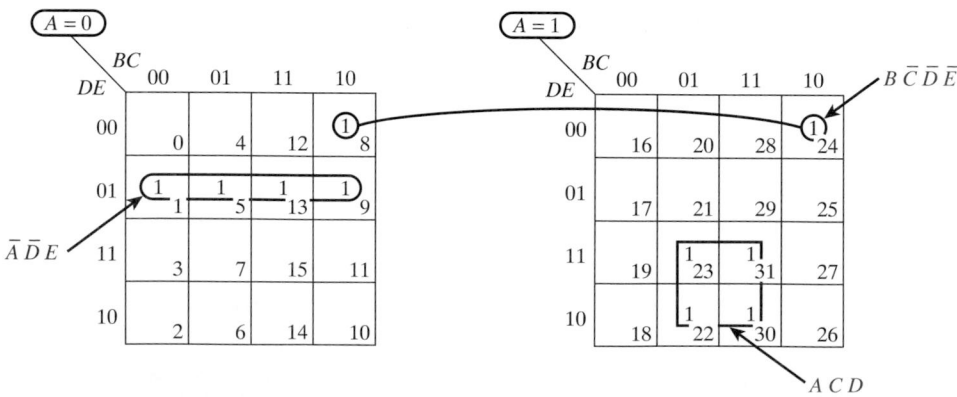

FIGURE 5.23   Black Box Design: Five Inputs

Sum of Products: Fours: m_1, m_5, m_9, $m_{13} = \overline{A}\,\overline{D}\,E$, m_{22}, m_{23}, m_{30}, $m_{31} = A\,C\,D$. Pair: m_8, $m_{24} = B\,\overline{C}\,\overline{D}\,\overline{E}$. The result is $f = \overline{A}\,\overline{D}\,E + A\,C\,D + B\,\overline{C}\,\overline{D}\,\overline{E}$.

•••

Another method useful for working with a large number of inputs is the ***Quine-McCluskey method.*** This technique does not involve Karnaugh maps, but instead concentrates on grouping zeros and ones together in certain ways to find a minimal set of input variables. Figure 5.24 demonstrates the steps involved in solving Example 5.14 using

FIGURE 5.24   Using the Quine-McCluskey Method

$f = m(1, 5, 8, 9, 13, 22, 23, 24, 30, 31)$

1	00001
5	00101
8	01000
9	01001
13	01101
22	10110
23	10111
24	11000
30	11110
31	11111

1	00001
8	01000
5	00101
9	01001
24	11000
13	01101
22	10110
23	10111
30	11110
31	11111

1	00001
8	01000
5	00101
9	01001
24	11000
13	01101
22	10110
23	10111
30	11110
31	11111

1/5	00X01
1/9	0X001
8/9	0100X
8/24	X1000
5/13	0X101
9/13	01X01
22/23	1011X
22/30	1X110
23/31	1X111
30/31	1111X

(*a*) Group according to number of 1s   (*b*) Compare terms and reduce

1/5	00X01
1/9	0X001
8/9	0100X
8/24	X1000
5/13	0X101
9/13	01X01
22/23	1011X
22/30	1X110
23/31	1X111
30/31	1111X

1/5/9/13	0XX01
8/9	0100X
8/24	X1000
5/13	0X101
9/13	01X01
22/23/30/31	1X11X

(*c*) Find the prime implicants

Prime Implicant	1	5	8	9	13	22	23	24	30	31	
0XX01	✓	✓		✓	✓						← Must have
0100X			✓	✓							
X1000			✓					✓			← Must have
0X101		✓			✓						
01X01				✓	✓						
1X11X						✓	✓		✓	✓	← Must have

(*d*) List minterms covered by prime implicants

$$\begin{array}{cccc} & ABCDE & ABCDE & ABCDE \\ f = & 0XX01 + & X1000 + & 1X11X \end{array}$$

$$f = \quad \overline{A}\overline{D}E \quad + \quad B\overline{C}\overline{D}\overline{E} + \quad ACD$$

(*e*) Find the equation

Quine-McCluskey. Initially, the binary codes for each minterm (there are ten) are grouped according to the number of ones they contain. This is illustrated in Figure 5.24(a). Next, the codes from one group are compared with those in the next group. As usual, only patterns that are different by one variable may be reduced. For example, the "1" and "5" patterns (00001 and 00101) make a reduced pattern of 00x01, where "x" means *don't care.* These comparisons take place in Figure 5.24(b) and Figure 5.24(c). The final set of six terms are called ***prime implicants.*** One or more of the prime implicant terms will be used in the solution.

In Figure 5.24(d) the minterms associated with each prime implicant are checked off. The 0xx01, x1000, and 1x11x terms are all required in the solution, since they specify unique minterm groups. For instance, the x1000 term is required, since it is the only way to include the "24" minterm in the solution. The other prime implicants (0100x, 0x101, and 01x01) are not needed, since the minterms they specify are already covered by the required terms.

Figure 5.24(e) shows the solution equation, the same one found via Karnaugh mapping in Example 5.14. You are encouraged to use the Quine-McCluskey method to solve some of the previous examples, to get a better feel for how it works.

The next example gives a problem and the answer. It is left to you to draw the Karnaugh maps and verify the result.

• •

EXAMPLE 5.15

Design a five-variable DBB for $f = m(0, 2, 3, 4, 8, 12, 14, 15, 16, 20, 24, 28, 30, 31)$.

SOLUTION

$f = \overline{D}\,\overline{E} + B\,C\,D + \overline{A}\,\overline{B}\,\overline{C}\,D$.

Duplicate your solution using the Quine-McCluskey method.
• •

5.7 PRACTICAL APPLICATIONS

In this section we will examine several applications of logic design that require us to use the techniques introduced in this chapter. You are encouraged to work each design fully.

Design of the BCD to Seven-Segment Readout Decoder/Driver

The seven-segment readout (Figure 5.25) uses light-emitting diodes arranged in a figure eight to represent digits. Such displays come with the LED anodes all tied together (common anode-type display) or with all cathodes tied together (common cathode-type display).

The segments are neatly identified with letters A–G. Often there is also a decimal point or colon as an additional indicator. Typically, such a display will require current-limiting resistors in series with each LED to protect the device from too much current. These are usually in the 150–220Ω range for use with a 5-V power supply. A more precise value would be determined from the data sheet for the readout device.

Since the information in a digital circuit is frequently BCD encoded, it is necessary to design a decoder that converts 4-bit BCD into the form the display requires. For instance,

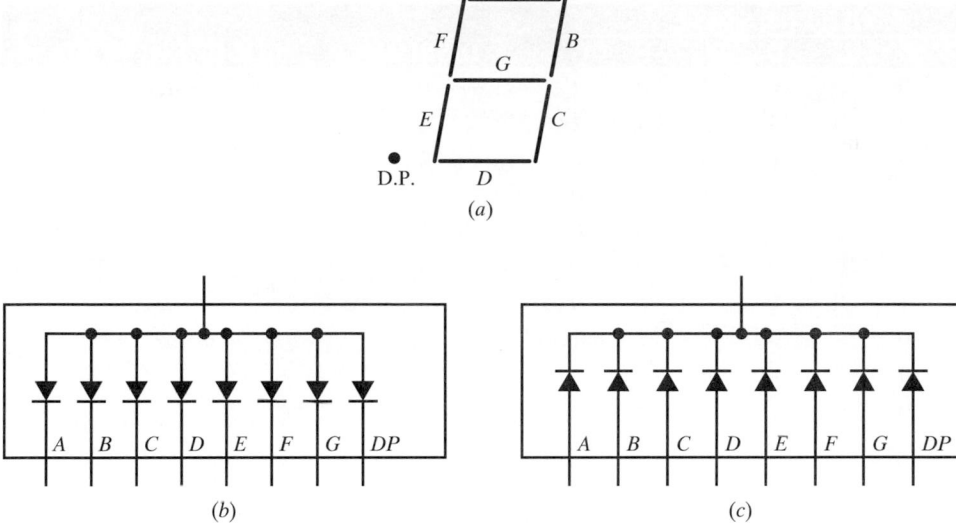

FIGURE 5.25 (*a*) **The Seven-Segment Display** (*b*) **Common Anode** (*c*) **Common Cathode**

if the BCD value is 0110 (six), then segments *C, D, E, F,* and *G* must be illuminated on the seven-segment readout. Figure 5.26 shows a 7490 decade counter driving a digital black box, which in turn drives the LED display. The DBB must provide a ground to the LED cathodes for the proper illumination of the display. Table 5.2 shows the truth table that the DBB must follow for correct operation of the display. We will suggest a DBB design for your consideration and study.

From Table 5.2 we observe that a seven-output DBB is required. The purpose of each output is to ground the appropriate segment at the proper BCD input. It is necessary then to use Karnaugh mapping to reduce each of the seven logic functions to build the circuitry for the digital black box. For practice, you should manually complete this prodigious task. However, we now present the 7447 BCD to seven-segment decoder driver that

FIGURE 5.26 Decade Counter Driving a Decoder

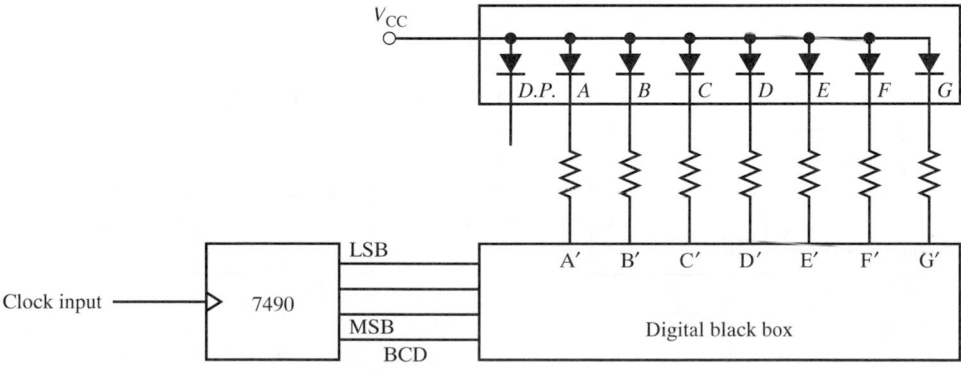

TABLE 5.2 Decoder Truth Table and Logic Equations

INPUT CODE				OUTPUT STATE							DISPLAY	LOGIC EQUATIONS*
d	c	b	a	A'	B'	C'	D'	E'	F'	G'		
0	0	0	0	0	0	0	0	0	0	1	0	$A' = m(1, 4, 6)$
0	0	0	1	1	0	0	1	1	1	1	1	$B' = m(5, 6)$
0	0	1	0	0	0	1	0	0	1	0	2	$C' = m(2)$
0	0	1	1	0	0	0	0	1	1	0	3	$D' = m(1, 4, 7, 9)$
0	1	0	0	1	0	0	1	1	0	0	4	$E' = m(1, 3, 4, 5, 7, 9)$
0	1	0	1	0	1	0	0	1	0	0	5	$F' = m(1, 2, 3, 7)$
0	1	1	0	1	1	0	0	0	0	0	6	$G' = m(0, 1, 7)$
0	1	1	1	0	0	0	1	1	1	1	7	
1	0	0	0	0	0	0	0	0	0	0	8	
1	0	0	1	0	0	0	1	1	0	0	9	

*$m(10, 11, 12, 13, 14, 15)$ are "don't care" states. The 7490 will not allow these states.

is available commercially to do the job, thus saving us the design of such a circuit. Figure 5.27 shows the 7447 driving a LED seven-segment readout.

Application of a BCD 4-bit count to the BCD input lines causes the proper digit to light on the display. The lamp test input is used to test the segments of the display. When lamp test is grounded, all segments are illuminated. The ripple blanking input ($\overline{RBI}$) causes the figure zero to be blanked (not illuminated) when a BCD zero (0000) is applied to the input. This is useful in multidigit displays in suppressing leading zeros of a number (e.g., 456, not 000000456). The ripple blanking output ($\overline{RBO}$) is used to drive the $\overline{RBI}$ line of an adjacent display.

Seven-segment displays have found very common usage in the digital and microprocessor industries. Since displays are frequently used in multiples, it is useful to observe the connection of several displays. Only one 7447-type decoder chip is necessary in this type of application. All the seven-segment lines (A'–G') are connected in common

FIGURE 5.27 7447 Driving a MAN1 Display

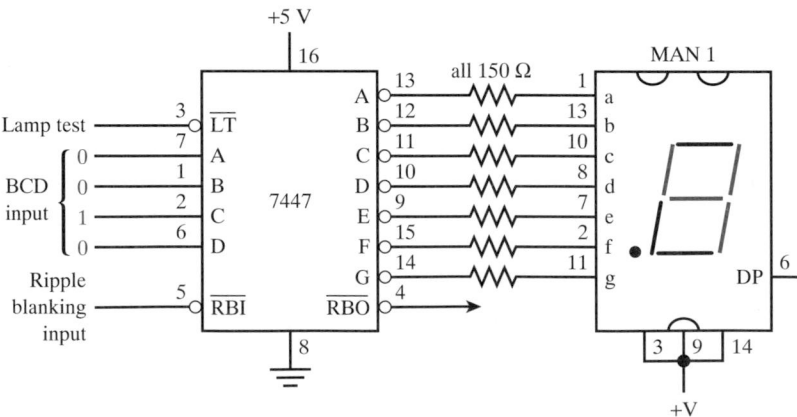

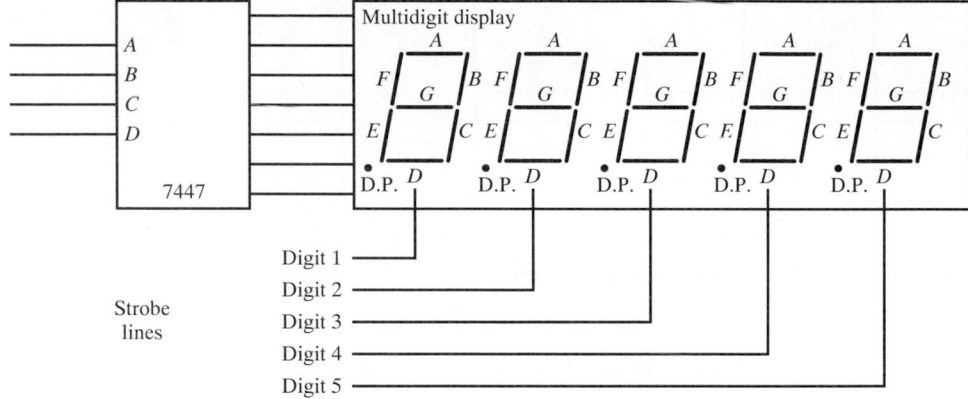

FIGURE 5.28   Using a Single Decoder to Drive Multiple Digits

as shown in Figure 5.28. The individual displays are strobed (pulsed) in sequence at a rate fast enough to elude detection by the human eye. The display will appear to be continuously illuminated if the displays are strobed (switched on) at a rate that exceeds the persistence of the human eye (approximately 30 pulses/sec). The data sent to the 7447 BCD inputs must of course be synchronized with the switching of the strobe lines. Since the strobe lines are the individual power lines for single displays, they must be able to supply sufficient current to illuminate the display. The timing diagram in Figure 5.29 shows a set of sequential pulses that could drive the five-digit display illustrated. As each digit is turned on, the appropriate seven-segment information is provided to that one digit. Since the other digits are not illuminated at this time, it does not matter that they get the seven-segment data too. This type of display is referred to as a *multiplexed* display. Each display shares the seven lines from a single decoder/driver circuit. This type of operation reduces the number of wires that are needed to operate a multidigit display.

The Design of a 2-Bit Comparator

A comparator is a digital circuit that examines the binary data on two sets of inputs, and outputs a signal indicating if the data values are equal. Figure 5.30 shows the operation of a 2-bit comparator. The EQU output goes high when the input numbers are the same. Since there are a total of sixteen input combinations, we can use a Karnaugh map to help

FIGURE 5.29   Strobed Digit Lines for a Multidigit Display

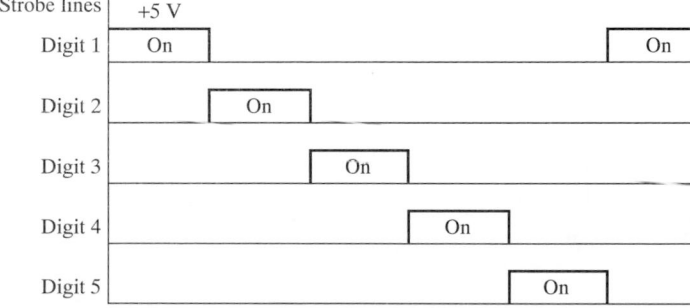

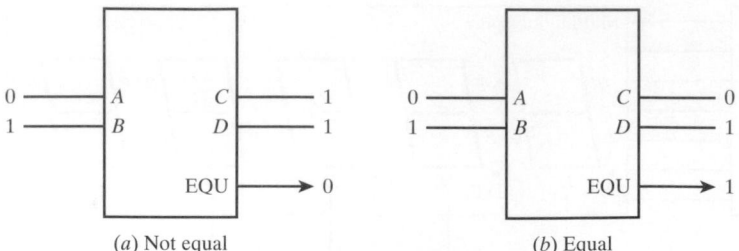

FIGURE 5.30   Operation of a Comparator

us with the design. A one must be placed in the map anywhere the row variables equal the column variables. The map and solution are shown in Figure 5.31. Without seeing the schematic, we can determine the gate cost from the equation. Four inverters, four 4-input AND gates, and a 4-input OR gate are required.

If we put a few Boolean identities and manipulations together, we are able to re-arrange and simplify the comparator equation significantly. Figure 5.32 shows how this is done.

By clever factoring, we are able to find exclusive NOR operations together with a single AND function. The three gates required for the 2-bit comparator provide a large savings over the original nine gates.

FIGURE 5.31   Karnaugh Map for 2-Bit Comparator

CD \ AB	00	01	11	10
00	1			
01		1		
11			1	
10				1

$$f = \overline{A}\,\overline{B}\,\overline{C}\,\overline{D} + \overline{A}B\overline{C}D + ABCD + A\overline{B}C\overline{D}$$

FIGURE 5.32   Rearranging the Comparator Equation

$$f = \overline{A}\,\overline{B}\,\overline{C}\,\overline{D} + \overline{A}B\overline{C}D + ABCD + A\overline{B}C\overline{D}$$

$$f = \overline{A}\,\overline{C}(\overline{B}\,\overline{D} + BD) + AC(BD + \overline{B}\,\overline{D})$$

$$f = (\overline{A}\,\overline{C} + AC)\,(\overline{B}\,\overline{D} + BD)$$

$$f = (\overline{A \oplus C})\,(\overline{B \oplus D})$$

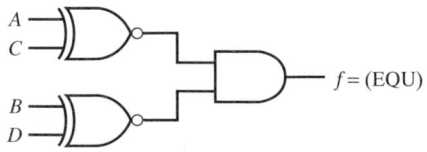

$f = (EQU)$

A full-function comparator also determines if one input number is greater than or less than the other. What type of logic circuit is required for this feature? Also, how could a 4-bit, or 8-bit comparator be made? Think of how you could *cascade* 2-bit comparators. Cascading involves connecting the output of one circuit to the input of another, in a daisy-chain format.

Designing a Code Translator

Table 5.3 shows the pairs of input and output patterns for a code translator. Three Karnaugh maps may be used to design the translator. Each map controls one bit of the output. Figure 5.33 shows the maps and their solutions. Notice the complexity of the equations for the B and C outputs. It may be worthwhile to convert the equations into all NAND or all NOR logic, to possibly save a few gates or packages. Verify for yourself that the package cost drops from four to two when this is done.

TABLE 5.3 Code Translator Operation

INPUT			OUTPUT		
A	B	C	A	B	C
0	0	0	1	1	1
0	0	1	1	1	0
0	1	0	1	0	0
0	1	1	1	0	1
1	0	0	0	0	1
1	0	1	0	1	1
1	1	0	0	1	0
1	1	1	0	0	0

FIGURE 5.33 Karnaugh Maps for Code Translator

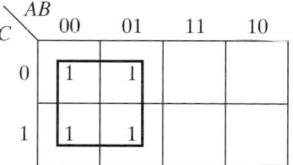

Output A $A_{\text{out}} = \overline{A}$

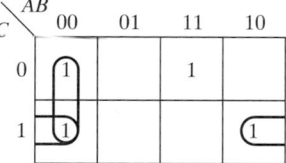

Output B $B_{\text{out}} = \overline{A}\,\overline{B} + \overline{B}C + AB\overline{C}$

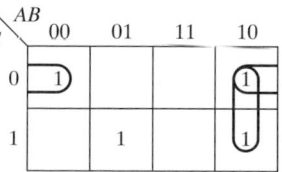

Output C $C_{\text{out}} = A\overline{B} + \overline{B}\,\overline{C} + \overline{A}BC$

5.8 AN INTRODUCTION TO PALS AND GALS

The *PAL** (programmable array logic) is a device that contains an assortment of internal gates that are connected *by the user* with a PAL programming system. The user (designer) writes a set of Boolean equations describing the logic behavior. A *PAL assembler* converts the equations into a *fuse map* that is used to program the PAL by selectively blowing internal fuses in an interconnection array. This technology is referred to as fusible-link technology. Figure 5.34 shows a simple logic circuit using fusible links. By selectively blowing the fuses, different logic functions can be created. By opening fuses F_2–F_4 and F_5–F_7, the output logic function becomes

$$\text{output} = I_1 + \overline{I_2}$$

By opening fuses F_2, F_3, F_5, and F_8, the output logic function becomes

$$\text{output} = I_1 \cdot \overline{I_2} + \overline{I_1} \cdot I_2$$

Many functions could be designed using this simple array by opening the appropriate fuses. Figures 5.35 and 5.36 show the resulting logic.

The PAL offers low-cost implementation of small logic functions as well as a simple, high-speed method for developing MSI logic functions. A PAL programmer can burn the inexpensive PAL from a set of simple Boolean equations. The PAL offers logic gates, counters, registers, comparators, and other functions typically used by the designer.

*PAL is a registered trademark of Monolithic Memories, Inc.

FIGURE 5.34   Simple Fusible Link Circuit

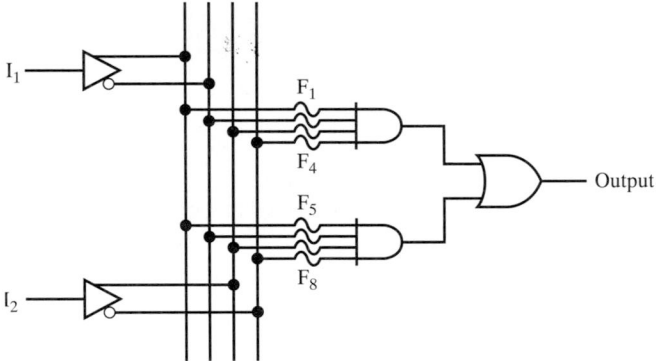

FIGURE 5.35   Simple Logic Function

$$I_1\,\overline{I_2} + \overline{I_1}\,I_2$$

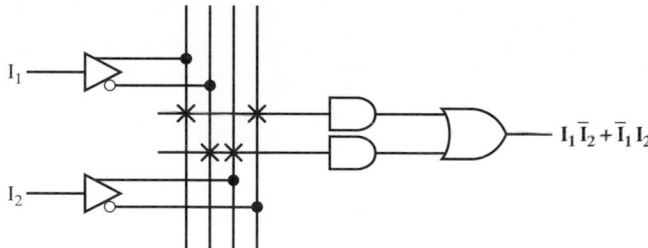

FIGURE 5.36 Implementation of Logic Function Using PAL

The PAL finds applications in memory decoders, video controllers, memory-mapped I/O, microprocessor interfacing, and security equipment. The PAL reduces hardware, increases speed, and decreases design and production time. We will learn more about PALs in Chapter 9.

Gate Array Logic (GAL)

A *gate array* is a large (> 1000) number of uncommitted transistors fashioned into logic cells. These cells can be turned into flip-flops, shift registers, counters, and more. Instead of designing a printed circuit board using TTL logic or a microprocessor-based system, you can have your entire logic requirement implemented in a "logic array." The advantage is speed not available in the microprocessor. In a microprocessor, each instruction requires time to execute; in a gate array, hard-wired logic is faster. It used to be very expensive to design a specific TTL-based system. The microprocessor made it more general and easier to change. However, certain applications require speed that the microprocessor cannot yet produce. Hard logic in an inexpensive form is now provided with the gate array.

The trick is that gate array logic does not become economically feasible until the 2000-gate level. Typical quantities range from 2000 to 200,000 arrays for a typical application. A logic array or gate array usually has an application requiring greater than 1000 gates. Too many PALs would be required to support the design. Gate arrays can have upward of 100 inputs and outputs. There can be 75 to 7500 gates per chip. Each chip performs a complete logic function. Gate arrays operate to 1 nsec/gate and 1 mW/gate. They are available in bipolar (TTL), CMOS, and ECL. Some of the literature suggests an "economic opportunity envelope" of 2000 to 200,000 units. In a gate array, the manufacturer of such devices connects the logic cells together in accordance with the purchaser's specifications. A single chip then replaces a standard PC board design using TTL- or microprocessor-based parts. Time and money are saved and a high-speed product is the result. It is a matter of designing the interconnects between standard "cells." In this way the gate array represents an economic alternative to traditional design.

5.9 TROUBLESHOOTING TECHNIQUES

Whenever you work with a Karnaugh map, it pays to study the map for a while before drawing the circles that group the ones together. Examine the map in Figure 5.37. There is a group of four ones in the center of the map. It is tempting to want to circle the group of four, but doing so actually *increases* the logic required. The other four ones in the map

CD \ AB	00	01	11	10
00			1	
01	1	1	1	
11		1	1	1
10		1		

FIGURE 5.37   A Tricky Map

can only be grouped in pairs, so there must be four pairs of ones no matter what we do. Circling the group of four is not necessary in this case, since all ones will be accounted for when the four pairs are circled. The point of this example is to show that blindly circling the biggest group of ones you can find is not always the best route to take.

Even when a Karnaugh map is used correctly, there may still be some simplification possible in the Boolean equation. This requires you to have a good understanding of the basic logic functions and Boolean properties covered in Chapter 4. We have already seen examples where an equation ended up in the form $f = A\overline{B} + \overline{A}B$, and was further simplified to $f = A \oplus B$, the exclusive OR function (a one-gate solution). A good designer knows how to spot these basic forms.

It is also a good idea to keep track of the package cost while working on a design. For instance, a new circuit is developed for a lab exercise. There are twenty-eight gates, mostly inverters, NAND gates, and OR gates. A total of nine ICs are required. Before breadboarding the circuit, look for ways to use spare gates to implement other logic functions. Two spare inverters and a spare NAND gate can make an OR function. Maybe there are nine OR gates in the design, and one can be eliminated using the spare gates. This allows an entire IC to be taken out of the design (two 7432s provide eight OR gates). This indicates that knowing how to apply DeMorgan's theorem is also a good addition to a designer's toolbox.

SUMMARY

Digital black box design is a very useful tool in the development of large-scale digital circuitry. While the numbers of inputs to a DBB can vary and become very large, simple methods are available to reduce the logic functions necessary to produce a valid output. These methods are Boolean reduction, Karnaugh mapping, and the Quine-McCluskey method. All serve to eliminate extra inputs or logic states. In all methods, similar terms are kept while different terms are eliminated.

STUDY QUESTIONS

General

1. Reduce each equation using Boolean reduction.
   (a) $f = A\,B + A\,\overline{B} + B\,C + \overline{B}\,\overline{C}$
   (b) $f = A\,B\,C + A\,\overline{B}\,C + \overline{A}\,\overline{B}\,C$
   (c) $f = A\,\overline{B}\,C + \overline{A}\,\overline{B}\,C + \overline{A}\,\overline{B}\,\overline{C}$

2. Show that each of the following can be reduced to fewer needed gates just by factoring.
   (a) $f = A\,\overline{B}\,\overline{C} + \overline{A}\,B\,\overline{C}$
   (b) $f = W\,\overline{X}\,\overline{Y} + W\,X + W\,Y$
   (c) $f = A\,\overline{B}\,\overline{D} + A\,\overline{B}\,\overline{C} + A\,\overline{B}\,D$

3. Design a three-input DBB using both sum of products and product of sums. Reduce the results using Boolean reduction.
   (a) $f = m(1, 3, 4, 6)$
   (b) $f = m(1, 3, 7)$
   (c) $f = m(2, 3, 5)$

4. Convert the results of Question 3 to both all-NAND and all-NOR logic using DeMorgan's theorem.

5. Use a three-variable Karnaugh map to design the circuitry for $f = m(0, 1, 4, 6, 7)$. Be certain to use both S/P and P/S methods.

6. Use a four-variable Karnaugh map to implement each of the following functions.
   (a) $f = m(0, 1, 4, 5, 6, 7, 10)$
   (b) $f = m(2, 4, 6, 10, 12, 14)$
   (c) $f = m(2, 3, 8, 9, 12, 13, 14)$

7. Write the simplest Boolean equation for the following Karnaugh maps.
   (a) Figure 5.38

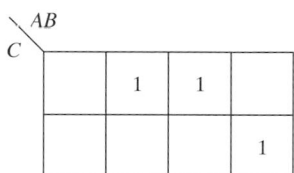

FIGURE 5.38   For Question 5.7(a)

(b) Figure 5.39

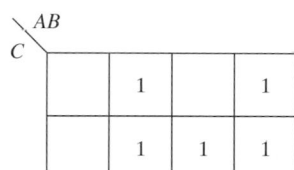

FIGURE 5.39   For Question 5.7(b)

(c) Figure 5.40

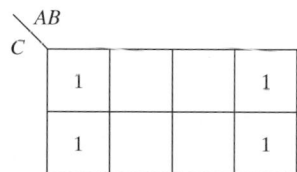

FIGURE 5.40   For Question 5.7(c)

8. Repeat Question 7 for the following maps:
   (a) Figure 5.41

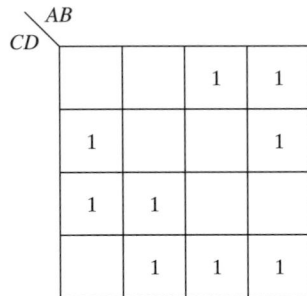

FIGURE 5.41 **For Question 5.8(a)**

(b) Figure 5.42

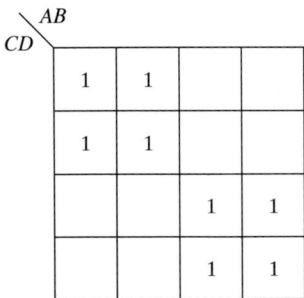

FIGURE 5.42 **For Question 5.8(b)**

(c) Figure 5.43

FIGURE 5.43 **For Question 5.8(c)**

(d) Figure 5.44

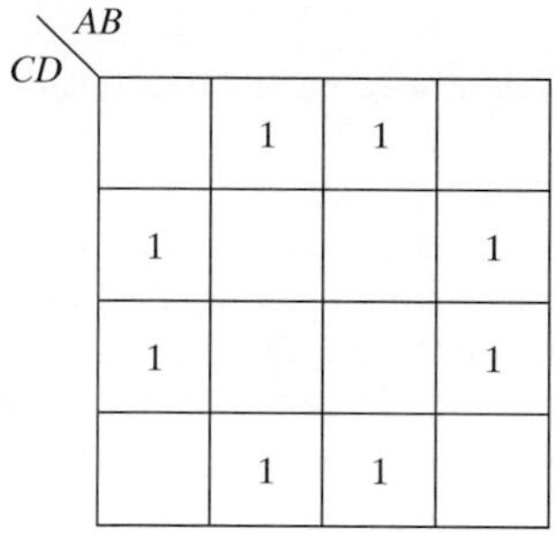

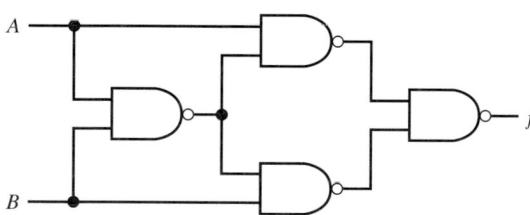

FIGURE 5.44 **For Question 5.8(d)**

9. Draw the circuitry necessary to implement the following functions in simplest form: $f = m(0, 1, 7, 15)$.

10. Convert the result of Question 9 into all-NOR logic.

11. How can an inverter be made out of a NOR gate? How can an OR gate be made from a group of NAND gates? Prove with Boolean algebra.

12. Use Boolean algebra to prove that the circuit in Figure 5.45 behaves like an exclusive OR gate.

FIGURE 5.45 **For Question 5.12**

13. Explain why $\overline{A} + A$ is always true (a one).

14. Show an example of how factoring a Boolean equation reduces the number of logic gates.

15. Factor this equation: $f = \overline{A} B C + \overline{A} \overline{B} \overline{C} + \overline{A} \overline{B} C + A B C$.

16. What are the before-factoring and after-factoring gate costs in Question 15?

17. What are the before and after package costs in Question 15?

18. How does gate cost affect package cost?

19. What is the basic technique behind the Quine-McCluskey method?

20. Find the reduced Boolean equation for each function using the Quine-McCluskey method:
    (a) $f = m(0, 1, 6, 7)$
    (b) $f = m(3, 7, 11, 15)$
    (c) $f = m(1, 2, 3, 8, 9, 15, 20, 22, 26, 31)$

21. What is the logic function of the Karnaugh map in Figure 5.37?

22. Draw a six-input Karnaugh map (use three inputs on each side). Label all row and column bit values and the truth table line numbers in all 64 map locations.

23. Explain how spare gates can be used to reduce the package cost.

24. What types of logic functions can be simulated by three spare NAND gates and one spare NOR gate?

25. Does a Karnaugh map always give the simplest possible Boolean function?

Electronics Workbench

26. Electronics Workbench contains a virtual instrument called the "Logic Converter." Its symbol looks like Figure 5.46(a). Clicking on the symbol opens up the instrument, which is shown in Figure 5.46(b). To design a digital black box, first click on the number of inputs you require (*A*, *B*, and *C* have been clicked in Figure 5.46(b) for a three-input DBB).

(*a*) Logic converter symbol

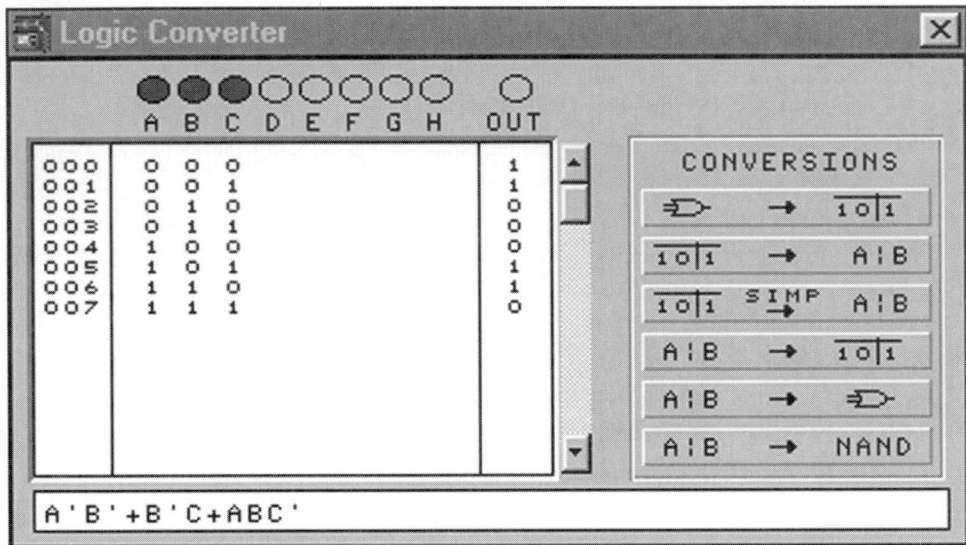

(*b*) Instrumental details

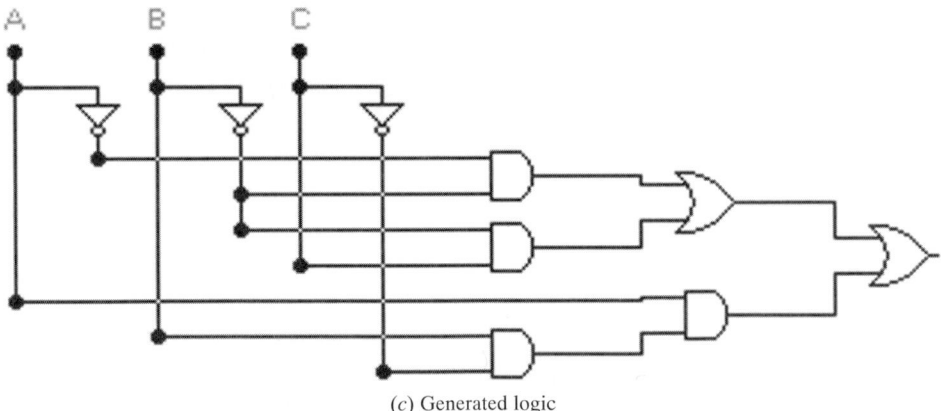

(*c*) Generated logic

FIGURE 5.46   For Question 5.26

Next, click on any bit in the Out column and enter the 0 or 1 value for that line of the truth table. Then, to find the Boolean equation for the truth table, click the conversion button marked "SIMP" and read the equation in the small window at the bottom of the dialog box.

To automatically generate a digital circuit from the equation, click the fifth button (second from bottom) and you will get a circuit similar to that shown in Figure 5.46(c). The logic converter does not provide the simplest hardware circuit because only two-input gates are used.

Use the Logic Converter to solve the truth tables in Figures 5.1 and 5.9.

27. Use the Logic Converter to solve for the equations needed in Study Questions 15 and 20.

Programmable Logic

28. How does a PAL reduce the number of ICs necessary for a logic function?

29. What is a fuse map?

30. What is a PAL assembler?

6

FLIP-FLOPS

When finished with this chapter, you should be able to:

1. Identify and describe the operation of the following 1-bit memory devices: $\overline{R}$-$\overline{S}$ latch, type D flip-flop, J-K flip-flop.

2. Read and interpret the truth table for standard 7400 series flip-flops.
3. Draw a timing diagram for a flip-flop.
4. Identify the basic shift register connection and describe its operation.

Keep the following questions in mind and try to answer them when you have completed the chapter.

1. How does a flip-flop store a zero or a one?
2. What do SET and RESET mean?

3. What do synchronous and asynchronous mean?
4. What is an illegal state in a flip-flop?
5. How are combinations of flip-flops able to form counters?
6. How does an edge-detector work?

6.1 INTRODUCTION

The purpose of this chapter is to acquaint you with flip-flops and circuits composed of flip-flops (such as counters and shift registers). ***Flip-flops*** are used in digital electronic circuits to store temporary results, to divide frequencies, and to rotate binary words, to name a few applications. While counters and shift registers are very useful, to understand them you must first be familiar with the operation of the flip-flop.

6.2 LEVEL-SENSITIVE INPUTS VS. EDGE-SENSITIVE INPUTS

We encounter two different types of flip-flops: those with level-sensitive enable inputs and those with edge-sensitive clock inputs. Figure 6.1 provides the timing details. Both types of flip-flops can be used for the same purposes; only the system timing and control circuitry are different, such as data setup and hold times, and other parameters we will encounter. An example is shown in Figure 6.2, where a digital sequencer cycles between four states, and an arithmetic pipeline starts a new operation every negative edge. The sequencer performs four non-overlapping operations in a repetitive manner. Each operation is performed when its control signal is low. The arithmetic pipeline performs operations in parallel, transferring data between independent logic units on every negative edge of the system clock.

Try to keep these differences in mind as you study the flip-flops presented in the following sections.

FIGURE 6.1 Two Types of Flip-Flop Control Signals

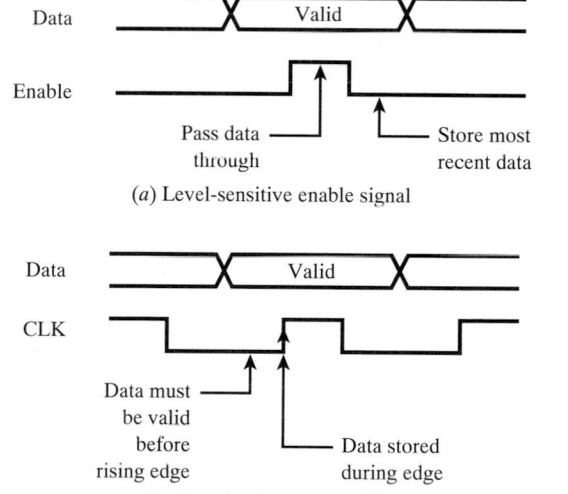

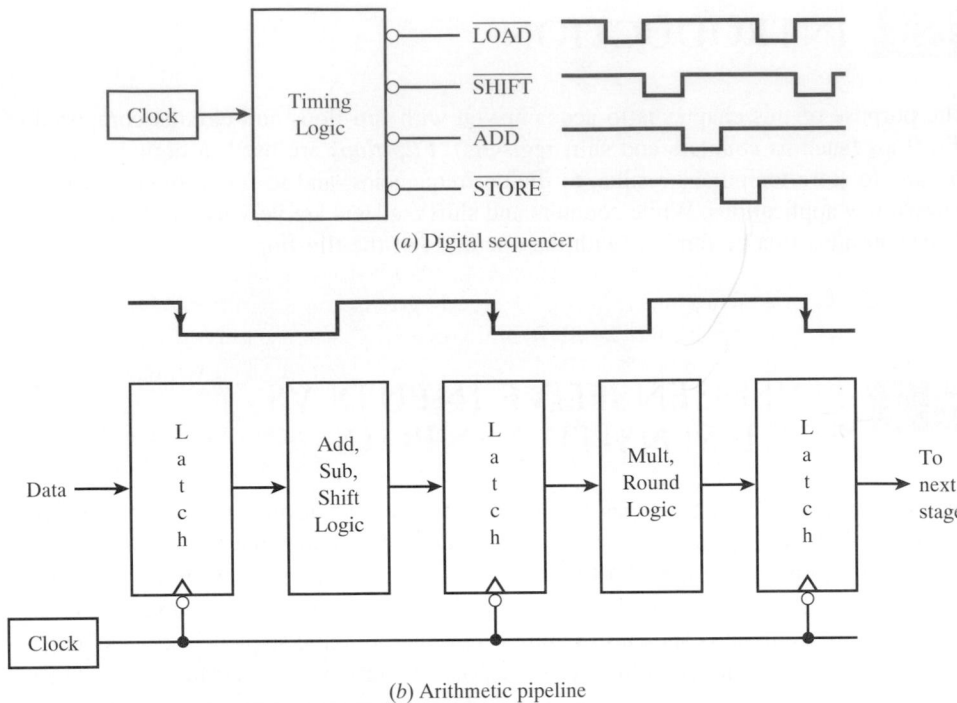

(a) Digital sequencer

(b) Arithmetic pipeline

FIGURE 6.2 Comparing Two Digital Systems

6.3 THE R-S LATCH

In Chapter 4 we saw how a switch was debounced by a simple circuit employing two-input NAND gates. Reproduced here in Figure 6.3, this circuit, called a **DC Latch,** is more commonly known as an $\overline{R}$-$\overline{S}$ latch. The term **latch** refers to the nature of the circuit, which "latches" into a 0 level or a 1 level until told to change. In an $\overline{R}$-$\overline{S}$ latch, there are two-inputs, $\overline{R}$ and $\overline{S}$, and two outputs, Q and $\overline{Q}$. The bars over the $\overline{R}$ and $\overline{S}$ inputs indicate that the reset and set functions occur when either $\overline{R}$ or $\overline{S}$ is taken low. To reset the latch, place

FIGURE 6.3 Latch Used as a Switch Debouncer

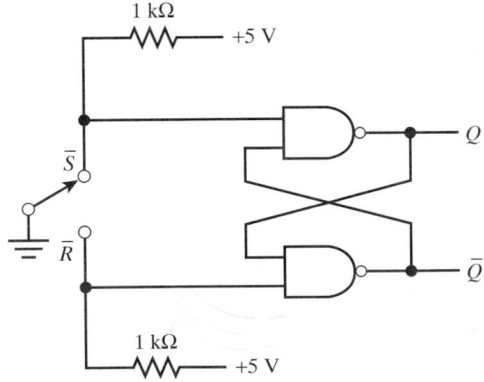

a zero on $\overline{R}$. To set the latch, place a zero on the $\overline{S}$ input. When not being used, $\overline{R}$ and $\overline{S}$ are normally left high.

The set and reset states of the latch are defined as follows:

① When Q is high the latch is "SET."

② When Q is low the latch is "RESET."

The latch can be used as a 1-bit memory device to remember one bit of a binary number. "SET" can be used to represent a stored one, "RESET" can be used to represent a stored zero. Since the latch has only these two states, it is ideal for storing a binary bit.

To store a one in an $\overline{R}$-$\overline{S}$ flip-flop we simply ground the $\overline{S}$ input. The $\overline{S}$ input is then returned high, and the flip-flop is "SET" and a one is considered to be stored. To store a zero in an R-S flip-flop we ground the $\overline{R}$ input. The $\overline{R}$ input is then returned high, and the flip-flop is "RESET" and a zero stored.

As we will see, all latches and flip-flops have some kind of input and almost always have the Q and $\overline{Q}$ outputs. By Q and $\overline{Q}$ we means that the outputs are complements of each other—when one output is a zero, the other is a one, and vice versa. Therefore, if the latch is SET, Q is 1 and $\overline{Q}$ is 0.

It is very useful to have access to both outputs when designing digital circuits because other TTL gates often require both signals. The outputs of a latch are *never* the same (either both one or both zero)! It is important to remember that this condition never occurs in latch circuitry. If two 1s or two 0s are found, the latch in use is operating incorrectly.

Let us take a look at the function table for the $\overline{S}$-$\overline{R}$ latch (Figure 6.4(a)) and see if we can gain an understanding of its operation. You will notice that when the $\overline{S}$ and $\overline{R}$ inputs are both high (H), the output indicates "No Change." This means that the output Q is in the same state or level when both inputs are high as the state it was in when the inputs were different. The Electronics Workbench circuit in Figure 6.4(b) shows what happens

FIGURE 6.4 $\overline{S}$-$\overline{R}$ **Latch**

$\overline{S}$	$\overline{R}$	Q	$\overline{Q}$
0	0	Illegal	
0	1	1	0
1	0	0	1
1	1	No change	

(*a*) Truth table

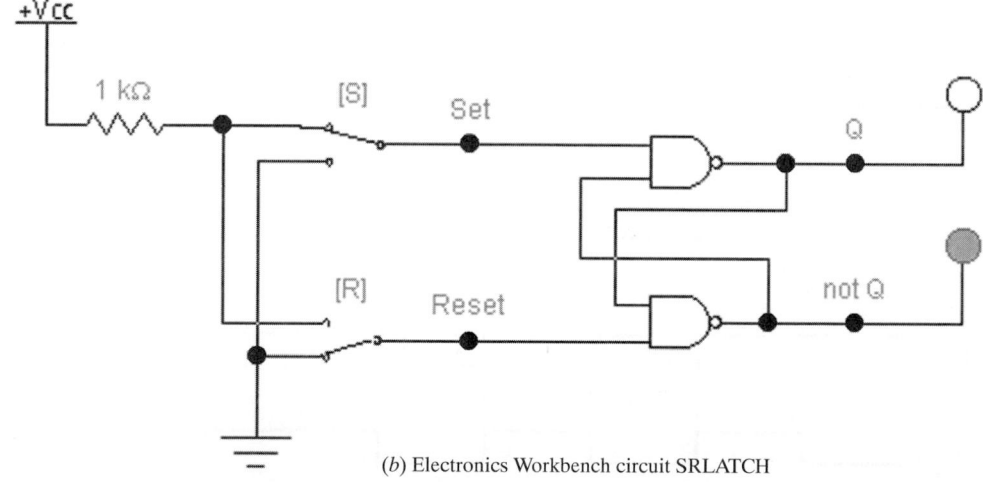

(*b*) Electronics Workbench circuit SRLATCH

when the $\overline{R}$ input (Reset) is low. Use the S and R keys to control the inputs. The outputs will follow the truth table.

To prove this, let's trace through the operation of a single $\overline{S}$-$\overline{R}$ latch, assuming the following initial conditions: The $\overline{S}$ and $\overline{R}$ inputs are at zero and one, respectively. From the truth table we know that the Q output should be high. Figure 6.5 shows this.

We know from Chapter 4 that the output of a NAND gate is a one whenever one or both inputs are low. In our example the $\overline{S}$ input is driving the input of NAND gate A low, producing a high at the output (which is also the Q output). This output also feeds one input of NAND gate B. Since the other input ($\overline{R}$) is also a one, the output of the gate becomes zero (this output is $\overline{Q}$). From this *steady state* condition, let's change the $\overline{S}$ input back to a logic "1" and see what happens (see Figure 6.6). Since we had a zero on input two of NAND gate A, changing input one ($\overline{S}$) from a zero to a one does not affect the output (Q). This is an example of the "No Change" principle. If we had used the $\overline{R}$ input instead, the Q output would have stayed low, just as the truth table predicts.

One last point before we move on to the next type of flip-flop. Notice that the first line of the truth table shows $\overline{S}$ and $\overline{R}$ both low, and the output as "Illegal." Both Q and $\overline{Q}$ will be high at the same time during the "Illegal" condition. We cannot predict which one will go low when a valid input state is restored. This condition exists in most latches and is a result of the DC characteristics of the device. Most digital circuits are designed to ensure

FIGURE 6.5 Operation of an $\overline{S}$-$\overline{R}$ Latch

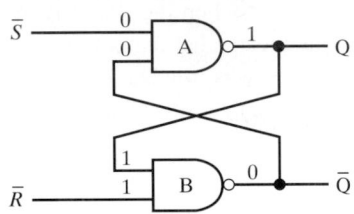

FIGURE 6.6 $\overline{S}$-$\overline{R}$ Latch with Both Inputs High

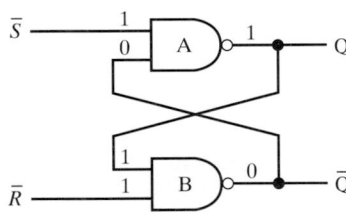

FIGURE 6.7 Sample $\overline{S}$-$\overline{R}$ Latch Timing

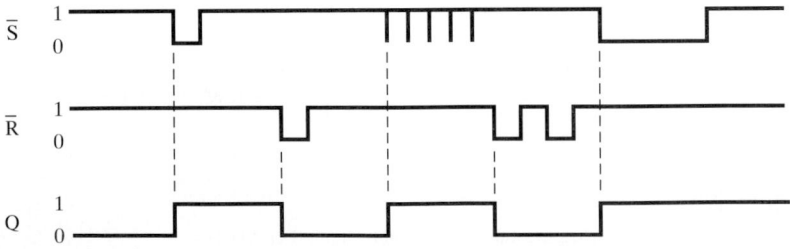

that they will never enter that condition. In other words, the circuit driving the latch will never apply two low inputs to the $\overline{S}$-$\overline{R}$ latch.

Another way to look at the $\overline{S}$-$\overline{R}$ latch is through a timing analysis. Figure 6.7 shows how signals on the $\overline{S}$ and $\overline{R}$ inputs affect the Q output. Notice that Q goes high or low the instant $\overline{S}$ or $\overline{R}$ goes low. The $\overline{S}$-$\overline{R}$ latch is level sensitive, so Q will remain in its current state until changed by activity on the $\overline{S}$ and $\overline{R}$ inputs.

6.4 TYPE D FLIP-FLOP (AND LATCH)

Before discussing the D flip-flop (and D latch), let us look at a few improvements to the basic $\overline{S}$-$\overline{R}$ latch. The outputs of the $\overline{S}$-$\overline{R}$ latch are controlled by the $\overline{S}$ and $\overline{R}$ inputs. Any time these inputs change, the output may be affected.

There are times when we would like to ignore the activity on the $\overline{S}$ and $\overline{R}$ inputs. In effect, we want to have control over when $\overline{S}$ and $\overline{R}$ are allowed to update the output of the latch. Figure 6.8 shows how two NAND gates added to the $\overline{S}$-$\overline{R}$ latch provide an additional control input called EN (for enable). When EN is low, the activity on S and R are ignored. The output remains in its current state.

When EN is high, the S and R inputs are allowed to make changes to the output state. The output state at the time EN goes low again is stored and does not change until EN goes back high at a later time. An additional consequence of adding the EN signal is that the S and R signals operate on different logic levels from the $\overline{S}$ and $\overline{R}$ inputs of the $\overline{S}$-$\overline{R}$ latch. When EN is high the NAND gates act like inverters, turning S into $\overline{S}$, and R into $\overline{R}$.

Now, why bother with the "Illegal" state at all? For that matter, if we get rid of the "No Change" state, there are only two lines in the truth table that perform any work on Q and $\overline{Q}$. Figure 6.9 shows how an inverter is used to limit the patterns received by S and

FIGURE 6.8 Adding an Enable Input to the $\overline{S}$-$\overline{R}$ Latch

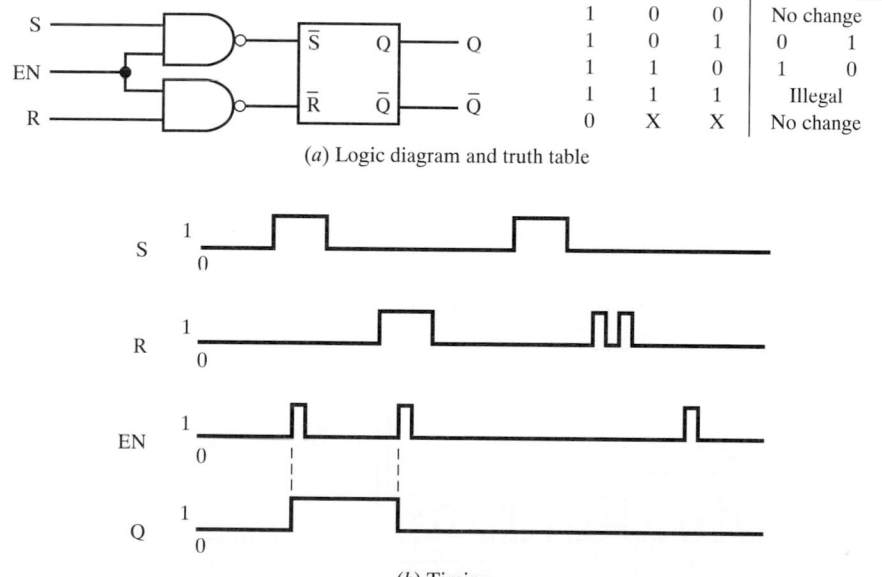

EN	S	R	Q	$\overline{Q}$
1	0	0	No change	
1	0	1	0	1
1	1	0	1	0
1	1	1	Illegal	
0	X	X	No change	

(a) Logic diagram and truth table

(b) Timing

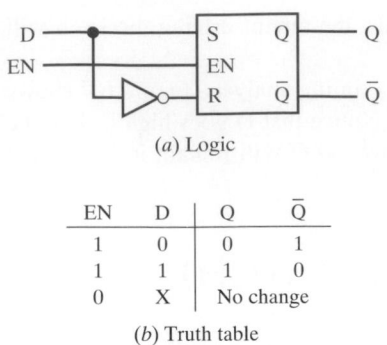

(a) Logic

EN	D	Q	$\overline{Q}$
1	0	0	1
1	1	1	0
0	X	No change	

(b) Truth table

FIGURE 6.9   Adding a Data Input to the Enabled S-R Latch

R. When the D input is low, Q will go low when EN is active. If D is high, Q goes high when enabled. This type of latch is used to store a zero or a one only and is called a D-type latch. Figure 6.10 shows the pinpoint of the 7475 Quad D latch. Two 7475s can be used to make an 8-bit latch. A pre-packaged 8-bit D latch (also called an *octal latch*) is found in the 74373. This device is shown in Figure 6.11. Note that there are no $\overline{Q}$ outputs available on the 74373. This is one reason why in some cases you may want to use two 7475s instead of one 74373.

The enable input adds some degree of additional control over the S-R and D latches. The problem with the enable input is that, if left high too long, many different S and R state changes may be received. It would be better if the active enable signal only lasted for a few

FIGURE 6.10   7475 Quad D Latch

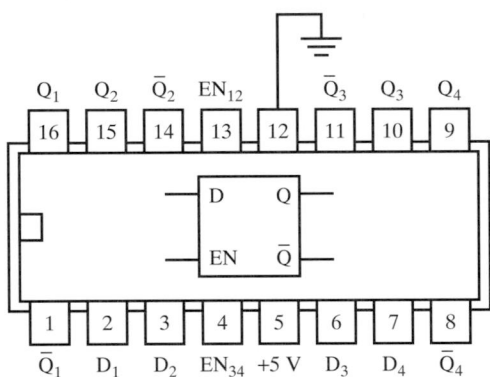

FIGURE 6.11   74373 Octal D Latch

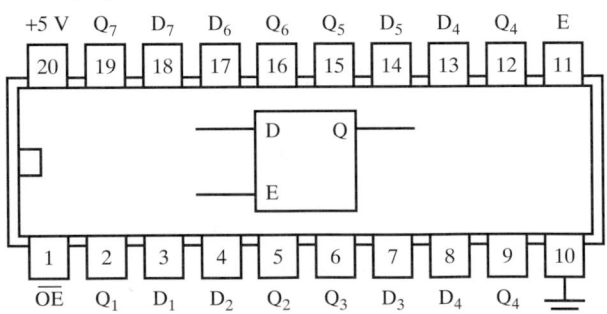

nanoseconds, so that a quick snapshot of the S and R inputs is all that is possible. This is essentially what happens when we use a ***clocked flip-flop.*** The flip-flop changes state only during a clock edge (rising or falling). A simple ***edge-detector circuit*** is needed between the new CLK input and the old EN input. This is shown in Figure 6.12. The edge-detector circuit depends on the internal timing of the inverter and the AND gate to generate its short output pulse (the EN signal). In this case the data on D is transferred to the Q output on the rising edge of the CLK signal.

Figure 6.13 shows how the inverter-AND combination generates a short positive pulse during a rising edge. The gate delay of the inverter forces the CLK signal to reach the inputs of the AND gate at different times. For a few nanoseconds both inputs are high, providing the brief high-going output pulse. Similar circuits are used for negative-edge triggering.

An operation that uses a clock line to control the flow of data into the flip-flop is called ***synchronous.*** That is, nothing happens until there is a change in the clock input.

FIGURE 6.12   Adding a CLK Input to the D Latch

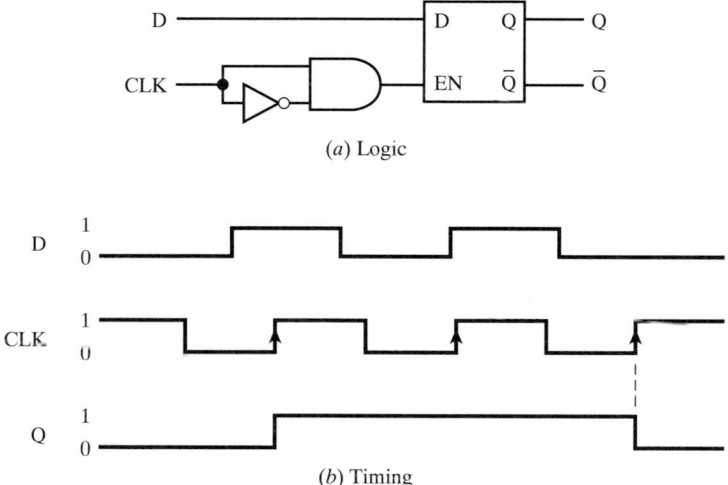

FIGURE 6.13   Positive Edge-Detector

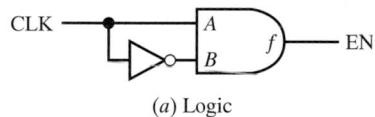

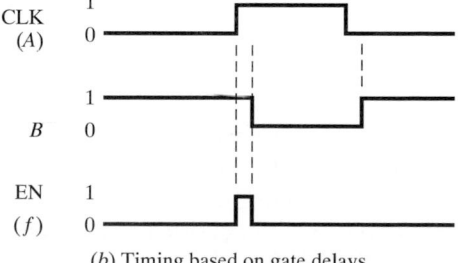

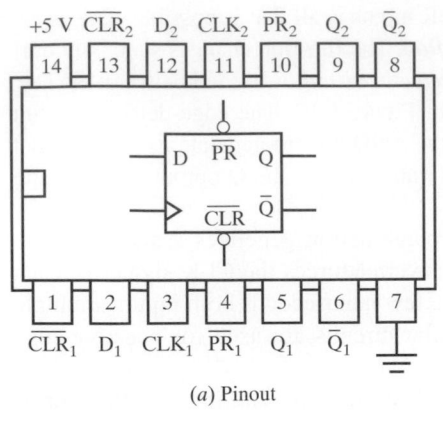

(a) Pinout

CLK	D	$\overline{PR}$	$\overline{CLR}$	Q	$\overline{Q}$
↑	0	1	1	0	1
↑	1	1	1	1	0
X	X	0	0	Illegal	
X	X	0	1	1	0
X	X	1	0	0	1
0	X	1	1	No change	

(b) Truth table

FIGURE 6.14   7474 Dual D Flip-Flop

If we look at the function table for a 7474 type D flip-flop (see Figure 6.14), we will see this principle more clearly.

Two other inputs to the D flip-flop are preset ($\overline{PR}$) and clear ($\overline{CLR}$). The circles at the inputs indicate that the operations occur when the inputs are taken low. To clear (reset) the flip-flop, ground the $\overline{CLR}$ input and return high. To preset (set) the flip-flop, ground the $\overline{PR}$ line and return high. These inputs cause instant change in the output, regardless of the state of the CLK input, and are called ***asynchronous inputs.*** The clock input (CLK) is shown with the standard symbol for a clock input, a triangle. This clock input responds to a positive edge. Other flip-flops respond to negative edges and have a circle on the CLK input, indicating that a high-to-low transition causes a trigger.

Lines 1 and 2 of the truth table show that the output Q will follow the data on input D when we receive a low-to-high (↑) transition on the clock line. Because this is a flip-flop, the data is also stored until we change it, so when the clock line goes low again, the outputs stay the same (see bottom line of the truth table). By using synchronicity we can control the flow of data into the flip-flop. This synchronous method of operation allows us to design circuits in which flip-flops are used to divide digital signals and shift binary words.

··

EXAMPLE 6.1

The 7474 is used as a "divide-by-2" circuit in the digital circuit shown in Figure 6.15. We will apply a square wave to the clock input and watch the Q and $\overline{Q}$ outputs. If we assume that the flip-flop is "*reset*" when we begin (or Q is low), the timing diagram for this simple circuit will look like Figure 6.16. During the first low-to-high transition of the clock,

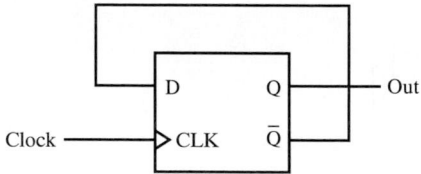

FIGURE 6.15 A 7474 Connected as a Divide-by-2 Circuit

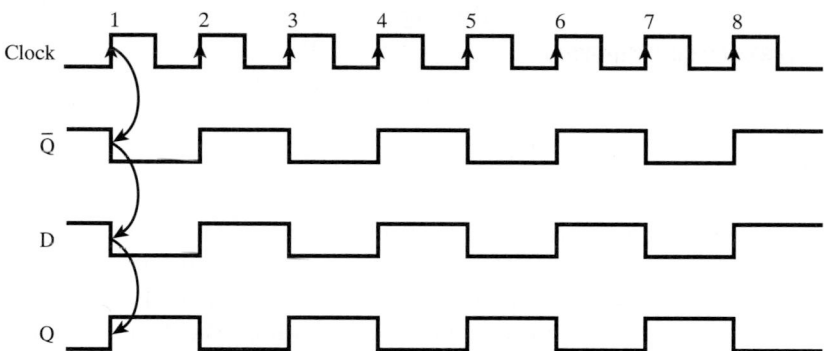

FIGURE 6.16 Timing Waveforms for 7474 Divide-by-2 Circuit; Q̄ Is High Because Q Is Low to Begin

a one gets clocked into the flip-flop (this is because D is tied to Q̄). When this happens, Q goes to one and Q̄ goes to zero. These outputs will remain this way until the next low-to-high transition on the clock input.

When the next transition does occur, a zero gets clocked into the flip-flop. By doing so, the Q and Q̄ outputs change back to zero and one again. We are now back where we started. If you take a close look at the timing diagram, you will notice that low-to-high transitions on the Q output occur once for every two low-to-high transitions on the clock input; thus we have a divide-by-2 circuit.

••

Conditions in the flip-flop are affected at any time by the $\overline{PR}$ and $\overline{CLR}$ inputs, regardless of what state the clock is in. A look at the truth table will show this. During normal operation both the $\overline{PR}$ and $\overline{CLR}$ inputs are high. This is the normal mode of operation. However, to set the flip-flop (Q is high) at any time, merely ground the $\overline{PR}$ line. In line with this, to clear the flip-flop at any time, pull the $\overline{CLR}$ line low. Remember, the outputs of the flip-flop will change immediately when the $\overline{PR}$ and $\overline{CLR}$ inputs are changed, regardless of the clock or data inputs. This is further illustrated in the truth table by the letter X, meaning we "don't care" what is happening on the clock and data inputs, we simply set Q to a logic "1." As in the S̄-R̄ latch, there is also an illegal state in the 7474 when both $\overline{PR}$ and $\overline{CLR}$ are low. Care must be taken to avoid this condition. We finish this section with a look at the 74374 Octal D flip-flop shown in Figure 6.17.

The 74373 and 74374 are useful in microprocessor-based circuitry, since the processor bus signals come in multiples of eight bits.

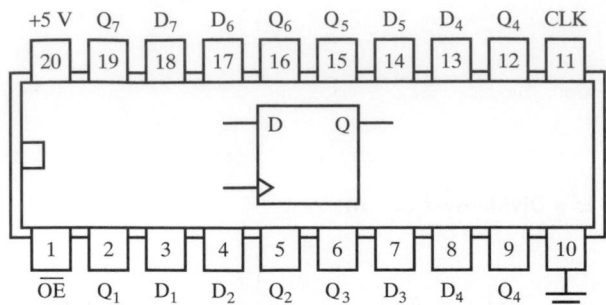

FIGURE 6.17   74374 Octal D Flip-Flop

6.5 THE J-K FLIP-FLOP

The last flip-flop we will look at, the type *J-K flip-flop* (see Figure 6.18), comes in a standard 16-pin DIP (7476) and has a few more functions than the type D flip-flop. Like the other devices we have studied, the J-K flip-flop has two outputs, Q and $\overline{Q}$. It also has $\overline{PR}$ and $\overline{CLR}$ inputs that operate just as they do in the type D flip-flop. The three exceptions in the J-K flip-flop are the J, K, and clock inputs.

The first thing to notice is the inverting circle on the clock input. We know that this means that we are inverting the input. By doing this, we will need a high-to-low (↓) transition (negative transition) on the clock input to operate the flip-flop.

FIGURE 6.18   7476 Dual J-K Flip-Flop

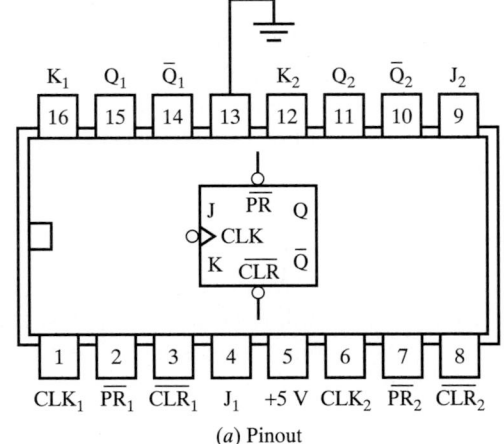

(a) Pinout

CLK	J	K	$\overline{PR}$	$\overline{CLR}$	Q	$\overline{Q}$	
↓	0	0	1	1	No change		
↓	0	1	1	1	0	1	(clear)
↓	1	0	1	1	1	0	(set)
↓	1	1	1	1	Toggle		
X	X	X	0	0	Illegal		
X	X	X	0	1	1	0	
X	X	X	1	0	0	1	

(b) Truth table

Second, let us consider the J and K inputs. They are slightly different from the $\overline{\text{S-R}}$ inputs and the D input on the 7474.

When both J and K are low, the flip-flop does not change state. The logic levels present at the Q and $\overline{Q}$ outputs *never* change, as long as both inputs are low. No amount of clocking the flip-flop will produce a change (assuming the $\overline{\text{PR}}$ and $\overline{\text{CLR}}$ inputs are not used and left high).

When J and K are different—that is, when they complement each other—data from J and K transfer straight through to Q and $\overline{Q}$ on the *falling edge* of a clock pulse. It is this mode of operation that enables us to make devices called shift resisters.

Last, when both J and K are high, the outputs **toggle** on each successive clock pulse. For example, Q and $\overline{Q}$ may be set to one and zero, respectively. On the next clock pulse, they toggle, or switch to zero and one. On the second pulse, they switch back to one and zero, and so on. This operation was accomplished in the 7474 D flip-flop by feeding the $\overline{Q}$ output back into the D input. In the J-K flip-flop, this is not necessary. To divide by 2, we simply tie both J and K high, enabling the toggle mode.

We will use J-K flip-flops to make counters and shift registers in the next two sections.

6.6 SHIFT REGISTERS

Another important device used in digital circuits is the ***shift register.*** As we have seen in Chapter 2, shift registers can be used to multiply numbers (or at least to aid in the multiplication process). We have many other uses for them, however. We use them to control the transmission of information from computers to electronic displays and printers. They also were used in early video games and pinball machines for sequencing lights. All in all, the shift register is a very important device in digital electronics.

Simple Shift Register

The shift register shown in Figure 6.19 is made up of J-K flip-flops. To operate it, we simply place the data to be shifted on the *serial* input, and look at the four outputs as we clock the data into and through the circuit. Keep the truth table for the 7476 in mind as we discuss this circuit. Assuming that all flip-flops are reset to begin with, the timing diagram in Figure 6.20 shows what happens in our simple 4-bit shift register as we change the input and look at the output. The B, C, and D outputs are all shifted one bit at a time to the right of the A output. This is the primary operation of the shift-register. As we will find in our next shift register, the user has a choice of shifting these 4 bits right or left.

The basic shift register has a serial data input at which the bits are applied one at a time during the loading process. We will see later that data may be loaded into another

FIGURE 6.19 Simple 4-Bit Shift Register

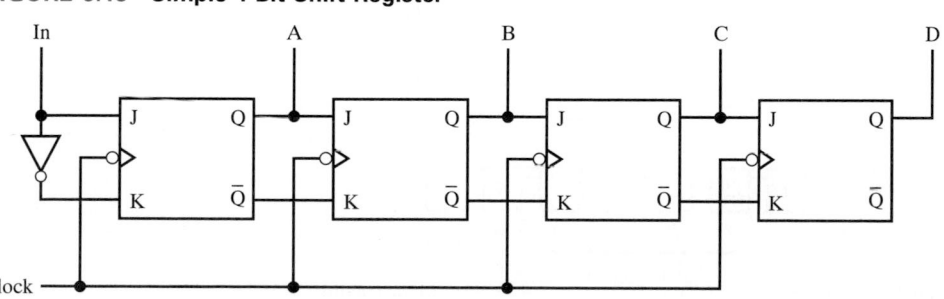

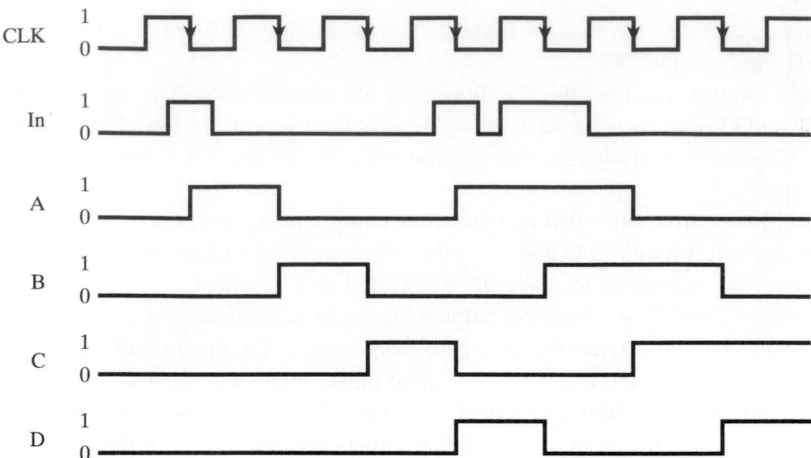

FIGURE 6.20 Output Waveforms of a 4-Bit Shift Register

type of register in parallel. In a parallel load shift register, the inputs to each flip-flop are all available at once. The register is loaded in one operation, all bits at once.

This basic shift register also has a serial output (the right-most flip-flop's Q output). To see all the stored data, it must be clocked through the right-most flip-flop and observed one bit at a time. Many shift registers have all the Q outputs available, and this is a parallel output shift register. Basically, *serial* means one bit at a time, whereas *parallel* means all bits at the same time.

Shift registers come in a variety of styles: serial in, serial out, parallel in, parallel out, and shift left/shift right. Parallel inputs use a *load* line to enter the data.

The 7495 4-Bit Right/Left Shift Register

In the 7495 we have the capability of shifting the 4 bits right or left in the shift register. Figure 6.21 shows a pinout of the 7495. In the operation of the shift register we place the 4-bit word on the A, B, C, D inputs. Then, by clocking the shift register either right or left (through the use of the mode control input), we obtain our data off of the A, B, C, D outputs. A shift register of this kind would be very useful if we were

FIGURE 6.21 7495 4-Bit Right/Left Shift Register

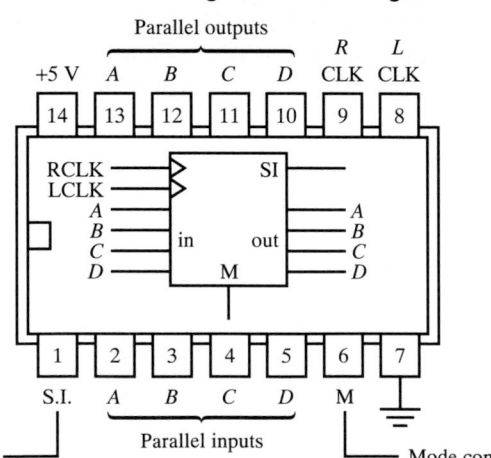

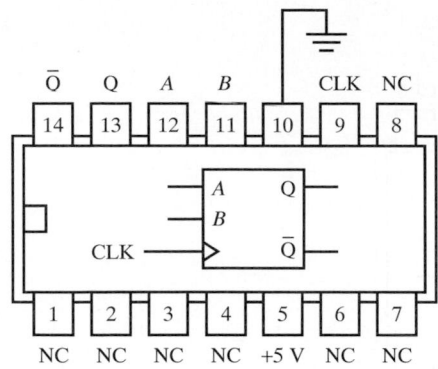

FIGURE 6.22 **7491 8-Bit Shift Register**

designing a TTL circuit to perform the shift operations for our BCD multiplication and division routines.

The 7491 8-Bit Shift Register

In microcomputers, we are frequently dealing with 8-bit words or bytes, so instead of using two 7495 shift registers, we look up a substitute that can handle 8 bits in one DIP. One integrated circuit that we might find is the 7491 8-bit shift register shown in Figure 6.22.

In this integrated circuit, we place the data to be shifted on the A and B inputs, and clock the shift register. After eight clock pulses, we will begin to see the data on the Q and $\overline{Q}$ outputs. For reasons that may not be apparent at this point, we can use this shift register to obtain a *delay*. It sometimes becomes necessary to delay a pulse from arriving at its destination for a short while (as in the case of the carry bit in some binary adders). To learn how to achieve such a delay, take a look at Example 6.2.

EXAMPLE 6.2

Use a 7491 to make an 80-μsec delay circuit.

SOLUTION

It will take eight clock pulses to get our signal out of the shift register. This is indicated in Figure 6.23. So, if we want to wait 80 μsec for our signal, and it takes eight clock pulses to get our signal out of the shift register, each clock pulse should take 80/8 or 10 μsec!

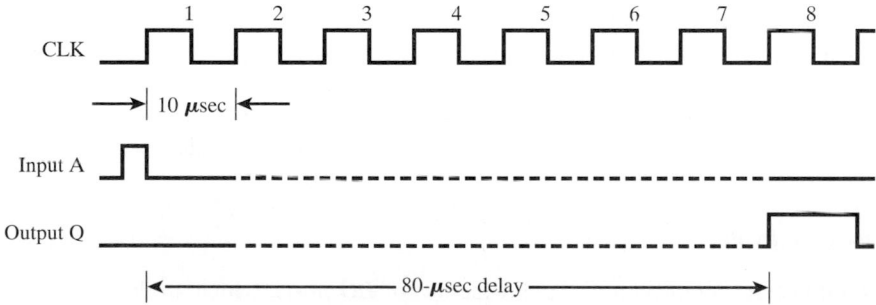

FIGURE 6.23 **The 7491 as an 80-μs Delay Circuit**

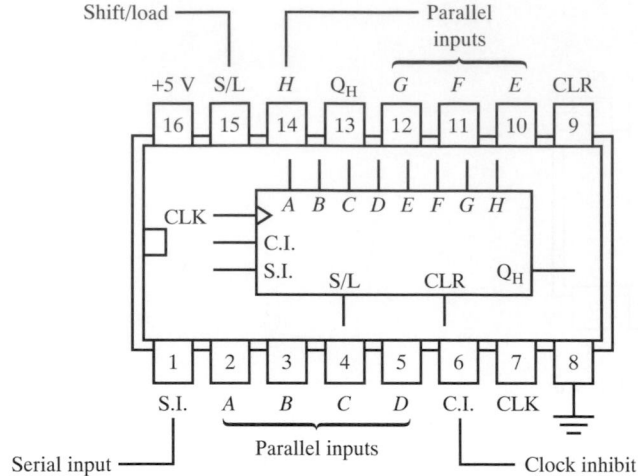

FIGURE 6.24   74166 8-Bit Parallel Load Shift Register

The 74166 8-Bit Parallel Load Shift Register

The last shift register we want to look at is the 74166 8-bit parallel load shift register. The 74166 enjoys widespread use in video display logic for sending out video information to television monitors. An image may be stored in binary in a microcomputer's memory.

Timing circuits take 8 data bits from the computer's memory and place them at inputs A through H of the shift register. The shift/load input is then toggled to get the data into the shift register. Once this has been done, the data bits are clocked out one by one and mixed with other video signals to become pixels on the television screen.

Figure 6.24 shows a pinout of the 74166. The parallel-to-serial nature of the 74166 makes it suitable for other operations, such as those involved in serial data transmission (Chapter 12).

6.7 PRACTICAL APPLICATIONS

In Chapters 7 and 8 we will see how flip-flops are used to construct many different kinds of counters. For now, let us look at a number of other useful applications for them.

Pushbutton Controller

In Figure 6.25 a J-K flip-flop is used to control a RUN signal to an external circuit. Two pushbuttons are used to operate the flip-flop. When the "Start" button is pressed, a logic zero is applied to the normally high $\overline{PR}$ input. This sets the flip-flop (RUN goes high). When the "Stop" button is pressed, the $\overline{CLR}$ input is active, and the flip-flop is cleared (RUN goes low). Any flip-flop that contains asynchronous preset and clear inputs can be used in place of the J-K flip-flop.

LED Output Port

Many microprocessor architectures support the use of *I/O ports,* which are hardware devices that allow eight or more bits of information to be exchanged between the processor and the outside world. Figure 6.26 shows one way to make an 8-bit output port. The 74373

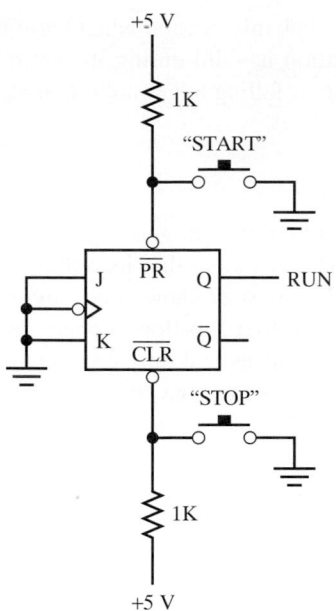

FIGURE 6.25   Pushbutton Control Circuit

FIGURE 6.26   8-Bit LED Output Port

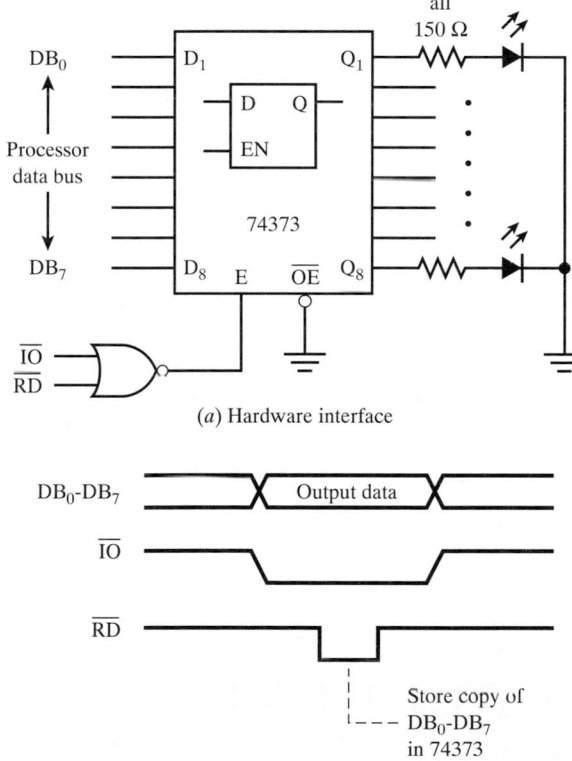

(a) Hardware interface

(b) Timing

Octal D latch is used to save a copy of the processor's data bus information when the $\overline{IO}$ and $\overline{RD}$ signals are active. Typically, the data bus information is valid during the entire interval of time when both controls signals are low. If rising or falling edges are required, change the 74373 to a 74374.

The 4-Bit Ring Counter

The digital sequencer of Figure 6.2(a) employed a 4-bit *ring counter* to generate the indicated waveforms. A ring counter is actually a shift register wired so that its output is fed back to its input, allowing data to circulate in a ring. Figure 6.27 shows one way to make a 4-bit ring counter using J-K flip-flops. Notice that the first flip-flop is preset by the $\overline{CLEAR}$ signal, while the others are all cleared. This guarantees that only one output is high to begin. Now, since the ring counter is actually a shift register, the single one being stored simply circulates from stage to stage with each CLK pulse. Figure 6.28 shows a screen shot of the Electronics Workbench RING circuit. The Space bar is used to clear the counter. You may wish to connect the logic analyzer to view the output waveforms.

The 3-Bit Johnson Counter

A variation of the ring counter is the *Johnson counter.* As shown in Figure 6.29, the Johnson counter has its output signals cross-connected back into the input. This causes the output sequence to toggle back and forth between groups of ones and groups of zeros. Note that each

FIGURE 6.27   4-Bit Ring Counter

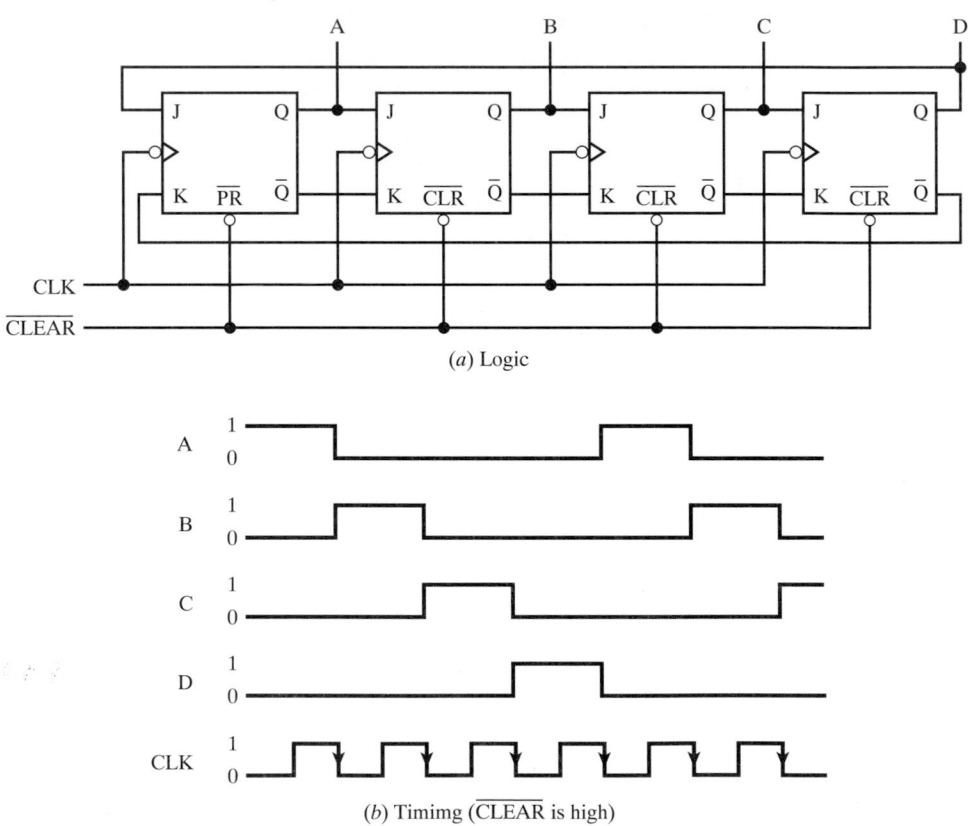

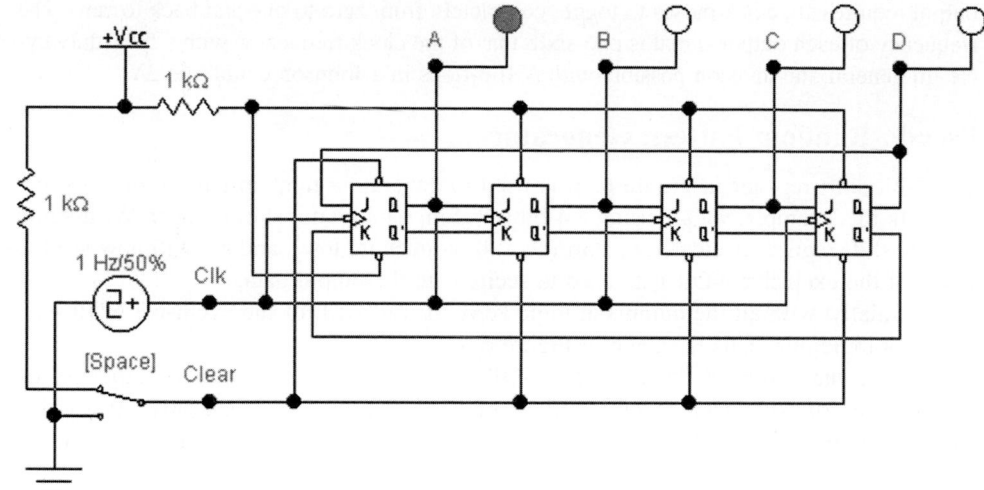

FIGURE 6.28   EWB Counter Circuit RING

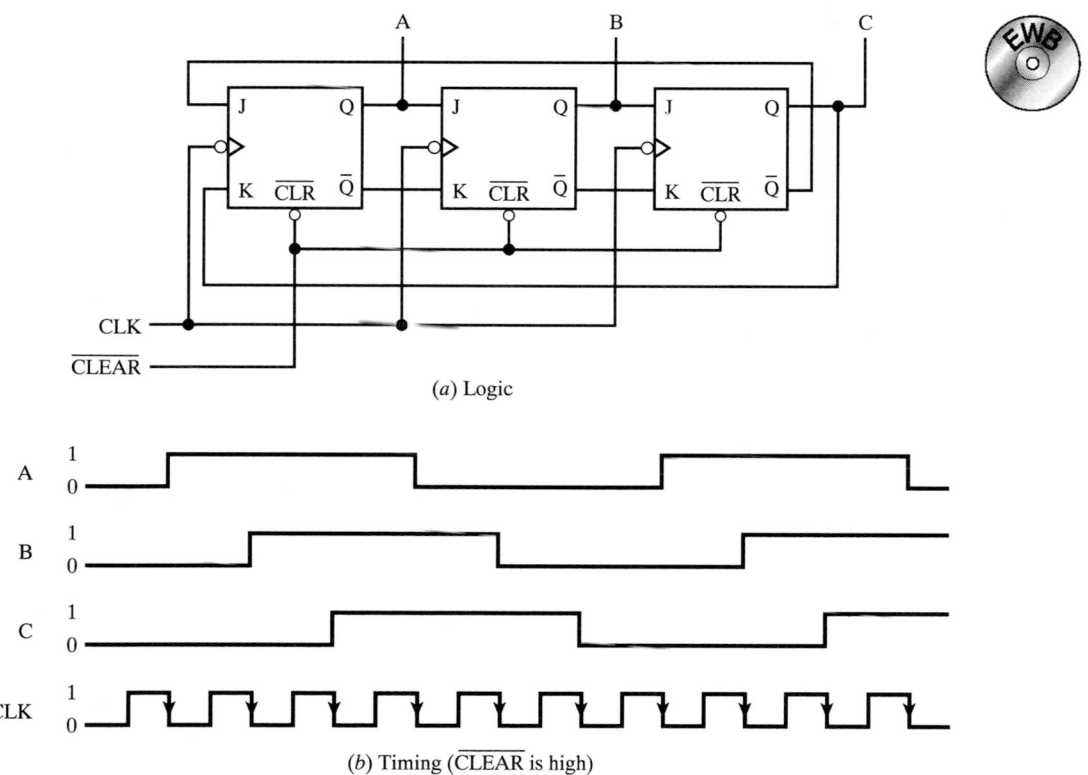

(*a*) Logic

(*b*) Timing ($\overline{\text{CLEAR}}$ is high)

FIGURE 6.29   3-Bit Johnson Counter

output requires six clock pulses to toggle completely from zero to one and back to zero. The frequency of each output signal is one-sixth that of the clock frequency, with a 50% duty cycle. In general, the division possible with N flip-flops in a Johnson counter is $2N$.

Pseudo-Random Pattern Generator

The 4-bit shift register in Figure 6.30 is used to generate a random pattern of ones and zeros. In this example, we are using a 4-bit serial in, parallel out shift register. We assume that the shift register is cleared to start (i.e., all outputs are low), and we will now see the effect of the exclusive NOR gate used to recirculate the output data.

Initially, with all the outputs at logic zero, the output E of the exclusive NOR gate will be a logic one. This is shown in Figure 6.31.

Since the output of the exclusive NOR gate is connected to the data input of the shift register, this one gets loaded into the register during the first clock pulse. When this happens, output A goes to a logic one, with B, C, and D unchanged. Figure 6.32 illustrates this pattern.

FIGURE 6.30 **Four-Bit Pattern Generator**

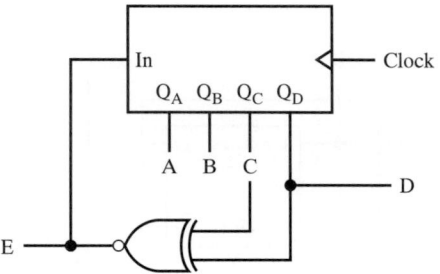

FIGURE 6.31 **Initial State of Shift Register**

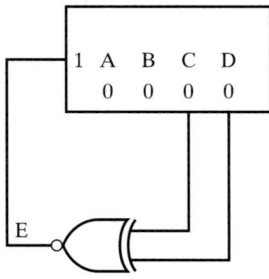

FIGURE 6.32 **Shift Register after First Clock Pulse**

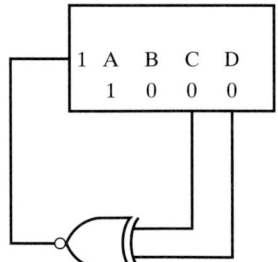

C/C++ HELPER

The SHIFTREG program on the companion CD-ROM is designed to simulate the operation of a serial-in, parallel-out shift register that may be up to 32 bits long. The first bit of the shift register is bit 0 and the last is bit 31. For a four-bit shift register, bits are numbered from 0 to 3.

The pseudo random sequence generated by the circuit in Figure 6.30 is found by executing SHIFTREG in this way:

```
C> SHIFTREG
Enter the number of bits ----> 4
Enter first output to use ---> 2
Enter second output to use --> 3

 1: 0 0 0 0
 2: 1 0 0 0
 3: 1 1 0 0
 4: 1 1 1 0
 5: 0 1 1 1
 6: 1 0 1 1
 7: 1 1 0 1
 8: 0 1 1 0
 9: 0 0 1 1
10: 1 0 0 1
11: 0 1 0 0
12: 1 0 1 0
13: 0 1 0 1
14: 0 0 1 0
15: 0 0 0 1
```

Compare this with the contents of Table 6.1.

SHIFTREG is useful for quickly determining the best outputs to use for a desired sequence.

Since the C and D outputs did not change, the data input remains at logic one $(0 \oplus 0 = 1)$. This in turn gets loaded into the shift register during the second clock pulse. (see Figure 6.33).

Now, after two clock pulses, the shift register outputs are 1 1 0 0. With one more clock pulse, they will be 1 1 1 0. Notice that outputs C and D are now different. The exclusive NOR of one and zero is zero! This zero gets clocked in on the fourth clock pulse. This sequence goes on and on and produces an output like the one shown in Table 6.1.

This sequence repeats every 15 clock pulses. We know this to be true because the shift register (on the fifteenth pulse) goes back to 0 0 0 0. From here the sequence repeats again.

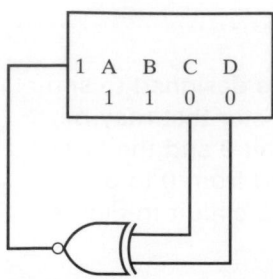

FIGURE 6.33   Shift Register after Two Clock Pulses

**TABLE 6.1   Pseudo-Random Sequence
for a 4-Bit Shift Register**

0 0 0 0	(0)	Initial State
1 0 0 0	(8)	1
1 1 0 0	(12)	2
1 1 1 0	(14)	3
0 1 1 1	(7)	4
1 0 1 1	(11)	5
1 1 0 1	(13)	6
0 1 1 0	(6)	7
0 0 1 1	(3)	8
1 0 0 1	(9)	9
0 1 0 0	(4)	10
1 0 1 0	(10)	11
0 1 0 1	(5)	12
0 0 1 0	(2)	13
0 0 0 1	(1)	14
0 0 0 0	(0)	15

By connecting the exclusive NOR inputs to different shift register outputs, different sequences can be made (they may not all be 15 patterns long, though). The decimal equivalents of each pattern are shown to illustrate the pseudo-random nature of the output.

Circuits like these (only containing more bits; hence, longer sequence lengths) are combined with summing amplifiers to make digital noise sources, music machines, and code scrambling devices.

Registered PAL Outputs

Figure 6.34 shows one type of *registered* PAL output. The D-type flip-flop represents a one-bit register that can be examined and modified through the connections made within the PAL. The sum-of-products output of the programmable AND-OR array provides the

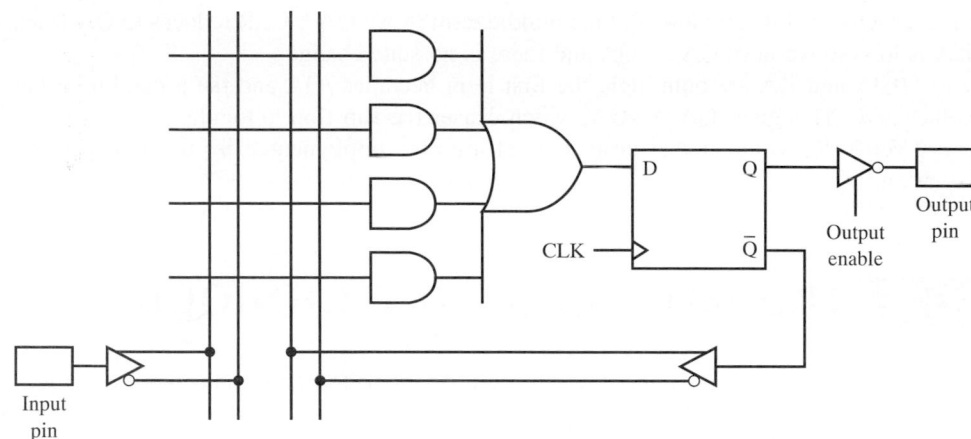

FIGURE 6.34 **Registered PAL Output**

input signal (D) to the flip-flop. Notice that the Q output is inverted before being applied to the output pin. Keep this important fact in mind as you examine the following sets of PAL equations.

D-Type Flip-Flop Equations

Let us define three PAL inputs, DA, PRE, and CLR. DA is the data input, PRE sets the flip-flop when high, and CLR clears the flip-flop when high. The PAL output /QA is defined as well (thus making the Q output of the flip-flop QA). The / indicates inversion. The equation for the D-type flip-flop is as follows:

QA := DA * /CLR   ;QA equals DA when CLR is low or zero when CLR is high

   + PRE   ;QA equals one when PRE is high

Bear in mind that the /QA output will be low when QA is high, and vice versa.

J-K Flip-Flop Equations

Using PAL input signals JA, KA, PRE, and CLR, and output /QA, the operation of a J-K flip-flop is simulated via this equation:

QA := JA * /QA * /CLR   ;JA and KA control QA when CLR is low

   + /KA * QA * /CLR   ;Otherwise, QA is low when CLR is high

   + PRE   ;QA equals one when PRE is high

This equation take a little time to understand. With JA and KA, four functions are possible (no change, clear, set, and toggle). Imagine that both JA and KA are low. This forces the JA * /QA * /CLR term low. But the middle term /KA * QA * /CLR reduces to QA when KA is low, so we have QA := QA and there is no state change.

If JA and KA are both high, the first term becomes /QA and the second term becomes zero. This gives QA := /QA, which causes the flip-flop to toggle.

Verify for yourself that clear and set are also implemented by the J-K flip-flop equation.

6.8 TROUBLESHOOTING TECHNIQUES

The very nature of the operation of the flip-flop (or latch) forces us to be aware of its special timing requirements. In a high-speed digital circuit, care must be taken to properly establish *setup* and *hold* times for the flip-flops or latches. As Figure 6.35 shows, data at the input to the flip-flop must be valid for a brief setup time (t_S) and must not change after the clock edge for the duration of the hold time (t_H). Setup and hold times are typically in the nanosecond range, with the hold time often specified at zero ns. Setup and hold times are one of the limiting factors in how fast a flip-flop may be clocked. For circuits that use combinational logic to control the S-R, J-K, or D inputs, it may be necessary to determine the propagation delay through the circuit, to guarantee that the setup and hold times are maintained. Failure to do so may result in a *race condition,* where the inputs change state at the same time as the clock edge, and the flip-flop has to guess what the input was.

The preset and clear inputs of a flip-flop also have control over what goes on at the output, as much as the other input signals do. When setting up a flip-flop, is it acceptable to leave the PR and CLR inputs unconnected if you are not using them? Flip-flops from one manufacturer may pull these unconnected inputs high internally and everything will be fine. Flip-flops from a different manufacturer may work entirely differently, or not at all, unless a pullup resistor is attached to the unused inputs to pull them to a logic one level. It is a good design practice to always connect an unused input to a pullup resistor (or to ground if a logic zero is required). Do not depend on the device to do it for you.

One last point regarding flip-flops: know which edge you are using. Often, a design is prevented from working simply because the clock input is being fed the wrong edge. Its operation may be intermittent, its setup and hold times being made and broken. If necessary and within the timing requirements, insert an inverter in series with the clock signal to change its phase. This will provide the opposite edge required.

FIGURE 6.35 Data Setup and Hold Times

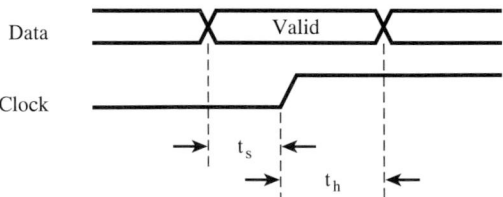

SUMMARY

In this chapter we examined the operation of latches, flip-flops, and shift registers. A latch changes state when its enable input goes to its active level. A flip-flop changes state when the appropriate edge is applied to its clock input. Shift registers move entire groups of bits to the left or right, and are used in serial/parallel conversions. Some typical flip-flop applications were presented, including counting and pattern generation.

STUDY QUESTIONS

General

1. Explain the meaning of a low-to-high transition as applied to a clock pulse.
2. What is the meaning of a circle on the input to a flip-flop?
3. What is the difference between a synchronous input and an asynchronous input?
4. Describe in words the operation of the J and K inputs on a J-K flip-flop.
5. What is the difference between a latch and a flip-flop?
6. Show how to make an S-R latch from two NOR gates.
7. What does it mean when we say a flip-flop is set?
8. Describe in your own words the operation of the simple 4-bit shift register.
9. How can an 8-bit shift register be used to produce a 240-msec delay?
10. Determine the truth table for the flip-flop shown in Figure 6.36.

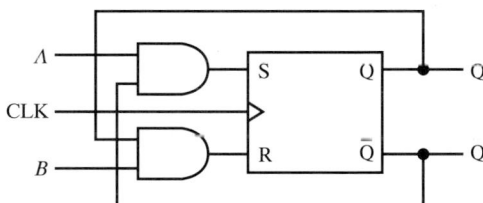

FIGURE 6.36   For Question 6.10

11. Design a circuit that generates a pulse when a negative edge is applied.
12. Show how two type D flip-flops can be used to divide a digital signal frequency by 4.
13. How does the propagation delay of a flip-flop work to our advantage when used inside a shift register?
14. Show how a 5-bit shift register can be made out of J-K flip-flops.
15. Show how an 11-bit shift register can be made out of an 8-bit shift register and some flip-flops.
16. If the clock line on a shift register is 10 MHz, what is the time between bits at the output?
17. Show how a second signal called DONE could also be used to stop the system (RUN equals zero) in the flip-flop push button controller (Figure 6.25).

18. Design a 3-bit ring counter. Verify its operation via the truth table for the flip-flop you used in the design.

19. Show how a J-K flip-flop can be wired to mimic the operation of a D flip-flop.

20. A 1-KHz clock is applied to a 5-bit Johnson counter. What is the frequency of any output?

21. What are the output sequences for each shift register in Figure 6.37? Assume all outputs of the shift register are initially low.

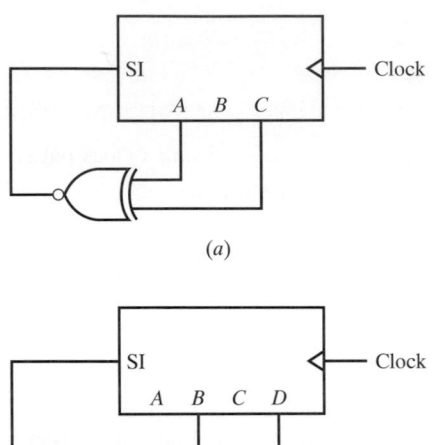

(a)

(b)

FIGURE 6.37   For Question 6.21

22. What is a race condition?

23. Design a 3-bit shift register using type D flip-flops.

24. What does the timing diagram for the circuit in Figure 6.38 look like? Assume all Q outputs are initially low.

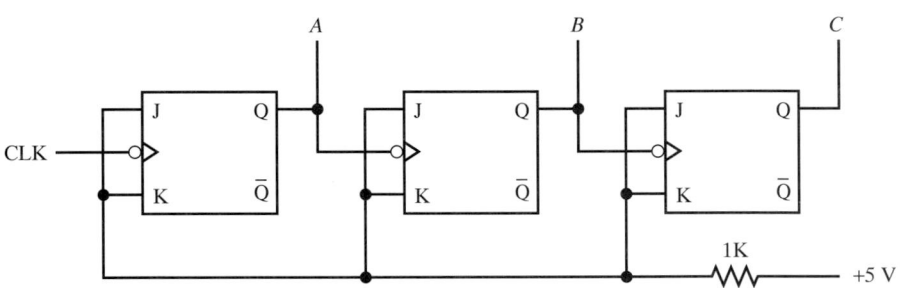

FIGURE 6.38   For Question 6.24

25. Is it possible to connect two level-sensitive latches so that they operate as if they were a single edge-sensitive device?

Electronics Workbench

26. Is it possible to simulate the operation of the edge detector in Electronics Workbench? Explain the results of your attempt.

27. Verify your answers to questions 18, 20, and 24 using Electronics Workbench.

Programmable Logic

28. Write the PAL equations required for a 4-bit shift register. Input signals are CLR and SI (serial input). Output signals are D0 through D3. Use D-type flip-flop equations.

29. Modify the PAL equations from Question 28 so that SI is fed from an internal exclusive-NOR function. Inputs to the exclusive-NOR are D2 and D3.

30. Write the PAL equations for the 3-bit ring and Johnson counters using J-K flip-flop equations.

7

COUNTERS

INSTRUCTIONAL OBJECTIVES

When finished with this chapter you should be able to:

1. Explain how flip-flops are used to make a counter.

2. State the differences between a ripple counter and a synchronous counter.

3. Compare a BCD counter with a binary counter and know when to use each.

4. Design a counter-based frequency division circuit.

SELF-EVALUATION QUESTIONS

Keep the following questions in mind and try to answer them when you have completed the chapter.

1. How fast can a ripple counter be clocked?

2. How fast can a synchronous counter be clocked?

3. How many states does a modulo-8 counter have?

4. How are large counters made from smaller counters?

5. What role does counting play in frequency division?

7.1 INTRODUCTION

In this chapter we extend the role of flip-flops into the world of counters. *Counters* come in many flavors: binary, BCD, modulo-*N,* and up/down to name a few. We will discuss the difference between ripple counters and synchronous counters, and see where counters are applied in digital circuitry. A number of standard counter packages will be studied.

7.2 USING FLIP-FLOPS TO MAKE A COUNTER

In Chapter 6 we saw a few examples of how flip-flops were used to make counters. These were the ring and Johnson counters. Basic shift-register connections among the flip-flops were used. In order to count in binary, or decimal, the flip-flops must be connected differently. Usually, flip-flops in a counter are set up so that their outputs toggle on selected clock pulses. Figure 7.1 shows how this is done. The type D flip-flop requires us to feed the $\overline{Q}$ output back to the D input. This forces the output to go to the opposite state with each clock pulse. To check this principle, imagine that the D flip-flop powers on into a clear state (Q equals zero). This means that $\overline{Q}$ is high, which in turn makes the D input high. The next clock pulse will clock the high level into the D flip-flop, causing Q to go high and $\overline{Q}$ to go low. Now a zero is waiting to be clocked into the D flip-flop on the next clock pulse. Thus, every two clock pulses the Q output goes from low to high and back low.

The J-K flip-flop makes it even easier for us, since toggle mode is selected by pulling the J and K inputs high. The toggling is a natural part of any counting sequence, as you will see when we examine truth tables and timing diagrams.

7.3 BINARY RIPPLE COUNTERS

When individual flip-flops are cascaded, as they are in Figure 7.2, we have what is called a binary *ripple counter.* Since three flip-flops are used to make the counter, we refer to it as a 3-bit counter. The 3-bit ripple counter has an output sequence shown in Table 7.1. The output wraps around from 111 to 000 on the eighth clock pulse.

Notice that the Q output of the first J-K flip-flop acts as a clock for the second flip-flop. When Q toggles from high to low, the second flip-flop also toggles. The second flip-flop controls the third flip-flop in the same way. Larger counting sequences are generated by adding more flip-flops. *Binary counters* have a maximum count of $2^N - 1$, where N is the number of flip-flops.

FIGURE 7.1 **Using Toggle Mode for Counting**

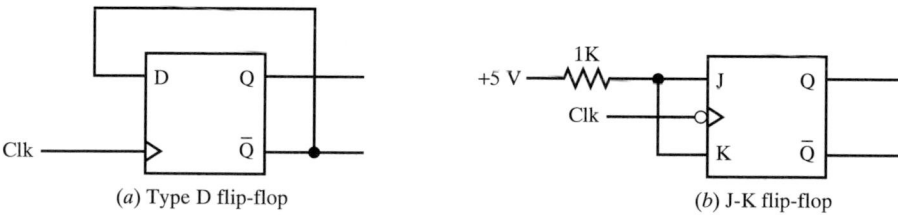

(*a*) Type D flip-flop   (*b*) J-K flip-flop

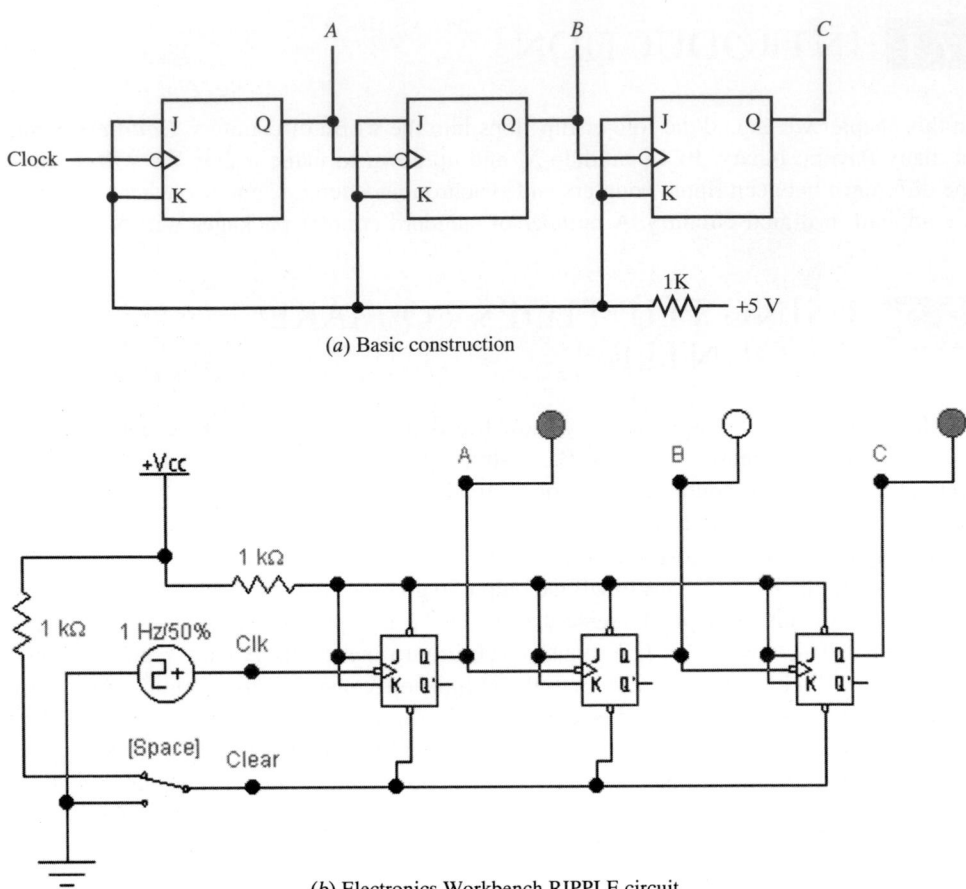

(a) Basic construction

(b) Electronics Workbench RIPPLE circuit

FIGURE 7.2   **3-Bit Binary Ripple Counter**

TABLE 7.1   **3-Bit Binary Counter Sequence**

STATE	C	B	A	
0	0	0	0	
1	0	0	1	
2	0	1	0	
3	0	1	1	
4	1	0	0	
5	1	0	1	
6	1	1	0	
7	1	1	1	(then back to 000)

The *ripple* term comes from the fact that the outputs do not all change state at the same time. To illustrate this point, imagine that the counter in Figure 7.2(a) has all three outputs high. The next clock pulse will roll the count over to 000, but not in the way you may think. First, the main clock pulse will cause the first flip-flop to toggle (after its internal propagation delay), sending output A low. This output is also the clock to the next flip-flop, which in turn toggles after its own propagation delay. The same is true for the third flip-flop, giving a total of three gate delays since the main clock pulse. That is how long it takes to get a valid count on the output (it goes from 111 to 011 to 001 to 000 as it rolls over). This total delay time determines how fast the counter can be clocked, since we have to wait for the output to stabilize before clocking it again. For example, if each flip-flop has a delay of 10 ns, the fastest the 3-bit ripple counter can be clocked is once every 30 ns, or a frequency of 33.3 MHz.

If the speed of the clock is not a concern, the ripple counter is a quick and easy way of getting the job done. If a clear input is added to reset the counter to zero asynchronously, the counter is even more useful. Figure 7.2(b) shows how the three-bit ripple counter is implemented in Electronics Workbench. Load the circuit file RIPPLE to examine its operation.

The 7493 4-Bit Binary Counter

Figure 7.3 shows a prepackaged binary counter, the 7493. Internally, there is a divide by two stage and a divide by eight stage. As shown in Figure 7.3(b), connecting the A output back to the B clock input is the normal way of operating the 7493, which counts from

FIGURE 7.3 7493 4-Bit Binary Counter

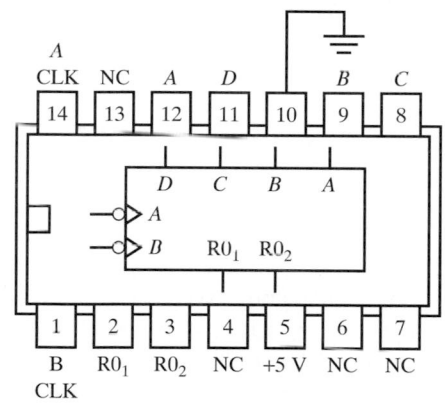

(*a*) Pinout

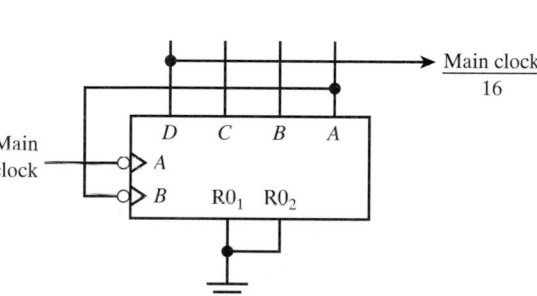

(*b*) Typical operation

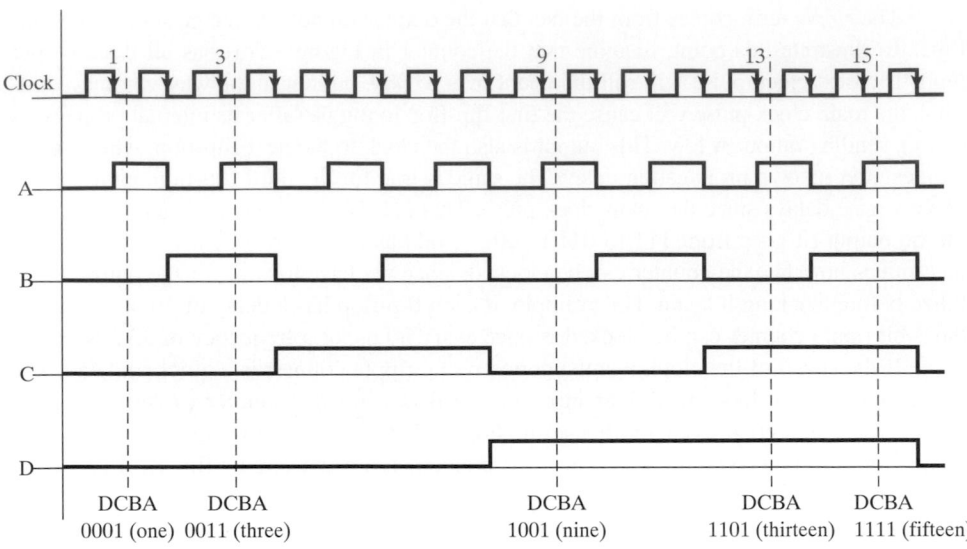

FIGURE 7.4 **The 7493 Timing Diagram**

0000 to 1111 (or divides by 16). The counter is automatically reset to 0000 whenever both R0 (reset) inputs are taken high. If one or both of the reset inputs are low, a normal 0000 to 1111 sequence will result.

Figure 7.4 shows a timing diagram for the 7493. Note that state changes occur on the falling edge of the clock input. The D output is the clock signal for larger counters that are made by cascading multiple 7493s together. We will examine this aspect in more detail in the applications section.

7.4 BCD RIPPLE COUNTERS

BCD ripple counters are similar to 4-bit binary ripple counters, except they reset after ten clock pulses, resulting in a 4-bit counting sequence from 0000 (0) to 1001 (9). This sequence is referred to as BCD, or *binary coded decimal.* The BCD counter resets when its output attempts to go to 1010 (10 decimal). This is accomplished by making use of the asynchronous clear inputs of the flip-flops in the counter. An AND gate can be used to determine when the B and D outputs are high (indicating a count of 1010, or 10 decimal) and clear the counter the instant it happens. So, in effect, the count rolls over from 1001 to 0000, as indicated in Table 7.2. BCD counters are used to make a variety of useful circuits, such as digital clocks, frequency counters, and timers.

The 7490 Decade Counter

Figure 7.5 shows the 7490 *decade counter,* a 14-pin DIP with a pinout similar to the 7493. The difference is that the 7490 contains two additional reset inputs ($R9_1$ and $R9_2$). If these inputs are both high, the output goes to 1001 (nine). This function does not serve much

TABLE 7.2 BCD Counter Sequence

STATE	D	C	B	A	
0	0	0	0	0	
1	0	0	0	1	
2	0	0	1	0	
3	0	0	1	1	
4	0	1	0	0	
5	0	1	0	1	
6	0	1	1	0	
7	0	1	1	1	
8	1	0	0	0	
9	1	0	0	1	(then back to 0000)

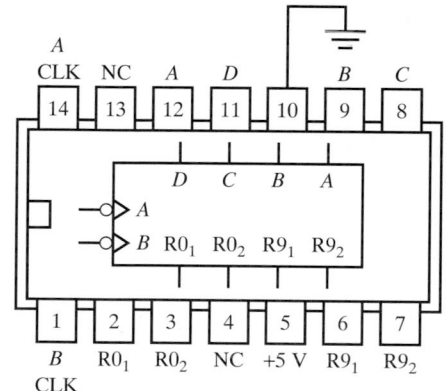

(a) Pinout

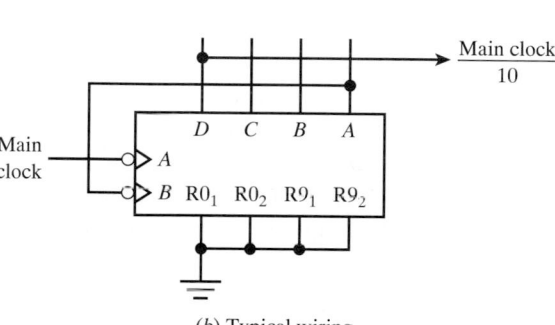

(b) Typical wiring

FIGURE 7.5   7490 Decade Counter

purpose, and so we typically ground these reset inputs to allow a normal zero to nine count.

The 7490 contains internal divide-by-two and divide-by-five counters that are cascaded by connecting the A output back to the B clock input. The output frequency of the 7490 (at D) is one-tenth the input frequency at the A clock input. The timing diagram for the 7490 is shown in Figure 7.6.

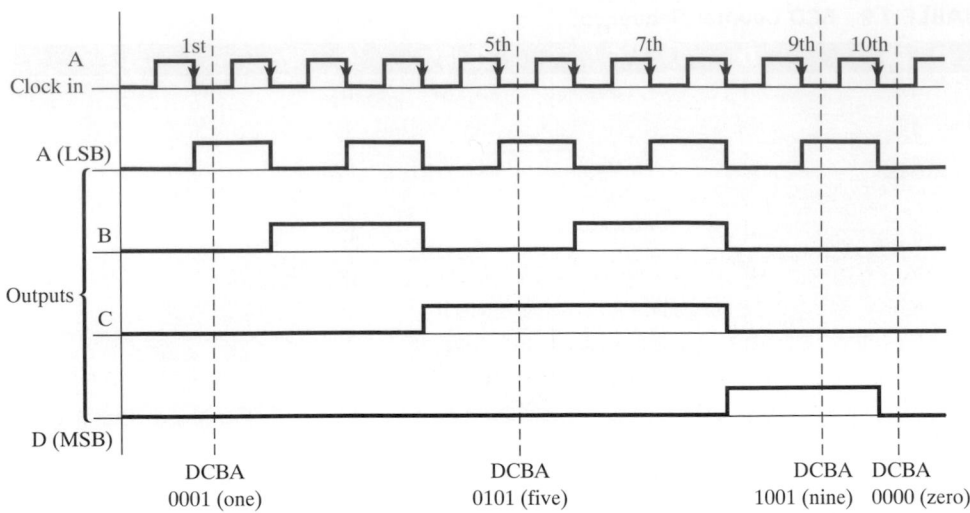

FIGURE 7.6   Timing Diagram for the 7490 Decade Counter

7.5   MODULO-N COUNTERS

It is possible to mimic the behavior of the 7490 decade counter using the 7493 binary counter. We just need to tell the 7493 to reset earlier than it normally would. Figure 7.7 shows how to connect the 7493 as a decade counter. The first binary pattern to have both the B and D outputs high is 1010 (ten). This pattern causes the 7493 to reset to zero within a few ns, so we effectively see 1001 roll over to 0000. This type of counter is called a modulo-10 counter. It counts by 10s. In general, a *modulo-N counter* counts by Ns, or has *N* different states from zero to $N - 1$. For example, a modulo-6 counter has six states: zero through five. We need to use a modulo-*N* counter when a prepackaged counter does not count the way we need it to. The basic task when designing a modulo-*N* counter is finding the correct reset circuit. In Figure 7.7 no logic is required for the reset circuit, only feedback connections from B and D to the reset inputs. Other values of *N* may require one or more logic gates to decode the reset pattern. For instance, for a modulo-11 counter we need to recognize the pattern 1011 (eleven) at the outputs. Somehow we must combine the three ones into a suitable reset signal to the R0 inputs. Figure 7.8 shows two ways to do this. The AND gates reduce the ones in the output to a signal reset signal.

FIGURE 7.7   Using a 7493 as a Decade Counter

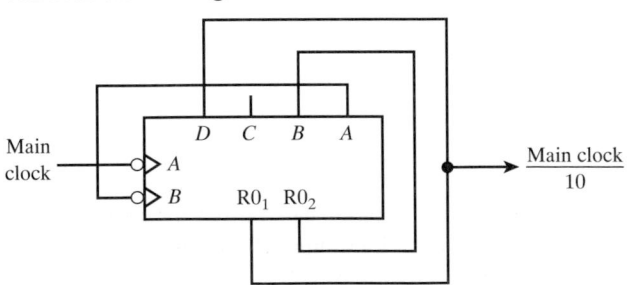

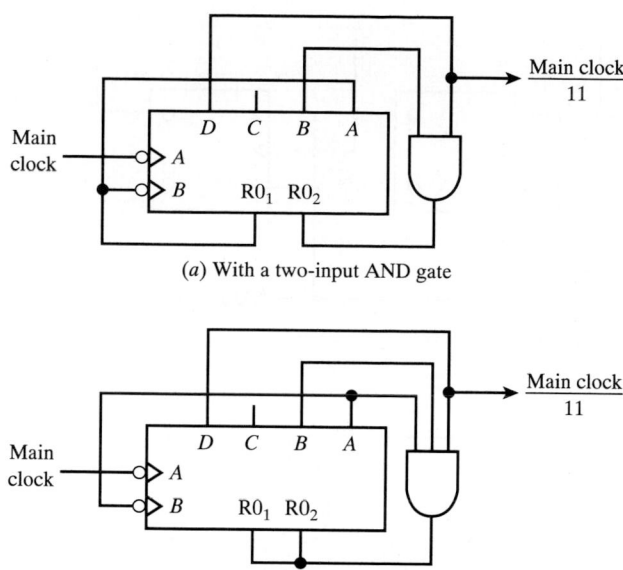

(a) With a two-input AND gate

(b) With a three-input AND gate

FIGURE 7.8 **Two Ways to Make a Modulo-11 Counter with a 7493**

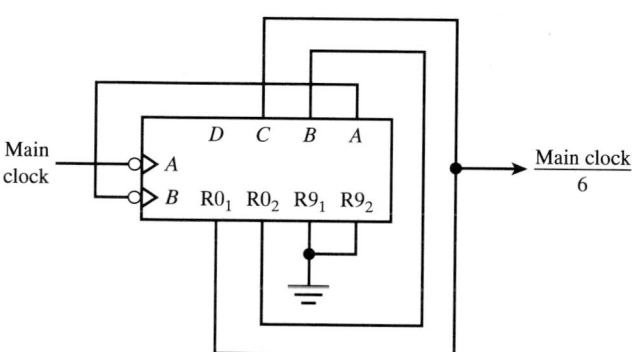

FIGURE 7.9 **Modulo-6 Counter Using the 7490**

One very useful modulo-N counter is the modulo-6 counter. This counter has six states, 0000 to 0101, and is ideal for use in digital clocks to keep track of 10s of seconds and 10s of minutes. Figure 7.9 shows how a 7490 decade counter is wired as a modulo-6 counter. The output state 0110 (six) causes the counter to reset back to 0000.

Modulo-N counters have a wide variety of applications. We will examine some of them in the applications section.

7.6 SYNCHRONOUS COUNTERS

The difference between a *synchronous counter* and a ripple counter is that in the synchronous counter all the flip-flops get clocked at the same time. Compare the 3-bit ripple counter in Figure 7.10(a) with the synchronous counter in Figure 7.10(b). In both cases

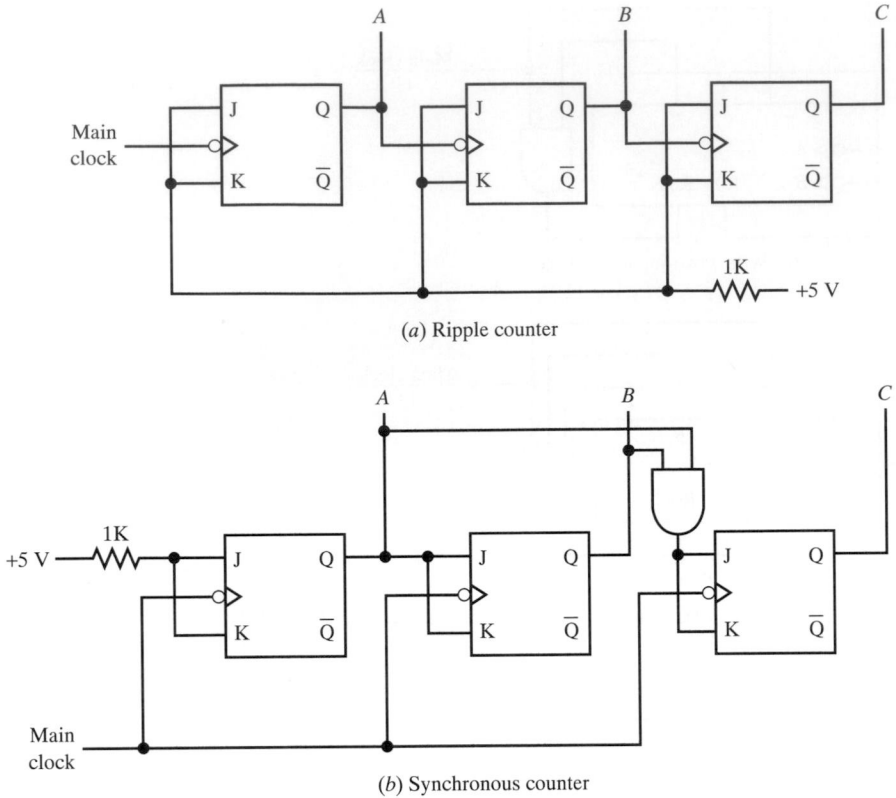

(*a*) Ripple counter

(*b*) Synchronous counter

FIGURE 7.10 **Comparing 3-Bit Counters**

the first stage (output A, the LSB) is hardwired for toggle. The next two stages of the synchronous counter are wired differently, with the second and third stages capable of two modes of operation: no-change and toggle. This should make sense, since the upper bits in the binary count will not change as fast as the LSB. Note the use of the AND gate on the inputs to the third flip-flop. The A and B outputs must both be high (counts three and seven) for the third flip-flop to toggle. The main advantage, even though we had to add a gate to the design, is that the synchronous counter may be clocked faster than the ripple counter. In Figure 7.10 the synchronous counter is three times faster than the ripple counter (best case). Theoretically, it may be clocked three times faster as well. Examine the following C/C++ Helper, where a 4-bit ripple counter and a 4-bit synchronous counter are examined.

We will examine the design method behind synchronous logic in detail in Chapter 8.

The 74192/74193 Synchronous Up/Down Counters

For a look at prepackaged synchronous counters, examine Figure 7.11, which shows the pinouts for the 74192 and 74193 counters. Both may count up or down, and may be preset to a specific count. The 74192 is a decade counter, the 74193 counts in binary.

C/C++ HELPER

The CSPEED program on the companion CD-ROM is used to determine the maximum speed a ripple counter or synchronous counter can be clocked. Two sample executions illustrate its operation.

```
C> CSPEED
Type of counter (R)ipple or (S)ynchronous ---> r
Number of bits in counter ---> 4
Enter the delay (in ns) per stage ---> 5

A 4 bit Ripple counter,
with a 5.0 ns delay per stage,
has a maximum clock frequency of 50.00 MHz.

C> CSPEED
Type of counter (R)ipple or (S)ynchronous ---> s
Number of bits in counter ---> 4
Enter the delay (in ns) per stage ---> 5

A 4 bit Synchronous counter,
with a 5.0 ns delay per stage,
has a maximum clock frequency of 200.00 MHz.
```

Bear in mind that these maximum clock frequencies may be reduced by additional logic used in the counters.

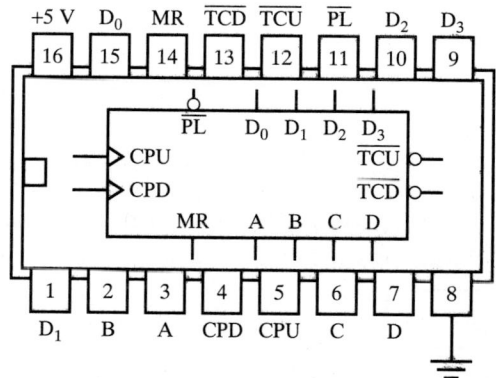

FIGURE 7.11   **74192/74193 Synchronous Up/Down Counters with Parallel Load**

The 74192 and 74193 each contain two clock inputs, one for counting up (CPU) and the other for counting down (CPD). Two *terminal count* outputs are provided to allow multiple 74192/74193s to be cascaded. These are $\overline{TCU}$ and $\overline{TCD}$. $\overline{TCU}$ goes low when the count is about to roll over to zero when CPU is received. $\overline{TCD}$ goes low when the count is zero and a CPD is received.

The master reset (MR) input is active when high and asynchronously clears the counter. The parallel load ($\overline{PL}$) input is also asynchronous, and loads the 4-bit count on the D_0 through D_3 inputs into the counter when low.

Figure 7.12 shows two ways a 74193 can be used to divide a signal's frequency by seven. In Figure 7.12(a) the 74193 is wired to count up and load a count of eight whenever the terminal count is reached. The unused CPD clock input must be tied high. When the counter reaches a count of 1111 it will automatically reset the count back to 1000 (eight). So, the counting sequence is 8, 9, 10, 11, 12, 13, 14, 15 (briefly, quickly resetting to 8). Since we do not count the 15 state there are a total of seven different states, which resulted from seven clock pulses on CPU. The frequency of $\overline{TCU}$ will be one seventh the frequency of CPU.

In Figure 7.12(b) the same division is accomplished by loading seven (0111) into the 74193 wired as a down counter. Now, when the terminal count of zero is reached, the counter is automatically reset back to seven. Once again, it will take seven clock pulses to roll the counter over.

To make larger counters, cascade multiple 74192/74193s by connecting the $\overline{TCU}$ and $\overline{TCD}$ outputs to the next stage's CPU and CPD inputs, respectively. In Chapter 8 we will see how a synchronous up/down counter is designed.

FIGURE 7.12   Two Ways to Divide 7 with a 74193

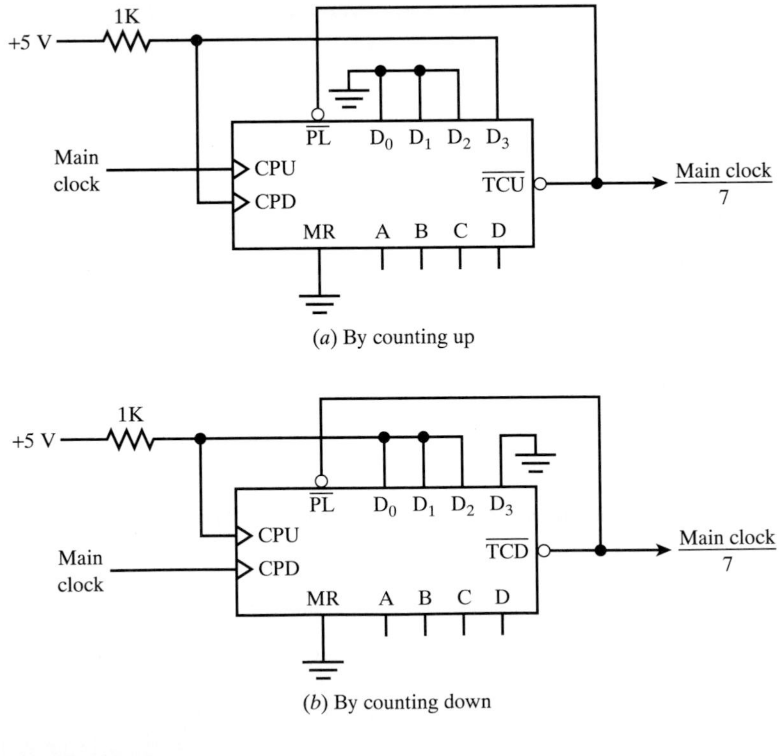

(*a*) By counting up

(*b*) By counting down

7.7 PRACTICAL APPLICATIONS

Let us look at several applications where counters play an important role.

Rotating Wheel Interface

Figure 7.13 shows a block diagram of a digital wheel interface. A rotating wheel contains 90 holes around the outer perimeter. A hole sensor sends a train of pulses to a rotation counter circuit that outputs a low-going pulse on the $\overline{REV}$ output when the wheel has gone around one complete revolution.

The problem is how to convert ninety pulses/revolution to one pulse/revolution. Figure 7.14 provides the solution in the form of two cascaded 7493 binary counters. The effective 8-bit counter is capable of dividing by 256; we need it to divide by 90. The binary pattern for 90 is 01011010 and represents the state of the output when we need a reset signal. The 4-input NAND gate normally outputs a logic one, but drops to zero when the four 1s show up at the output. This generates the required $\overline{REV}$ signal and also, via the inverter, resets the counters. When the NAND gate is normally high, the inverter provides the logic zero required by the reset inputs to allow counting.

You are encouraged to think about what types of divisions are possible with additional counters, with BCD counters, and with a mixed set of counters.

Microprocessor Program Counter

The operation of a typical microprocessor involves fetching an instruction from memory, decoding it, executing it, and repeating the process at the next location. Instructions are fetched from memory by supplying the memory system with the binary address of the de-

FIGURE 7.13   Wheel Interface

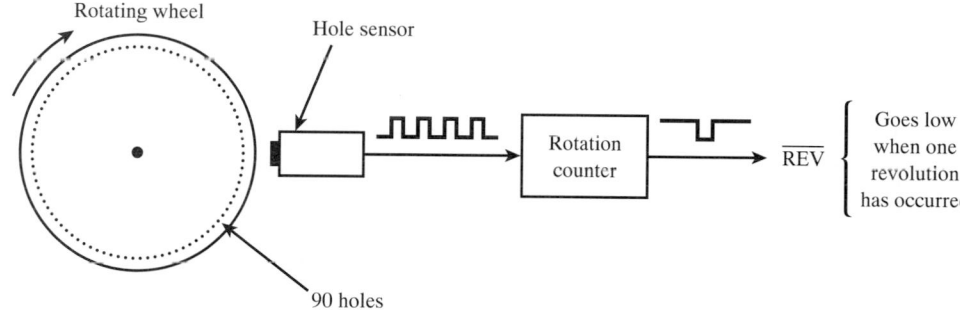

FIGURE 7.14   Rotation Counter Logic

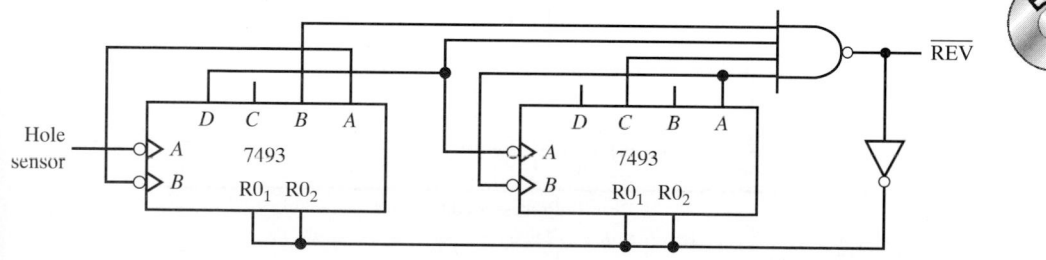

sired instruction. The address of the next instruction is stored in a register called the ***program counter.*** A 16-bit program counter is capable of addressing 64K memory locations.

Normally, as a program executes, the program counter simply increments by one after each instruction to prepare for the next instruction sequence. But some instructions such as JUMP, CALL (subroutine), and RET (return from subroutine) cause the program to change course. In these cases, a new address must be loaded into the program counter. Execution resumes at the new address pointed to by the program counter.

Figure 7.15 illustrates a block diagram of the program counter logic for a simple microprocessor. The program counter is composed of two 8-bit counters capable of parallel load. The counters are loaded with the address supplied by the JUMP-CALL-RET logic. A total of four 74193s could be used to implement the program counter by cascading them correctly.

Digital Clock

In the "old days," designing and building a digital clock was a big effort, as we will see, containing many individual counters, decoders, timing, and control logic. Today, a single chip contains all the functions necessary to maintain a 24-hour clock, 12-month calendar, and alarm and the logic to drive the displays. Let us examine the design of a digital clock anyway, for it will be a good review of the basics, and possibly will give us a look at some new tricks.

The basic component of the digital clock is the circuit module that contains all the logic required to maintain a single digit in the display. The basic module can be found in the Electronics Workbench circuit file 7SEG, and is shown in Figure 7.16. The 7490 counts from 0 to 9 over and over again. The decoder converts BCD into the appropriate seven-segment display control signals. This file is included to let you begin experimenting with counting circuitry, or as the beginnings of your own digital clock.

Two slightly different modules are actually used, a modulo-10 counter and modulo-6 counter. These are shown together in Figure 7.17. The modulo-6 counter is used for the 10s of seconds and 10s of minutes digits. The choice of display driver depends on the flavor of the seven-segment display (common anode or common cathode).

FIGURE 7.15   Simple Program Counter for a Microprocessor

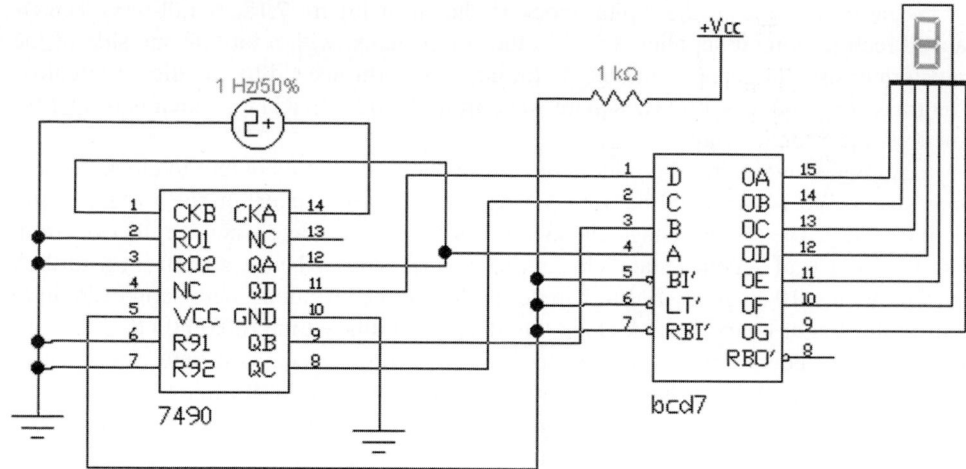

FIGURE 7.16   EWB Circuit 7SEG (Modulo-10 Counter)

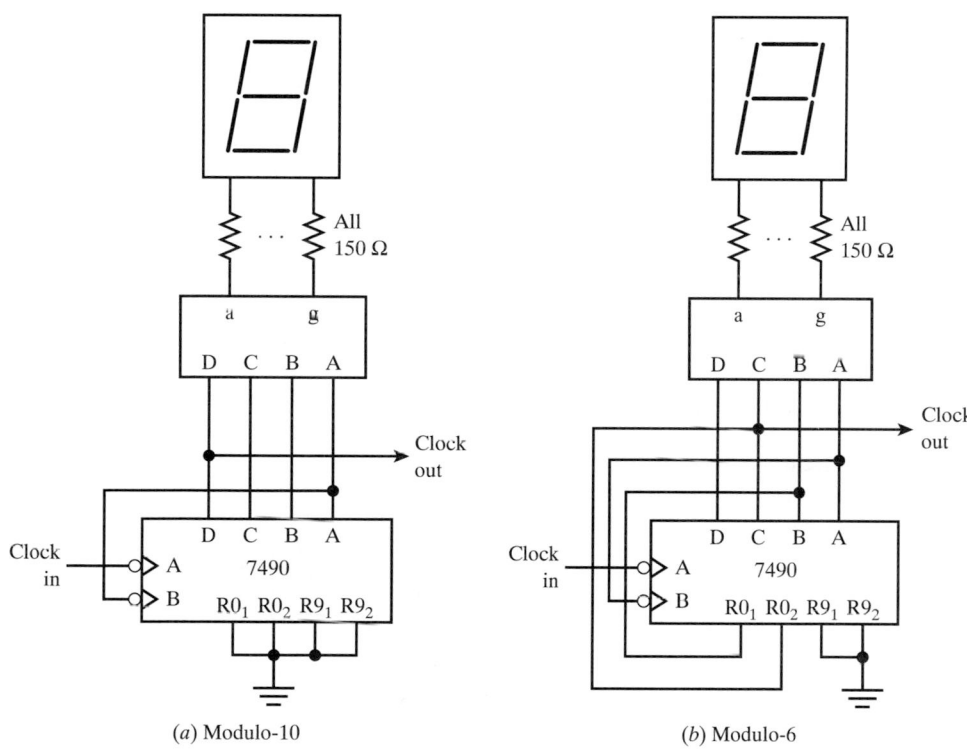

(*a*) Modulo-10   (*b*) Modulo-6

FIGURE 7.17   Modulo-10 and Modulo-6 Circuits

The basic logic of the digital clock is shown in Figure 7.18. A full-wave center-tapped rectifier circuit supplies +5 V to the components, with a tap off one side of the transformer used to generate the 60 Hz timing waveform needed by the digital circuitry. The transistor produces a 60 Hz square wave from the 60 Hz AC tap, which is further refined via the Schmitt trigger.

A divide-by-60 circuit produces one pulse per second at its output to clock the seconds stage. Two 7490s could be used (one as a modulo-6 counter) or some other equivalent circuitry. The seconds stage clocks the 10s of seconds stage when it rolls over from 9 to 0. The 10s of seconds stage clocks the minutes stage when it goes from 5 to 0. A special modulo-10 stage is used to handle the hours display, along with support logic to control the 10s of hours digit. A clock that displays military time (00:00:00 to 23:59:59) requires trivial hours logic to reset from 23 back to zero. A standard time clock, where 12:59:59 becomes 1:00:00 has more complex hours logic. A 7490 cannot be used, as it resets to zero, not one. In addition, the modulo-10 counter for the hours digit must reset at both 9 (when 10 o'clock comes around) and at 2 (going from 12 o'clock to 1 o'clock). A good solution might be to use a 74192 or 74193, with the appropriate parallel load values and reset logic.

We will return to our digital clock in Chapters 8 and 10, adding more features and redesigning others.

FIGURE 7.18 Digital Clock

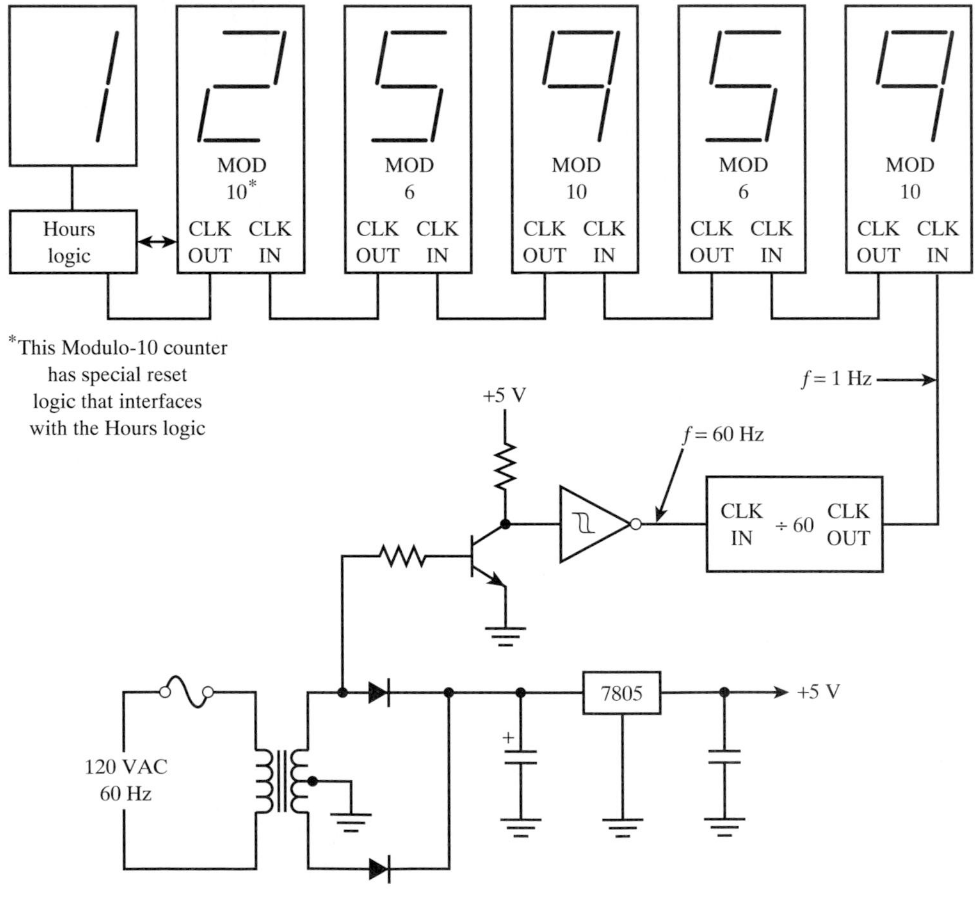

$f = 60$ Hz

$f = 1$ Hz

*This Modulo-10 counter
has special reset
logic that interfaces
with the Hours logic

3-Bit PAL Counter

Let us look at the design of a 3-bit synchronous counter (review Figure 7.10(b) for the details). Recall from Chapter 6 that the basic equation for a J-K flip-flop looks like this:

```
QA := JA * /QA * /CLR
    + /KA * QA * /CLR
    + PRE
```

Eliminating the PRE input gives us

```
QA := JA * /QA * /CLR
    + /KA * QA * /CLR
```

The first term is used for toggling and the second term for holding the same state.

The first bit in the counter will simply toggle with each clock pulse. The second two bits toggle under certain conditions and otherwise remain in the same state. The equations necessary for the 3-bit counter are as follows:

```
QA := /QA * /CLR                ; always toggle

QB := /QB * QA * /CLR           ; toggle if LSB is high
    + QB * /QA * /CLR           ; preserve state if LSB is low

QC := /QC * QA * QB * /CLR      ; toggle if lower two bits are
      high
    + QB * /QA * /QB * /CLR     ; preserve state otherwise
    + QB * QA * /QB * /CLR
    + QB * /QA * QB * /CLR
```

The equation for QC is complicated because there are many conditions that require QC to stay in the same state.

Being able to use counters within our PAL designs makes programmable logic even more attractive.

7.8 TROUBLESHOOTING TECHNIQUES

When working with counters, a number of points should be kept in mind when things do not go as planned and the counter is not working properly.

◆ Check power and ground connections. Sometimes a missing ground causes intermittent behavior or grossly distorted waveforms.

◆ For the 7490 and 7493, ground the R0 inputs (and the R9s on the 7490), apply a clock to the A-CLK input, and look at the A output. If there is no waveform, and the power connections are correct, the chip may be bad.

◆ When using all the outputs of the 7490 or 7493, remember to feed the A output back into the B-CLK input, or they will not count properly.

◆ Sometimes it will appear that the counter is bad, when in fact it is the circuit the counter is driving that has the problem. If possible, isolate the counter outputs from the rest of the circuit while testing.

◆ Check the data sheet for the counter. If the wrong logic level is used on the clear input, the counter will be stuck at zero. The same is true for other inputs as well. If you use the wrong logic level on the parallel load input the counter will be stuck at the loaded count.

◆ Be sure to use the proper edge on the clock input. The counter may appear to operate correctly, but will become unstable at high speeds.

◆ If possible, use a logic analyzer to examine all inputs and outputs of the counter. A relationship between two signals may be found, which can lead the way towards a solution. For instance, you may discover that the B and C outputs look exactly the same when viewed with the logic analyzer, even though you may not have noticed the similarity when probing individual signals with an oscilloscope. The cause of the problem may turn out to be a solder blob between two pins on the bottom of the circuit board or two loose wire-wrapped connections side by side.

In general, if all else fails you may want to try replacing the counter chip. If this does not solve the problem, turn the counter circuit over to someone else for fresh ideas. Eventually, the problem will be found. With only a few simple connections to make, it should never be impossible to get a counter functioning.

SUMMARY

In this chapter we examined the operation of ripple and synchronous counters, the BCD, binary, and modulo-N counting modes, and several applications involving frequency division. A number of standard TTL counter packages were illustrated as well.

STUDY QUESTIONS

General

1. How many flip-flops are required to count from 0 to 63?

2. Draw the schematic of a 5-bit ripple counter.

3. If the propagation delay for each flip-flop in a 4-bit ripple counter is 8 ns, what is the maximum frequency allowed on the clock?

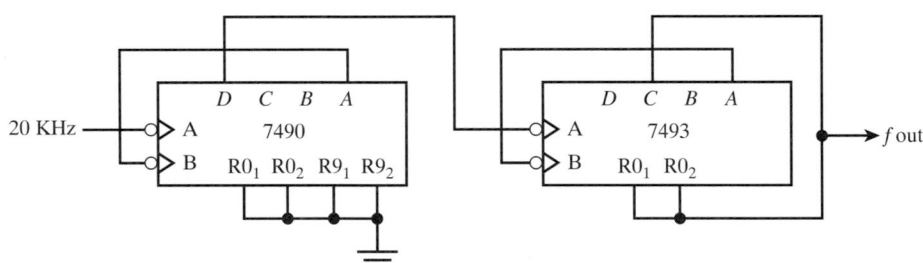

FIGURE 7.19 For Question 7.4

4. What is the output frequency of the circuit in Figure 7.19?

5. Design a divide-by-7 and a divide-by-13 counter using a 7493.

6. Sketch the timing diagram for the two counters designed in Question 5.

7. Determine the effect on the binary count of a 4-bit (7493) counter when the ABCD outputs are inverted ($\overline{A}\,\overline{B}\,\overline{C}\,\overline{D}$).

8. What division values are possible with one 7490? How about one 7493?

9. The input frequency to a 7490 is 15 KHz. What is the frequency of each output? Repeat

for the 7493.

10. Use two 7490 decade counters to make a divide-by-88 circuit.

11. Use two 7490s to make a divide-by-48 circuit.

12. A certain circuit will divide by 88 when its control line DIV is high, and by 48 when DIV is low. Modify the circuits from Questions 10 and 11 to divide by 48 or 88 depending on DIV; use only two 7490s and some support TTL gates.

13. What is the maximum count (or division) that can be obtained with each of the following counters?
    (a) Two 7490s   (d) Two 7493s
    (b) A 7490 and a 7493   (e) Three 7490s
    (c) A 7490 and a D flip-flop   (f) Four 7493s

14. Design a modulo-20 counter. Use a 7490 and a J-K flip-flop.

15. How is ripple eliminated in a synchronous counter?

16. What is the basic reason synchronous counters may be clocked faster than ripple counters?

17. Design a 74193-based modulo-9 counter. Show an up-counting circuit and a down-counting circuit.

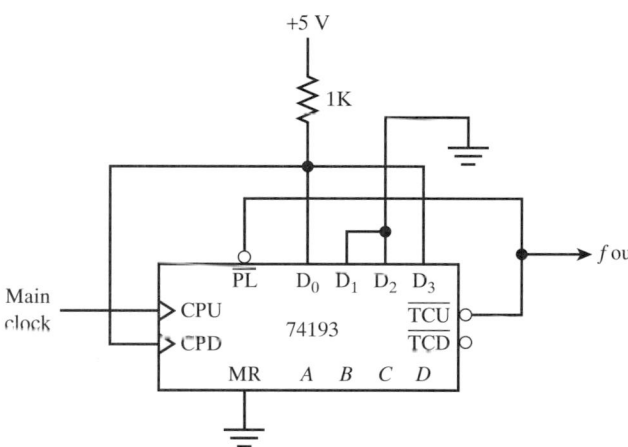

FIGURE 7.20 For Question 7.18

18. What is the counting sequence and output frequency for the counter in Figure 7.20?

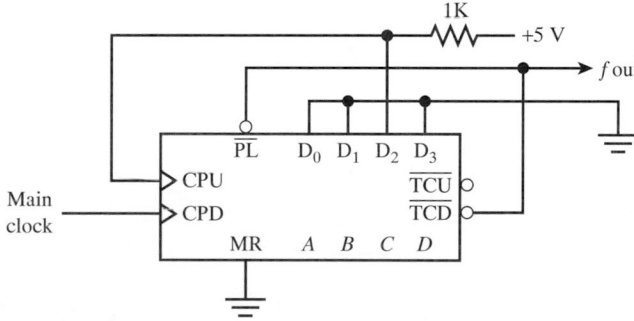

FIGURE 7.21 For Question 7.19

19. Repeat Question 18 for the counter in Figure 7.21.

20. Show how two 74193s can be cascaded to make an 8-bit up-counter.

21. Design a counter that starts at 7 and counts up to 13 before resetting.

22. Design the standard time hours logic for the digital clock. Four BCD bits must be output for the hours digit, and two bits for the 10s of hours digit. The logic should roll over from 23 to 0 (reset at 24).

23. Repeat the design of the hours logic for a standard 12-hour clock.

24. What is required to add a tenth-of-a-second digit to the digital clock?

25. Design a 16-bit counter that begins counting when a START button is pushed and stops counting, but does not reset, when a STOP button is pushed. Pressing START resets the counter before counting begins. Assume a 1 KHz clock is available for use.

Electronics Workbench

26. Compare the timing diagrams of a 3-bit ripple counter with a 3-bit synchronous counter using Electronics Workbench. Explain your results.

27. Verify your solutions to Questions 5, 9, 10, and 14 using Electronics Workbench.

Programmable Logic

28. Write the equations for a 4-bit synchronous counter. Build on the equations given for the 3-bit counter.

29. How can an additional control signal called HOLD be used to hold the counter in Question 28 in its current state (HOLD is high) or count normally (HOLD is low)?

30. How can an additional control signal called UP be used to select incrementing (UP is high) or decrementing (UP is low) in the counter from Question 28?

CHAPTER

8

SYNCHRONOUS LOGIC CIRCUIT DESIGN

INSTRUCTIONAL OBJECTIVES

When finished with this chapter you should be able to:

1. Describe the difference between synchronous logic and combinational logic.

2. Design a state machine from its state diagram or transition table.

3. Analyze a synchronous logic circuit to determine its transitions.

SELF-EVALUATION QUESTIONS

Keep the following questions in mind and try to answer them when you have completed the chapter.

1. What is a state diagram?

2. How are the J-K flip-flop excitations derived?

3. Why is a counter a state machine, but a state machine need not be a counter?

4. Why would a ROM be used in a state machine?

8.1 INTRODUCTION

This chapter takes a different look at logic design. Memory elements (flip-flops) are now added to combinational logic to make synchronous logic circuits, whose output depends on its input and its *last output*. The circuit must remember its previous state. None of this was necessary with plain combinational logic because the output followed the input after a short propagation delay. But synchronous logic will allow us to do much more than before, to do it faster, and to do it all at the same time.

8.2 SYNCHRONOUS LOGIC VS. COMBINATIONAL LOGIC

Figure 8.1 illustrates the difference between synchronous logic and combinational logic. The output of the combinational logic circuit in Figure 8.1(a) is a simple Boolean function of its input. The synchronous logic circuit in Figure 8.1(b) shows that its output depends on its current input *and* the current output. Furthermore, it is important to understand that the output of a synchronous logic circuit *does not change* until a clock pulse is applied. Even if the input signals are changed, the output will remain in its current state, changing only when a clock pulse is applied. Synchronous logic circuits are also referred to as *state machines,* since they go from state to state in a synchronous fashion and perform a specific function. All microprocessors are state machines.

The storage element of a synchronous logic circuit is a flip-flop (or set of flip-flops) that store the current state of the circuit. State changes occur when the flip-flops are clocked. Compare this to the combinational circuit, whose output changes after a brief gate delay whenever the input changes. Let us take a more detailed look at a synchronous logic circuit.

Figure 8.2 shows how two J-K flip-flops and an AND gate are used to make a 2-bit synchronous up counter. It is not necessary to use J-K flip-flops; other types will work as well, such as D flip-flops.

The level of the Count input determines what happens when a clock pulse occurs, along with the state of the A output. When Count is low, both flip-flops have zeros on their J and K inputs. This selects the "no change" condition for the next clock pulse. When Count is high, the A flip-flop is allowed to toggle, since its J and K inputs are high (following Count). Flip-flop B only toggles when the A output is high and Count is also high. It is important to see that controlling the state of the J and K inputs is the way to move from one state to the next.

FIGURE 8.1   Comparing Combinational Logic and Synchronous Logic

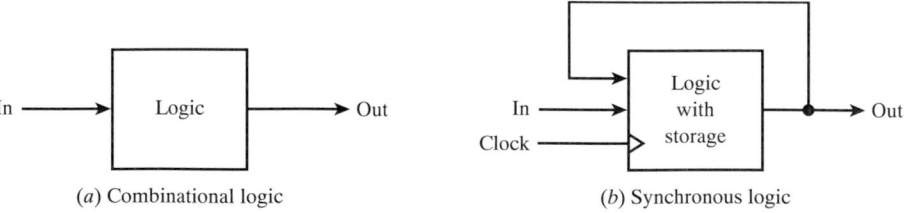

(*a*) Combinational logic   (*b*) Synchronous logic

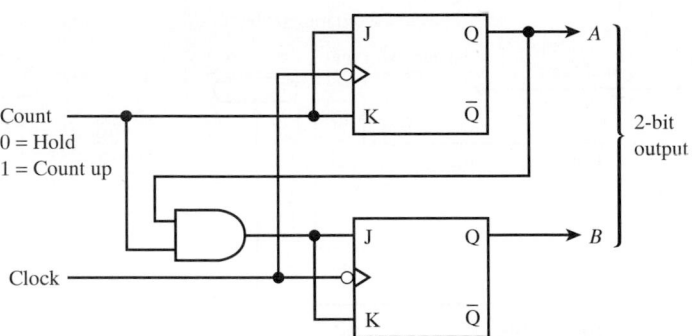

FIGURE 8.2   Simple Synchronous Logic Circuit (2-Bit Up Counter)

Designing a synchronous logic circuit is a matter of knowing how to set up the various J-K inputs of each flip-flop. In the next section we will examine the J-K flip-flop from a different viewpoint than we have previously.

8.3	TRUTH TABLES
	AND EXCITATION TABLES

8.3   TRUTH TABLES AND EXCITATION TABLES

We are accustomed to looking at the J-K flip-flop through its truth table, which is shown in Figure 8.3(a). Another valid way of representing the operation of the J-K flip-flop is through its *excitation table*. Figure 8.3(b) shows the excitation table for the J-K flip-flop. We are interested in what J and K logic levels are needed to get a desired change in the Q output of the flip-flop. For example, the first line of the excitation table indicates that Q is low before the clock pulse and also low after the clock pulse. To get Q to behave this way we must place a zero on the J input. The logic level of the K input does not matter (it can be low or high). During the design of a synchronous logic circuit we first determine how we want the Q outputs to change, and then figure out how to provide the necessary J and K levels.

Where does the excitation table come from? The truth table can help answer that question. Figure 8.4 shows how the first line of the excitation table is derived. Since Q is

FIGURE 8.3   Two Ways of Looking at the J-K Flip-Flop

J	K	Q	$\bar{Q}$	
0	0	No change		
0	1	0	1	(clear)
1	0	1	0	(set)
1	1	Toggle		

(*a*) Truth table

Q_{BEFORE}	Q_{AFTER}	J	K
0	0	0	X
0	1	1	X
1	0	X	1
1	1	X	0

where X means "don't care" (0 or 1)

(*b*) Excitation table

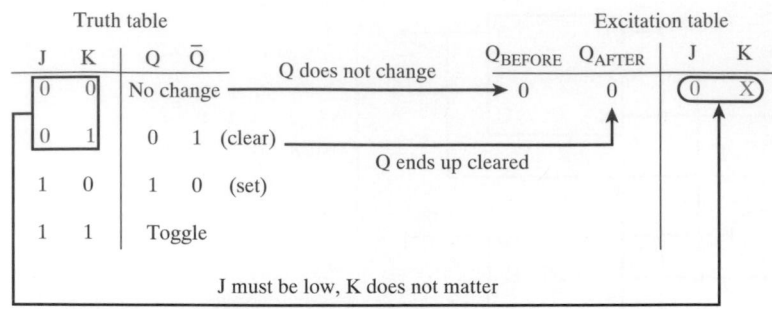

FIGURE 8.4 **Deriving the Excitation Table**

low before the clock pulse *and* after the clock pulse, the J-K flip-flop could have been set up for the "no change" condition or the "clear" condition. To select either of these two conditions it is necessary to place a zero on the J input prior to clocking. The K input can be low or high, and is listed in the excitation table as "x," for "don't care."

By a similar method, the other three lines of the excitation table are derived as follows:

◆ Line 2: Set or Toggle (1,x). Q is low before the clock pulse and high after.

◆ Line 3: Clear or Toggle (x,1). Q is high before the clock pulse and low after.

◆ Line 4: No change or Set (x,0). Q is high before and after the clock pulse.

Other flip-flops have their own specific excitation tables. It is useful to practice deriving these other excitation tables, since you may not always use J-K flip-flops in your designs.

8.4 STATE DIAGRAMS AND STATE TRANSITION TABLES

The operation of a synchronous logic circuit may be represented graphically through use of a ***state diagram*** that depicts how the circuit changes from state to state. Figure 8.5 shows the state diagram for a simple 2-bit up counter (no Count input). A state is

FIGURE 8.5 **State Diagram for a 2-Bit Up Counter**

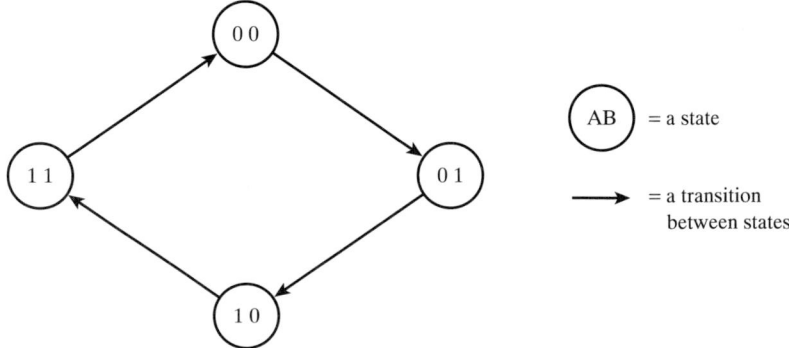

indicated by a circle enclosing a 2-bit state number. Each bit in the state number is actually the Q output of the flip-flops used in the actual state machine. So, the state 10 corresponds to Q = 1 on the A flip-flop, and Q = 0 on the B flip-flop.

The arrows indicate how the counter changes from state to state and are taken when a clock pulse is received. State 00 goes to state 01, state 01 to state 10, etc.

Figure 8.6 shows the addition of the Count input, which is used to hold the counter when low. The "Count = 0" arrows cause the current state to loop on itself, thus keeping the state of the machine constant. When Count equals one the states are allowed to change.

The information present in a state diagram can also be represented in a ***state transition table.*** The state diagram from Figure 8.6 is expressed in Table 8.1. Note that whenever the Count input is low, the "Next State" column contains the same state number as the "Present State" column. Only when Count equals one do we see the "Next State" values change.

In the next section we will see how the state transition table is used to determine the flip-flop excitations required to get the desired state transitions.

FIGURE 8.6   2-Bit Up Counter with Hold

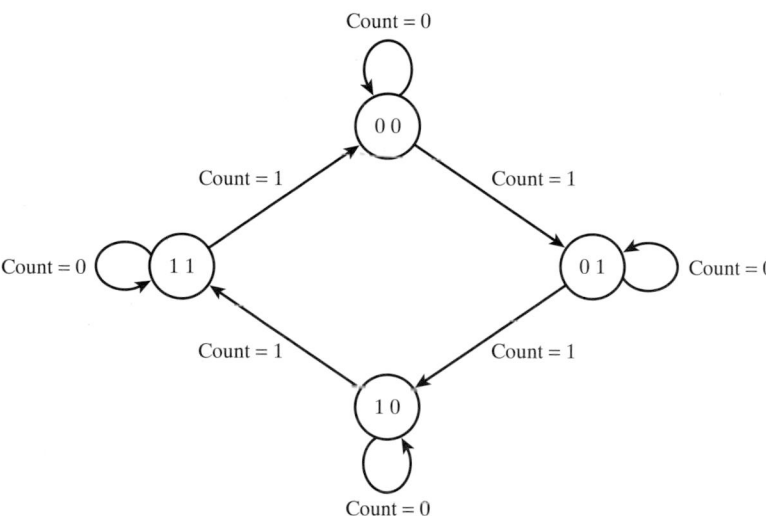

TABLE 8.1   State Transition Table for 2-Bit Up Counter with Hold

PRESENT STATE			NEXT STATE	
A	B	COUNT	A	B
0	0	0	0	0
0	0	1	0	1
0	1	0	0	1
0	1	1	1	0
1	0	0	1	0
1	0	1	1	1
1	1	0	1	1
1	1	1	0	0

8.5 DESIGNING A 2-BIT SYNCHRONOUS UP COUNTER

The state transition table for the 2-bit up counter must be examined to determine the required logic levels needed on the J and K inputs of each flip-flop. This process begins in Figure 8.7. There are four lines in the excitation table for the J-K flip-flop. The first line is for 0-0 transitions on the Q output. Figure 8.7 shows the first 0-0 transition in the map, and how the excitations for JA and KA (0,x) are found. We are looking for A and B levels that are zero before the clock pulse and zero after the clock pulse. Figure 8.8 has the rest of the 0-0 transitions filled in, along with the first 0-1 transition. These require 1,x on the J-K inputs.

Figures 8.9 and 8.10 show how the remaining excitations are found. Note that we have covered every possible way that an output can change from one state to the next.

FIGURE 8.7 Looking Up and Controlling State Changes (0-0 Transitions)

Present State			Next State		Flip-Flop Excitations			
A	B	Count	A	B	JA	KA	JB	KB
0	0	0	0	0	0	X		
0	0	1	0	1				
0	1	0	0	1				
0	1	1	1	0				
1	0	0	1	0				
1	0	1	1	1				
1	1	0	1	1				
1	1	1	0	0				

Q_{BEFORE}	Q_{AFTER}	J	K
0	0	0	X
0	1	1	X
1	0	X	1
1	1	X	0

FIGURE 8.8 Finding 0-1 Transitions

Present State			Next State		Flip-Flop Excitations			
A	B	Count	A	B	JA	KA	JB	KB
0	0	0	0	0	0	X	0	X
0	0	1	0	1	0	X		
0	1	0	0	1	0	X		
0	1	1	1	0	1	X		
1	0	0	1	0			0	X
1	0	1	1	1				
1	1	0	1	1				
1	1	1	0	0				

Q_{BEFORE}	Q_{AFTER}	J	K
0	0	0	X
0	1	1	X
1	0	X	1
1	1	X	0

Present State			Next State		Flip-Flop Excitations			
A	B	Count	A	B	JA	KA	JB	KB
0	0	0	0	0	0	X	0	X
0	0	1	0	1	0	X	1	X
0	1	0	0	1	0	X		
0	1	1	1	0	1	X		
1	0	0	1	0			0	X
1	0	1	1	1			1	X
1	1	0	1	1				
1	1	1	0	0	X	1		

Q_{BEFORE}	Q_{AFTER}	J	K
0	0	0	X
0	1	1	X
1	0	X	1
1	1	X	0

FIGURE 8.9 Finding 1-0 Transitions

Present State			Next State		Flip-Flop Excitations			
A	B	Count	A	B	JA	KA	JB	KB
0	0	0	0	0	0	X	0	X
0	0	1	0	1	0	X	1	X
0	1	0	0	1	0	X		
0	1	1	1	0	1	X	X	1
1	0	0	1	0	X	0	0	X
1	0	1	1	1			1	X
1	1	0	1	1				
1	1	1	0	0	X	1	X	1

Q_{BEFORE}	Q_{AFTER}	J	K
0	0	0	X
0	1	1	X
1	0	X	1
1	1	X	0

FIGURE 8.10 Finding 1-1 Transitions

There is no guessing involved, even though the excitation table entries are filled with "don't cares." Table 8.2 shows the final state transition table for the 2-bit synchronous up counter.

The next step is to find equations for the J and K inputs that are functions of the inputs (A, B, and Count). Figure 8.11 shows the way the JA equation is found. Recognize that the JA column in the excitation table is the output of a three-input digital black box. The logic levels in the JA column can be transferred into a Karnaugh map for solution.

When circling groups, use the "don't care" entries only if they help circle a larger group that contains actual one levels. In Figure 8.11 the don't care in m_7 is paired with the one in m_3 to get a simplified equation. Not using the don't care results in additional or more complex logic. It is not necessary to circle the group of four don't cares in the bottom row because they may be interpreted as zeros. Do you see how a don't care can be a zero and a one at the same time? This is the role played by the don't care in m_7.

TABLE 8.2 **Final State Transition Table**

PRESENT STATE			NEXT STATE		FLIP-FLOP EXCITATIONS			
A	**B**	**COUNT**	**A**	**B**	**JA**	**KA**	**JB**	**KB**
0	0	0	0	0	0	x	0	x
0	0	1	0	1	0	x	1	x
0	1	0	0	1	0	x	x	0
0	1	1	1	0	1	x	x	1
1	0	0	1	0	x	0	0	x
1	0	1	1	1	x	0	1	x
1	1	0	1	1	x	0	x	0
1	1	1	0	0	x	1	x	1

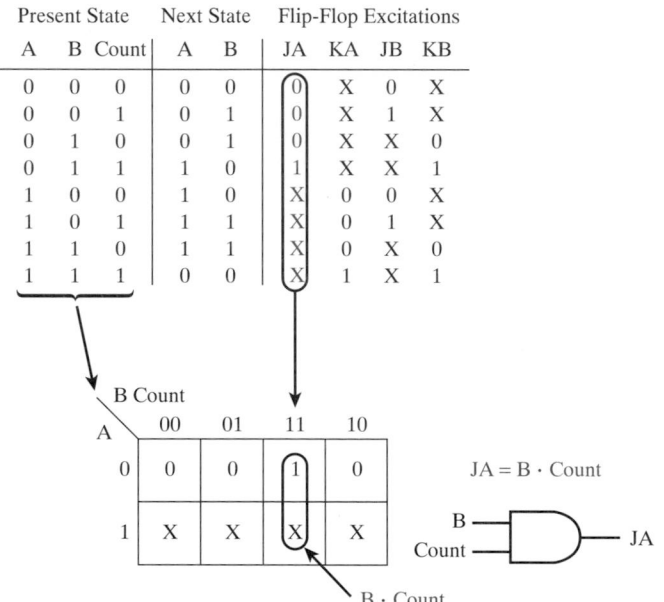

FIGURE 8.11 **Mapping a Single Column from the Excitation Table**

Figure 8.12 shows all four Karnaugh maps for the 2-bit counter. The Count input is part of each equation, helping to control what happens at each flip-flop. The schematic diagram for the counter is shown in Figure 8.13. Both flip-flops are clocked at the same time, as expected in a synchronous counter. The Count input, when low, applies 0s to the J-K inputs of flip-flop B, and via the AND gate, another set of 0s to flip-flop A. This guarantees that the counter will not change state during the clock pulse. When Count is high, flip-flop B is allowed to toggle with each clock pulse (the pattern normally seen on the LSB of a binary count). Flip-flop A only toggles when the output of flip-flop B is high (and Count is also high).

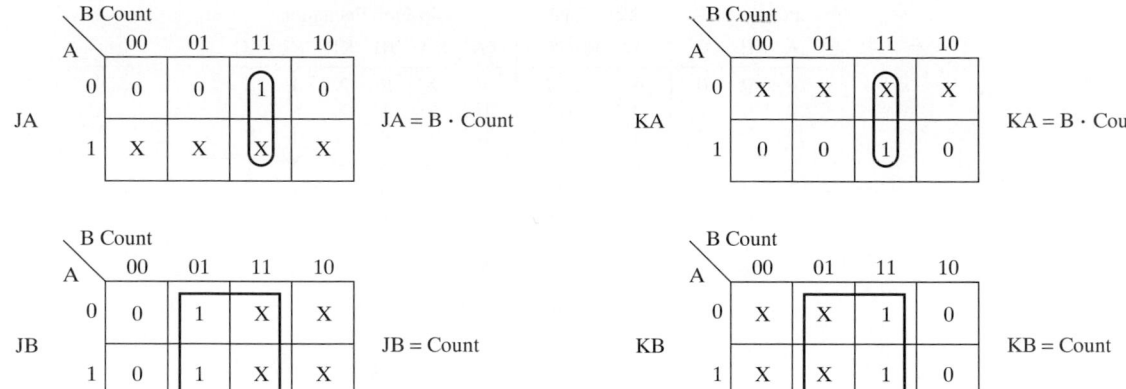

FIGURE 8.12 Complete Karnaugh Map Solution

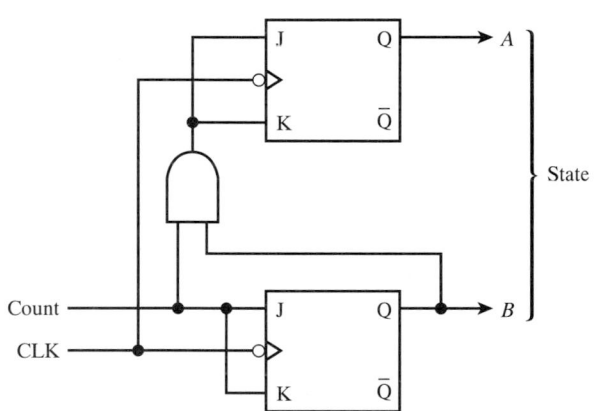

FIGURE 8.13 2-Bit Synchronous Up Counter with Hold

8.6 STATE MACHINES

A state machine, in general, is a synchronous circuit that has a finite number of states that transition from one to another according to prearranged rules. Here is a short list of uses for a state machine:

◆ Recognizing a pattern
◆ Performing arithmetic
◆ Controlling an industrial process
◆ Flashing holiday lights in different sequences
◆ Counting
◆ Dialing a telephone
◆ Transmiting/receiving serial data
◆ Controlling change in a vending machine

Let us look at two simple examples.

Present State			Next State			Flip-Flop Excitations					
A	B	C	A	B	C	JA	KA	JB	KB	JC	KC
0	0	0	0	0	1	0	X	0	X	1	X
0	0	1	0	1	0	0	X	1	X	X	1
0	1	0	0	1	1	0	X	X	0	1	X
0	1	1	1	0	0	1	X	X	1	X	1
1	0	0	1	0	1	X	0	0	X	1	X
1	0	1	1	1	0	X	0	1	X	X	1
1	1	0	1	1	1	X	0	X	0	1	X
1	1	1	0	0	0	X	1	X	1	X	1

(*a*) State transition table

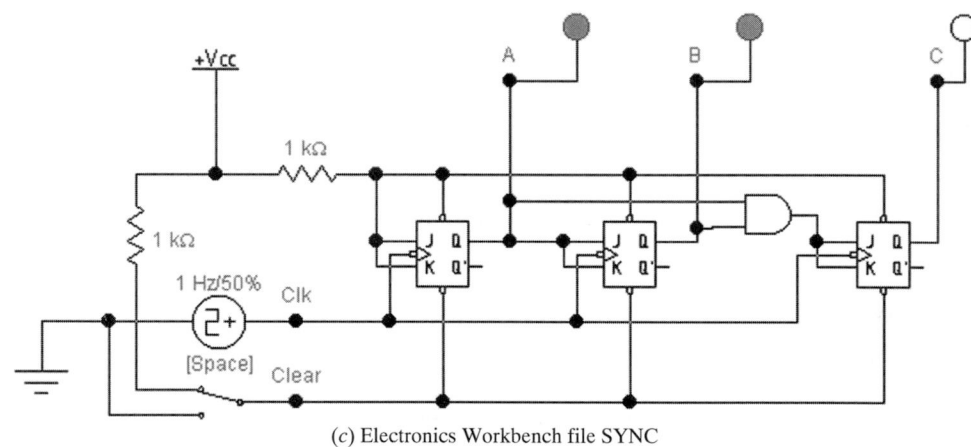

(*b*) Schematic

(*c*) Electronics Workbench file SYNC

FIGURE 8.14   3-Bit Synchronous Up Counter

The transition table for a 3-bit up counter is shown in Figure 8.14, along with the associated schematic. The Karnaugh maps are left for you to work out on your own. An Electronics Workbench version of the counter (saved as SYNC) is included to further illustrate its operation.

Flip-flop C (see Figure 8.14(b)) always toggles with each clock pulse. This is normal for the LSB of a binary count. Flip-flop B only toggles when Q_C is high. Flip-flop A only toggles when the first two flip-flops are both set.

The pattern at the output of a state machine does not have to resemble a binary count; it may wander all over the place according to the needs of the designer. This is

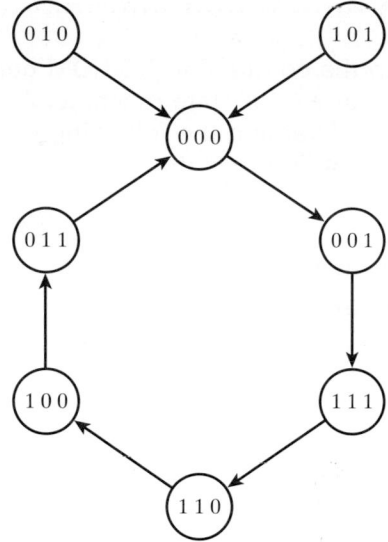

(*a*) Transition diagram

Present State			Next State			Flip-Flop Excitations					
A	B	C	A	B	C	JA	KA	JB	KB	JC	KC
0	0	0	0	0	1	0	X	0	X	1	X
0	0	1	1	1	1	1	X	1	X	X	0
0	1	0	0	0	0	0	X	X	1	0	X
0	1	1	0	0	0	0	X	X	1	X	1
1	0	0	0	1	1	X	1	1	X	1	X
1	0	1	0	0	0	X	1	0	X	X	1
1	1	0	1	0	0	X	0	X	1	0	X
1	1	1	1	1	0	X	0	X	0	X	1

(*b*) Transition table

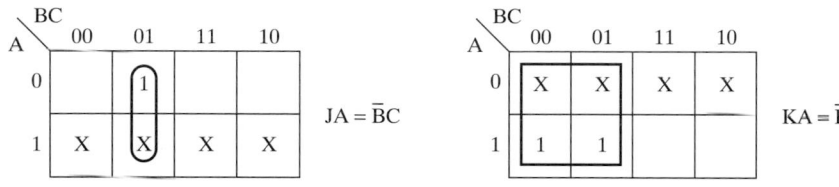

(*c*) Equations for first flip=flop

FIGURE 8.15 Pseudo-Random Sequence

illustrated in the state diagram shown in Figure 8.15(a). The counting sequence is *pseudo random,* not really random but appearing to be. The count goes like this: 0, 1, 7, 6, 4, 3. If the machine happens to power up into states 2 or 5, the first clock pulse will get the machine to state zero. The excitations required for three J-K flip-flops are shown in Figure 8.15(b). The Karnaugh maps and equations for flip-flop A are shown in Figure 8.15(c). The other equations for flip-flops B and C are left for you to work out on your own. To save time in a synchronous design, use the MACHINE program illustrated in the following C/C++ Helper.

····**C/C++ HELPER**·····································

The MACHINE program on the companion CD-ROM generates the required flip-flop excitations for a 3-bit state machine. All you need do is enter the state transitions. Examine the following execution, which uses the state transitions from Figure 8.15.

```
C> MACHINE
Enter the state transitions...
From state 0 to state --> 1
From state 1 to state --> 7
From state 2 to state --> 0
From state 3 to state --> 0
From state 4 to state --> 3
From state 5 to state --> 0
From state 6 to state --> 4
From state 7 to state --> 6
```

Present State			Next State			JK Excitations					
A	B	C	A	B	C	JA	KA	JB	KB	JC	KC
0	0	0	0	0	1	0	x	0	x	1	x
0	0	1	1	1	1	1	x	1	x	x	0
0	1	0	0	0	0	0	x	x	1	0	x
0	1	1	0	0	0	0	x	x	1	x	1
1	0	0	0	1	1	x	1	1	x	1	x
1	0	1	0	0	0	x	1	0	x	x	1
1	1	0	1	0	0	x	0	x	1	0	x
1	1	1	1	1	0	x	0	x	0	x	1

Verify for yourself that the J-K excitations are identical to those shown in Figure 8.15.

MACHINE saves you the trouble of having to compare present state and next state transitions and look up their excitations.

··

Handling Multiple Inputs

When additional inputs are added to a state machine, the number of possible combinations becomes too numerous to reduce using Karnaugh maps, and possibly even with the Quine-McCluskey method. Just adding one additional input to the 3-bit counter previously discussed doubles the number of next-state patterns. Figure 8.16(a) shows the state diagram for a modulo-6 counter that may count up or down depending on the $D/\overline{U}$ input. Even though there are only eight possible states (with states 6 and 7 designated as illegal

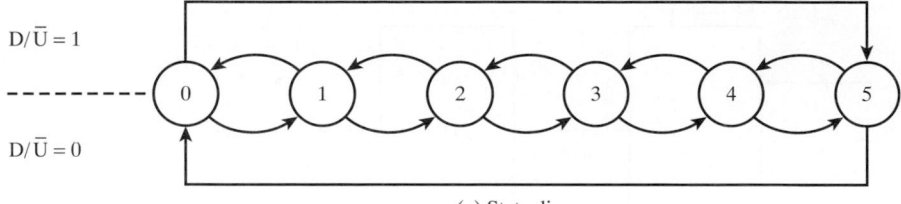

$D/\overline{U} = 1$

$D/\overline{U} = 0$

(a) State diagram

Present State				Next State		
A	B	C	$D/\overline{U}$	A	B	C
0	0	0	0	0	0	1
0	0	0	1	1	0	1
0	0	1	0	0	1	0
0	0	1	1	0	0	0
0	1	0	0	0	1	1
0	1	0	1	0	0	1
0	1	1	0	1	0	0
0	1	1	1	0	1	0
1	0	0	0	1	0	1
1	0	0	1	0	1	1
1	0	1	0	0	0	0
1	0	1	1	1	0	0
1	1	0	0	0	0	0
1	1	0	1	0	0	0
1	1	1	0	0	0	0
1	1	1	1	0	0	0

Illegal states
all return to 000
on first clock pulse

(b) State table

FIGURE 8.16 Synchronous Up/Down Modulo-6 Counter

for modulo-6 operation), there are sixteen different input combinations to check, as illustrated in Figure 8.16(b). A four-input Karnaugh map must be used to determine the equations for the J and K inputs. Even with some reductions, the logic required to implement the state machine may grow significantly as more inputs are added (32 combinations for 5 inputs, 64 for 6, and so on).

One way to eliminate the huge number of logic gates required is to store the transition information in a digital memory. Each location in the memory stores the data required to determine the next state and output pattern for the state machine. This memory is called a ***ROM,*** for read only memory. A ROM is programmed to output a specific pattern of data for each location. Once programmed, a ROM never forgets its data, even after power has been shut off and turned back on. A special type of ROM called an ***EPROM,*** for erasable-programmable read only memory, is often used for development purposes. Once the design is final, a ROM is used for manufacture.

Examine Figure 8.17 before continuing. The inputs to a ROM-based state machine are combined with the current ROM outputs to become the address used by the ROM to look up the new state information. Part of the data read out from the ROM specifies the new machine state (actually a new address in the ROM). The remainder of the ROM data become the output signals of the state machine. Specifically, the eight-bit ROM output is divided into three output bits and five state bits. Five state bits can select one of 32 locations. In addition, notice that the state machine has four inputs. This means that we have to store 16 groups of 32 state-output patterns to represent all the possible transitions. This amounts to a total of 512 locations. So, a 512-byte ROM is required.

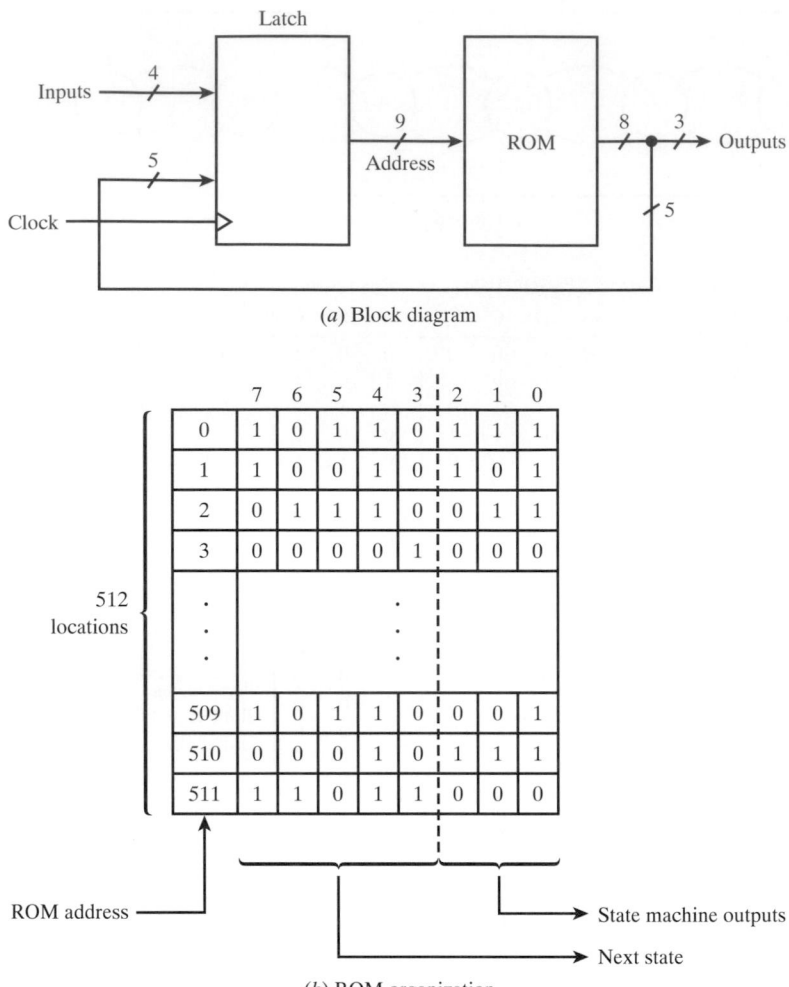

FIGURE 8.17 **ROM-Based State Machine**

A preloaded state memory is the basic idea behind the use of ***microcode*** in a microprocessor. The microcode ROM controls the operation of the instruction logic. Remember that the main benefit is that the single ROM may replace 10s or even 100s of logic gates. The speed of the machine is based on the access time of the ROM.

Imagine what you could do with a RAM-based state machine. Unlike a ROM, a ***RAM*** (random access memory) allows its patterns to be easily changed, storing them only as long as power is applied. By loading the RAM with a different set of patterns, you can actually change what the hardware does without changing the hardware.

8.7 ANALYZING A STATE MACHINE

You may be called upon to analyze a state machine to determine what the transitions are. It may have been designed by someone else, or you may just want to verify your own design by walking through it on paper. Figure 8.18 shows how this is done for a simple

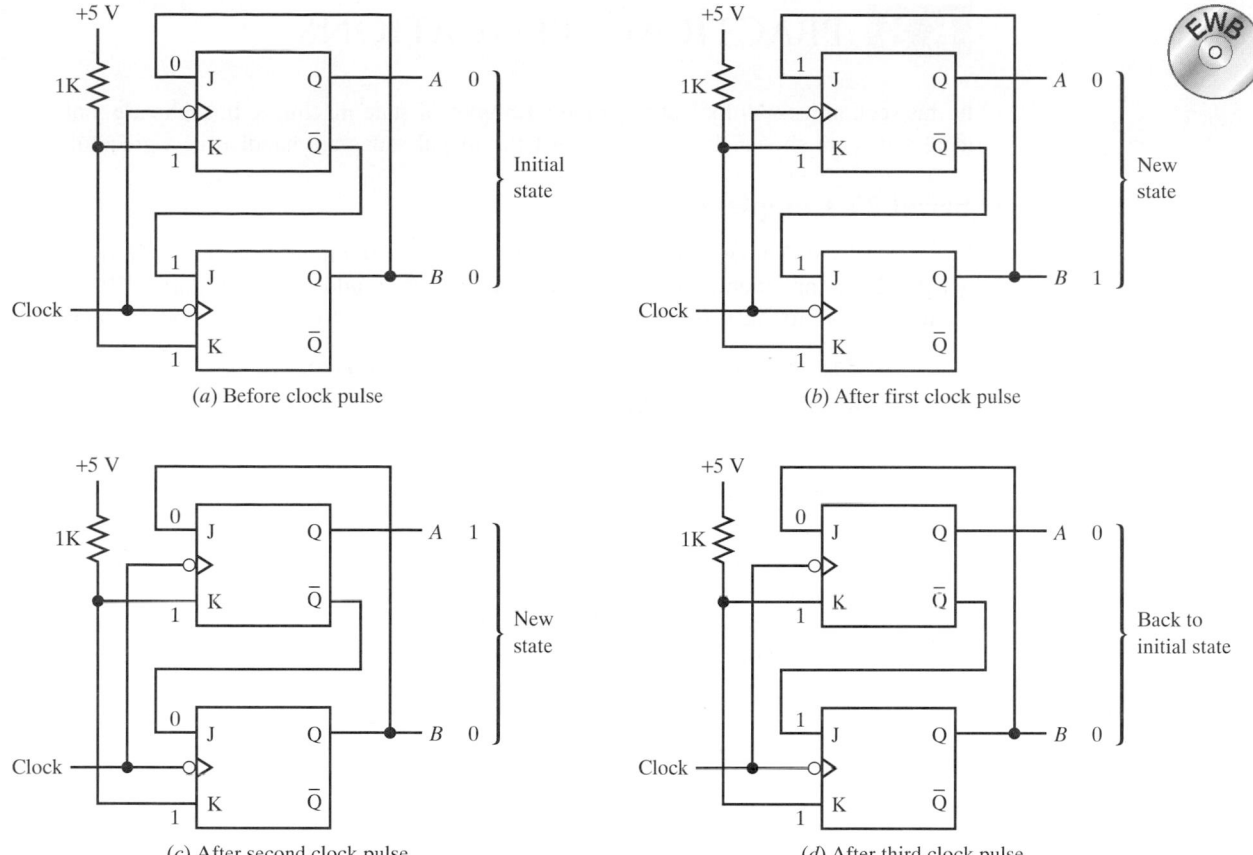

FIGURE 8.18   A 2-Bit State Machine

2 bit state machine. In Figure 8.18(a) we assume that both flip-flops start out clear. This is a perfectly reasonable assumption, for we could have just turned on power and the flip-flops may have come up that way.

The initial state on the A and B outputs, together with the pullup resistor, set up the J-K inputs so that flip-flop A is ready to clear and flip-flop B is ready to toggle, which it does on the first clock pulse. The new state is shown in Figure 8.18(b). The change in the B output now sets up flip-flop A for toggle on the next clock pulse. Since output A did not change (it is still zero), flip-flop B is still set up for toggle. When both flip-flops toggle on the second clock pulse, we get the results shown in Figure 8.18(c). Now the state machine's A and B outputs set up both flip-flops for the clear operation when the third clock pulse comes. This will get us back to our original state. The state sequence was 00, 01, 10, or 0, 1, 2.

What about state 11? Can we just ignore it since it did not show up in the analysis? No, that could lead to trouble. What if the flip-flops happen to come up both set? Maybe the machine never gets out of the 11 state. To see what would happen, assume it does start in the 11 state and see what the J-K inputs are set up for. In our current example, if A and B are both high, the new state will be 00 after the next clock pulse. The machine takes itself out of the illegal state to correct itself. This is a good thing to include in your design. Do not leave any holes in the state table.

8.8 PRACTICAL APPLICATIONS

In this section we will look at two more examples of state machines. In each case, only the legal states are shown. It is assumed that the illegal states are handled in a graceful way.

Serial 2's Complementer

Figure 8.19 shows how a binary number and its 2's complement are related. Normally, to get the 2's complement, we invert all the bits and then add one. In Figure 8.19, the 2's complement is found using a serial technique, one bit at a time. The process is simple:

1. Beginning with the LSB, keep all bits that are zero.
2. Keep the first one also.
3. Invert all the remaining bits.

A state diagram that implements this technique is shown in Figure 8.20(a). The terminology is slightly different on the transitions. We specify the input *and output* on each transition. Note from the state table in Figure 8.20(b) that the Out signal is only high when the machine is in state 01. This is easy to decode with a single AND gate, as illustrated in Figure 8.20(c). So, we design the machine from the state table like we usually do, determining the excitations for the two state flip-flops, and then add the AND gate to generate the

FIGURE 8.19 A Number and Its 2's Complement

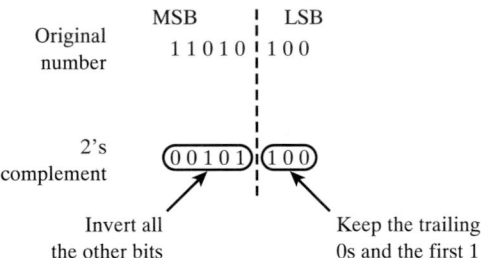

FIGURE 8.20 Serial 2's Complement Circuit

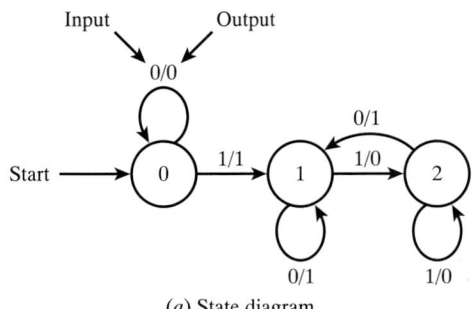

(*a*) State diagram

Present State		Next State	
AB	Data	AB	Out
00	0	00	0
00	1	01	1
01	0	01	1
01	1	10	0
10	0	01	1
10	1	10	0
11	0	00	X
11	1	00	X

(*b*) State table

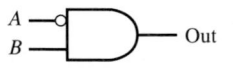

(*c*) Generating the serial output

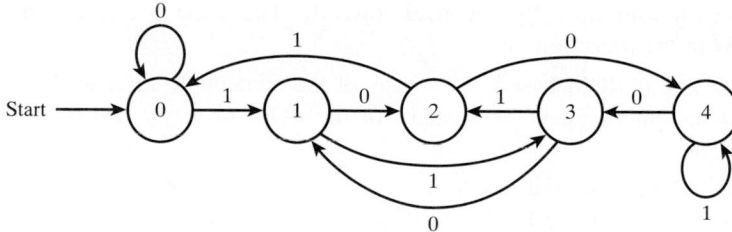

FIGURE 8.21 **State Machine That Finds Numbers Divisible by 5**

output signal. Though this technique takes longer than the parallel method of inverting all the bits and adding one, it requires a significantly smaller amount of hardware. Other operations, such as serial addition and subtraction, or comparison, can be done just as easily.

Pattern Recognition

Pattern recognition as applied to binary numbers means recognizing a pattern in the number. For example, the pattern we want to find might be all binary numbers that are evenly divisible by five. This is not as easy as it might appear, since 101, 1111, 11001, and 100011, and many other numbers, are all divisible by five but have nothing in common with each other in their binary representations.

Let us look at this problem from a different viewpoint. What are all the possible results we could get when dividing a number by five? What is left over? You should agree that either the number is divisible by five, and there is nothing left over, or the remainder is between one and four. This is the key point that allows us to find a simple solution to this problem. Figure 8.21 shows a five-state machine, with state zero indicated as the start state. Now, take a number like 1010 (ten). It is divisible by five. The machine is designed to return to state zero when the last bit of the input number is checked and the number is divisible by five. Otherwise, the state you end up in is the proper remainder. Trace the 1010 pattern through the machine. You should go from state 0 to 1, then from 1 to 2, then from 2 to 0, and finally from 0 back to 0. Try this with a number that does not work, like 111 (seven). The state sequence is 0 to 1 to 3 to 2. The machine ends in state 2 because 7 divided by 5 has two left over.

Since there are five states, three flip-flops must be used to keep track of the state number. You are encouraged to complete the design of the pattern recognizer. You may find patterns of your own while doing so.

8.9   TROUBLESHOOTING TECHNIQUES

If you are faced with the task of troubleshooting a synchronous circuit, you can do a number of things to obtain information about it.

◆ Using a logic analyzer, capture all of the inputs and outputs. Clock the logic analyzer probe from the state machine's clock. Use the waveforms to construct a state diagram, if possible, that you can compare with the original.

◆ Clock the circuit yourself. Use a bounceless pushbutton to provide a clean clock pulse. Clock the machine as fast or slow as necessary to enable you to see what is happening.

◆ In addition to the obvious steps of checking the wiring and required logic levels, go through the Karnaugh maps (or other reduction technique) and verify that the maps were filled in correctly and read correctly. The same is true for the J-K excitations in the transition table.

◆ If the state machine is ROM-based, get a copy of the contents to refer to when probing individual states. Read the contents of the ROM to verify it was programmed correctly.

◆ If you've designed a high-speed synchronous circuit, check the worst-case timing from the output back to the J-K inputs (through the excitation logic). Be sure that the setup and hold times for the flip-flops are within range.

The biggest synchronous logic circuit of them all is the microprocessor. Many of these tips apply to microprocessor troubleshooting as well.

SUMMARY

In this chapter we examined the properties of synchronous logic circuits. We saw how to use the J-K flip-flop in a different way, by supplying it with specific logic levels every clock pulse to control its output. We worked with state diagrams and transition tables and used Karnaugh maps to determine the logic equations required to control the flip-flops. A number of synchronous circuits were designed, including 2- and 3-bit counters.

STUDY QUESTIONS

General

1. What is synchronous about a synchronous circuit?
2. What makes a state machine different from ordinary combinational logic?
3. What is the difference between a truth table for a flip-flop, and an excitation table?
4. Derive the excitation table for a D flip-flop.
5. A state diagram contains seven states. How many flip-flops are required to keep track of the state number?
6. Repeat Question 5 for a state diagram containing 18 states.
7. Design the synchronous logic circuit for the state diagram shown in Figure 8.22.

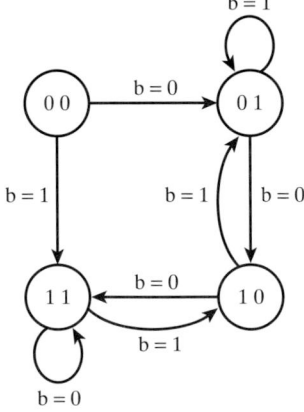

FIGURE 8.22   For Question 8.7

8. Design a synchronous modulo-10 counter.

9. Design a synchronous counter that counts from one to twelve.

10. Design a 3-bit binary counter with a HOLD input. The count is frozen when HOLD is high.

11. Design a 3-bit binary counter with a synchronous CLR input. The counter is cleared when CLR is high *and* the circuit is clocked.

12. Design a 3-bit shift register. How does the design differ from those covered in Chapter 6?

13. How do don't cares in the state table help us when reducing the Boolean equations?

14. Complete the design of the pseudo-random sequencer described in Figure 8.15.

15. Design a pseudo-random state machine for this sequence: 1, 15, 8, 3, 12, 6, 5, 4.

16. Complete the design of the modulo-6 up/down counter illustrated in Figure 8.16.

17. How many locations are required in the ROM of a ROM-based state machine given the following:

    ◆ The ROM output is 8 bits wide. Four bits are output signals, the other four are next-state bits.

    ◆ There are six inputs to the state machine.

18. Complete the design of the serial 2's complementer presented in Figure 8.20. Test your design on paper.

19. Design the division-by-5 state machine required by Figure 8.21.

20. Is the binary number 110111101 divisible by five? Use Figure 8.21 to prove your answer.

21. Repeat Question 20 for the number 10011010.

22. Design the synchronous logic required for both state diagrams in Figure 8.23.

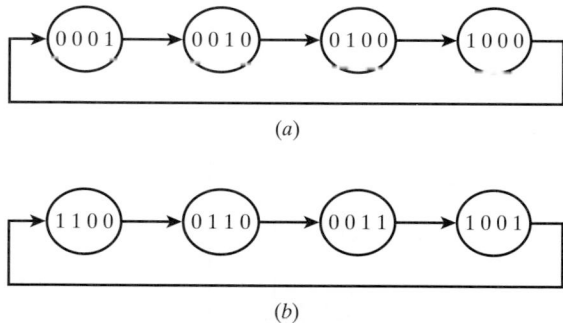

(a)

(b)

FIGURE 8.23   For Question 8.22

23. Determine the transition table for the circuit in Figure 8.24.

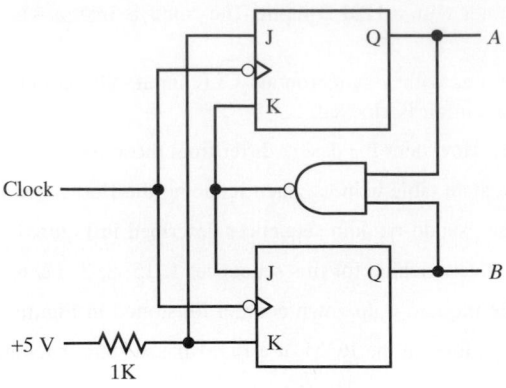

FIGURE 8.24 **For Question 8.23**

24. Repeat question 23 for the state machine in Figure 8.25.

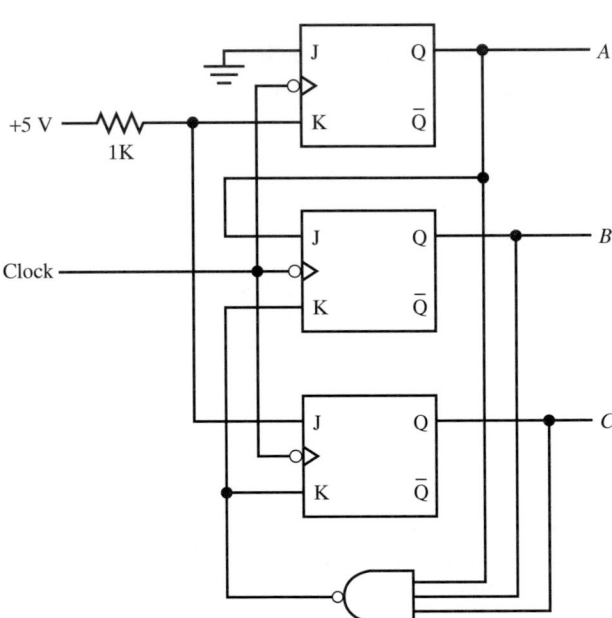

FIGURE 8.25 **For Question 8.24**

25. List five types of circuits that are actually state machines.

Electronics Workbench

26. Verify your solutions to Questions 7, 9, and 15 using Electronics Workbench.

27. Use Electronics Workbench to simulate the state machine shown in Figure 8.18.

Programmable Logic

28. Write the PAL equations for the state machines shown in Figures 8.15, 8.18, and 8.20.

29. Convert these excitation equations into PAL equations:

$JA = A + \overline{B}C$ $\qquad$ $JB = \overline{A}$ $\qquad$ $JC = \overline{B} + C$

$KA = 1$ $\qquad\qquad$ $KB = BC$ $\qquad$ $KC = \overline{AB}$

30. Explain how a 4-bit synchronous counter may be parallel loaded using PAL logic.

9

CIRCUIT DESIGN USING PROGRAMMABLE LOGIC

INSTRUCTIONAL OBJECTIVES

When finished with this chapter you should be able to:

1. List the benefits of using programmable logic.
2. Describe how a PAL is programmed, from the beginning.

3. List the various types of PALs.
4. Describe the purpose of a design entity.

SELF-EVALUATION QUESTIONS

Keep the following questions in mind and try to answer them when you have completed the chapter.

1. How does a PAL's part number describe its operation?

2. What is a PAL assembler for?
3. What is VHDL?
4. How can one PAL eliminate several ICs?

9.1 INTRODUCTION

The introduction of programmable logic into the digital design world provided many new directions for growth. Anyone who is good at digital design with basic logic gates can transfer that expertise to a programmable device with little effort. We need only write the Boolean equations (or draw a schematic or state diagram using an interactive tool), assemble them to create a fuse map, and burn the programmable device. Let us see how this is done.

9.2 WHY USE PROGRAMMABLE LOGIC?

The title of this section is an appropriate question. Why go to all the trouble to learn how to write PAL equations, use the assembler and PAL programmer, and spend money on design software? There are many answers. Programmable logic reduces the number of packages in a digital circuit. Breadboarding a new circuit takes less time, as there are fewer connections to be made. Changes in the circuit can be made by changing the logic equations and programming a new device. There are fewer signals to troubleshoot. PALs come in high-speed and low-power CMOS versions. In the long run, programmable logic is cheaper to use when manufacturing large quantities of electronic circuitry. A special security fuse can be blown in the PAL to prevent it from being read after programming. This provides a measure of security to a sensitive design.

Whatever your reasons for using it, programmable logic is here to stay, and improving every day. Already there are hundreds of different programmable devices, with more arriving every day. Getting a small exposure to them here in this chapter will be a good first step towards being a programmable-logic designer.

9.3 THE DESIGN PROCESS

A new digital circuit is designed for PAL operation using the following steps:

1. Specify the circuit. This may involve drawing a state diagram, making up a truth table, or writing the Boolean equations.

2. Choose the PAL needed for the circuit. This decision is based on many factors (whether we need a clock, registered outputs, or low power).

3. Translate the circuit specification into the appropriate form for the PAL assembler.

4. Assemble the PAL equations to create a fuse map or JEDEC file and generate simulation patterns. JEDEC stands for Joint Electronic Device Engineering Council. A JEDEC file is a standard way of representing PAL data.

5. Program the PAL and test it.

6. Put the PAL into the circuit and turn it on. Hopefully, it performs as desired. Several steps may need to be repeated if the PAL does not perform as intended.

It would be a good idea to use erasable PALs during the design/test phase, to reduce overall cost.

9.4 PAL BASICS

First, let us examine the naming convention used for PALs. Table 9.1 lists some typical PALs and their part number meanings. The number of inputs refers to the number of connections to the internal interconnection array, not the input signals to the chip. For example, in the 16R8, there are only 20 pins on the package, so there cannot be 16 individual inputs and 8 outputs. Instead, there are eight signal inputs to the chip (and thus the interconnection array), and eight additional inputs that are feedback signals from the eight internal registers of the 16R8. That is where the sixteen inputs come from. Figure 9.1 shows the logic diagrams of the PAL16L8 and PAL16R8 devices. Both require 20-pin DIPS, and are also available in 20- or 28-pin PLCC packages as well.

Table 9.2 lists a large portion of PALs that are currently available.

The simplified internal structure of a PAL is shown in Figure 9.2(a) on page 208. A set of AND gates and OR gates are available for constructing logic functions. The inputs to the AND gates and the OR gates are connected by selectively burning internal fuses at various intersection points, so that only the required connections remain. The AND-OR gate combination allows us to implement virtually any sum-of-products equation, making it ideal for designing combinational logic circuitry. PALs come with active high and active low outputs, so it may be necessary to use DeMorgan's theorem to convert the equation before using it.

Figure 9.2(b) shows what the PAL may look like after it is programmed. The "x" at each of the intersections indicates a connection has been made. The upper AND gate is connected to the inverted I_0 signal, and the I_1 signal. The second AND gate is connected to I_0 and the inverted I_1. The outputs of both AND gates are connected to the first OR gate. Figure 9.2(c) shows the equation implemented by the PAL.

To program the PAL, a PAL program must be written that identifies the signal names, the specific PAL, and the equations. A portion of the PAL program for the PAL in Figure 9.2 might look like this:

```
PAL10L8                          Pal Design Specification
XOR Function                     Alan C. Dixon
Broome Community College
A    B    /C   D    E    /F  NC  NC  NC  GND
NC  /FA  /FB  FC   FD   NC  NC  NC  NC  VCC
IF (VCC) FA = A * /B + /A * B
```

The first line specifies the part number used in the design. The assembler requires this to generate the appropriate programming data.

The fourth and fifth lines identify the signal names. Names preceded by a / (as in /C and /FA) are complemented (active low). The signal names are listed in the order of

TABLE 9.1   PAL Part Numbers and Meanings

PART	MEANING
10H8	10 inputs, active high, 8 outputs
16C1	16 inputs, complementary, 1 output
12L6	12 inputs, active low, 6 outputs
16R8	16 inputs, registered, 8 outputs
20X10	20 inputs, XOR registered, 10 outputs

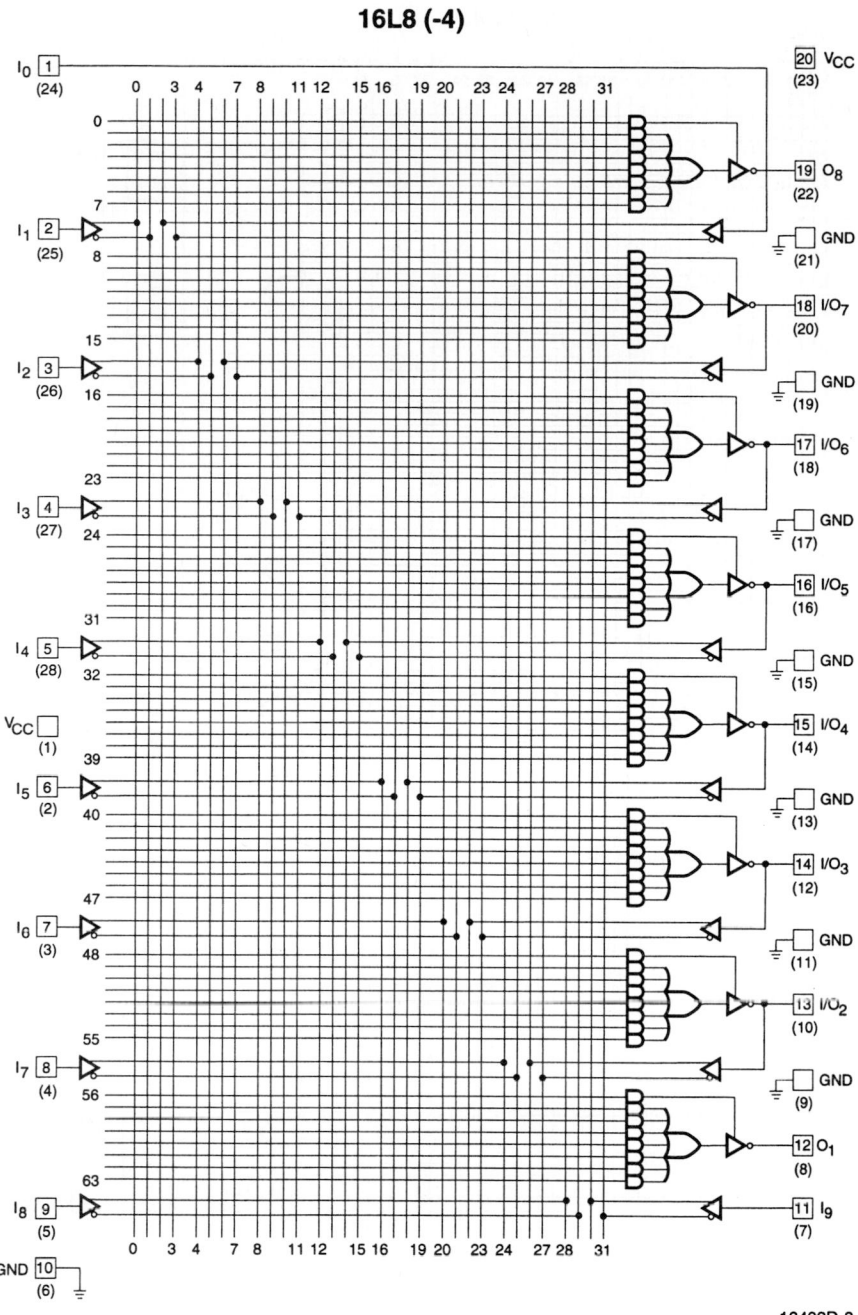

16492D-8

FIGURE 9.1(a) **PAL16L8 Logic Diagram (Copyright © 1996, Vantis. Reprinted with permission of copyright owner. All other rights reserved.)**

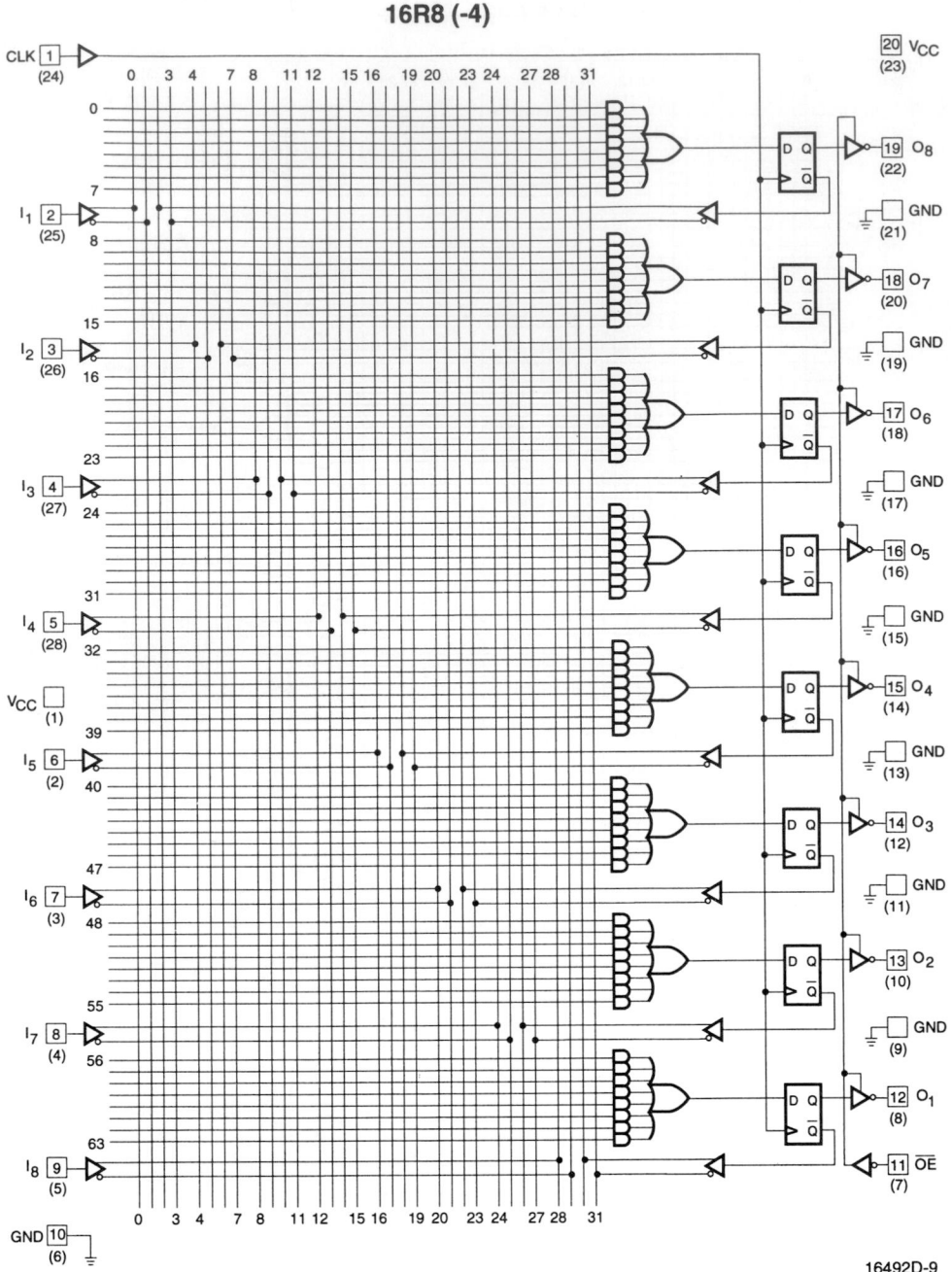

FIGURE 9.1(b) PAL16R8 Logic Diagram (Copyright © 1996, Vantis. Reprinted with permission of copyright owner. All other rights reserved.)

TABLE 9.2 PAL Packages

ACTIVE HIGH	COMPLEMENTARY OUTPUT	ACTIVE LOW	REGISTERED	XOR
10H8	16C1	10L8	16R8	20X10
12H6	20C1	12L6	16R6	20X8
14H4		14L4	16R4	20X4
16H2		16L2	20R8	
		12L10	20R6	
		14L8	20R4	
		16L6		
		18L4		
		20L2		
		16L8		
		20L8		
		20L10		

the pins on the device. Pin 1 is first (A), pin 10 last (GND) on line four. Pin 11 is first and pin 20 last on line five.

The IF statement is used to create the connections between the AND and OR gates. When the signal FA is high on the inside of the PAL, its counterpart /FA on the outside goes low.

Flip-flops included in a registered PAL use a slightly different notation:

```
Q7 := /Q7 * /LOAD + /D7 * LOAD
```

Here the := symbol indicates a flip-flop output.

Instead of having to write equations, some PAL assemblers allow you to draw a state diagram for a synchronous state machine, or use a truth table to describe the input/output relationships. This is a nice feature, since we do not have to write any Boolean equations. For example, the truth table for a BCD-to-seven-segment decoder may look like this:

```
[D,C,B,A] -> [A,B,C,D,E,F,G]
[0,0,0,0] -> [1,1,1,1,1,1,0]
[0,0,0,1] -> [0,1,1,0,0,0,0]
[0,0,1,0] -> [1,1,0,1,1,0,1]
   .
   .
   .
[1,0,0,1] -> [1,1,1,1,0,1,1]
```

The PAL assembler uses Boolean reduction to find the simplest logic circuit to implement the truth table.

9.5 VHDL

VHDL stands for VHSIC Hardware Description Language. **VHSIC** stands for Very-High-Speed Integrated Circuit, a research interest of the Department of Defense. VHDL is used to describe a digital system, its gates and their interconnections, and to aid in its simulation.

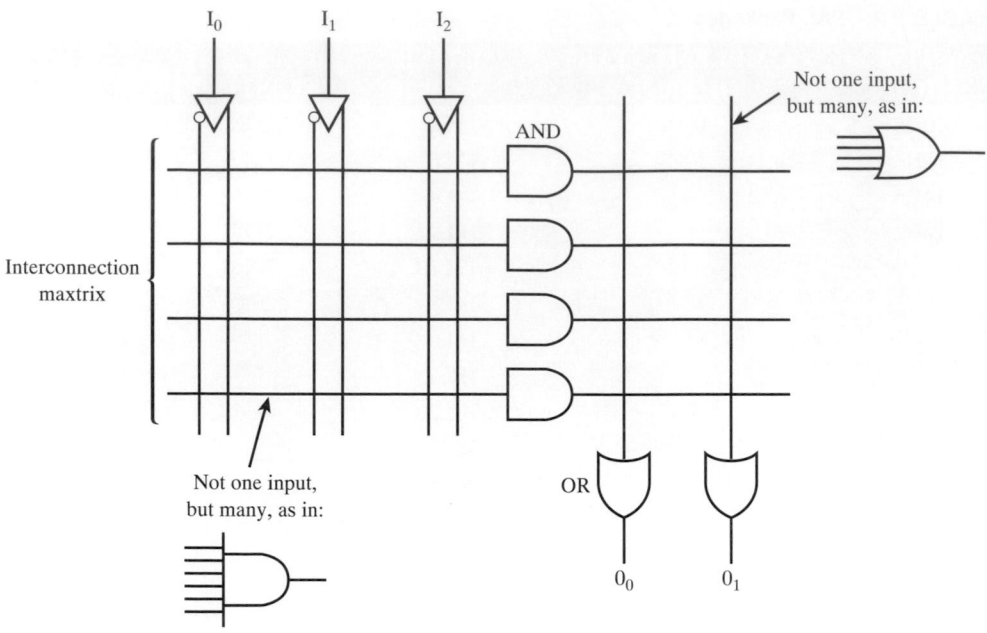

(a) Internal structure

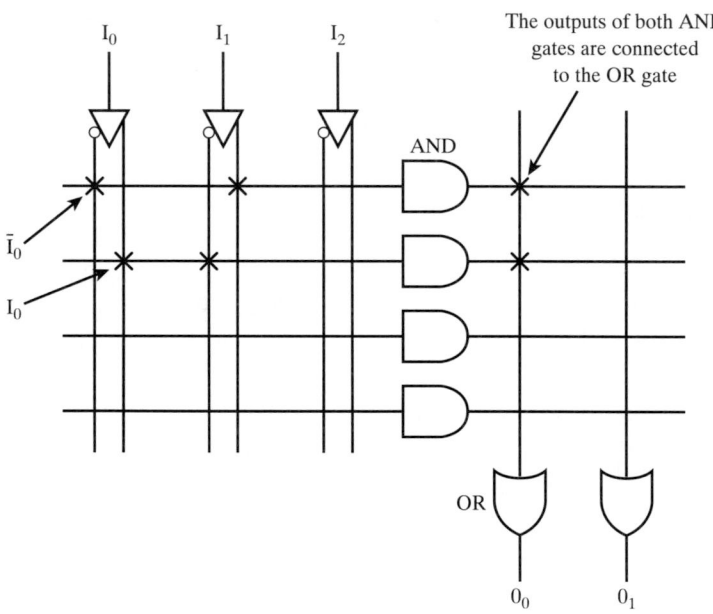

(b) Internal connections

$$0_0 = \bar{I}_0 I_1 + \bar{I}_0 I_1$$

(c) Equation

FIGURE 9.2   Internal PAL Circuitry (simplified)

Objects called *design entities* are used to specify the signals to a functional block. The half adder we are familiar with, which contains two inputs A and B and two outputs Sum and Carry, might be represented like this in VHDL:

```
entity HalfAdder is
port (A, B : in bit;
      Sum, Carry : out bit);
end HalfAdder;
```

All components in the system have a corresponding entity.

The operation of an entity is specified by additional statements, as in:

```
Sum <= (A xor B) after 4 ns;
Carry <= (A and B) after 4 ns;
```

Notice that the gate delay of the AND and XOR gates may be specified in the VHDL equations.

Designing with VHDL, or some other hardware description language, such as Verilog, is performed from a *top-down* approach, beginning with the basic overall operation of the system, and then adding more and more detail as the system is refined.

9.6 PRACTICAL APPLICATIONS

Let us look at two ways to use programmable logic.

68000 Memory Controller

The first application is for a PAL used as a controller for memory and I/O signals in a 68000 microprocessor-based system.

Figure 9.3 shows how a two- to four-line decoder is used to perform partial-address decoding in the 68000 system. Address lines A_{15} and A_{16} are used because they map the 68000's memory into convenient ranges. With both A_{15} and A_{16} low, the 74LS139 decoder will output a 0 on the $\overline{PROM}$ line; thus, $\overline{PROM}$ will be low whenever the 68000 addresses memory in the 00000 to 07FFF range. This allows us to use EPROMs as large as 16KB each for the monitor program. If A_{15} is high while A_{16} is low, the $\overline{RAM}$ signal will go low. This corresponds to an address range from 08000 to 0FFFF. If either $\overline{PROM}$ or $\overline{RAM}$ goes low, (as one will during a valid memory reference), the output of the AND gate will go low issuing $\overline{DTACK}$. The $\overline{SERIAL}$ signal is low whenever the 68000 addresses memory from 10000 to 17FFF, and this signal is used in the serial section and also to issue

FIGURE 9.3 Partial Address Decoder and $\overline{DTACK}$ Circuit

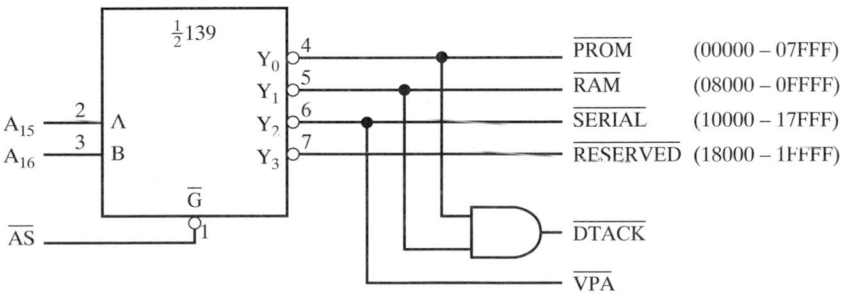

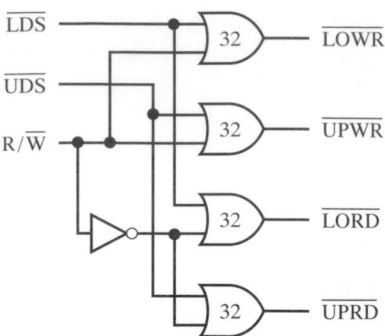

FIGURE 9.4   Memory Read/Write Logic

$\overline{\text{VPA}}$. Notice in Figure 9.3 that the enable input of the 74LS139 ($\overline{\text{G}}$) is connected to $\overline{\text{AS}}$. This ensures that the 74LS139 will work only during valid bus cycles (when $\overline{\text{AS}}$ is low).

Figure 9.4 shows how we generate the required memory read and write signals. We need separate read and write signals for both lower and upper addresses, to allow for byte operations in memory. By using only four OR gates and an inverter, we are able to generate these signals rather easily.

Although the parts requirements for the memory system are few, it is possible to further reduce the number of components needed, by adding programmable logic.

It is not uncommon that a single PAL can be programmed to replace the logic operation of five TTL ICs! The following example shows how a PAL may be used to generate some of the required memory section signals, thus replacing the circuitry of Figures 9.3 and 9.4.

Figure 9.5 shows the connections to a 10L8 PAL, a device that contains 10 inputs and eight programmed outputs. The inputs to the PAL are on pins 1 through 9 and 11. The four NC inputs are no-connects, inputs that are not needed for this application. The eight programmed outputs are on pins 12 through 19. The PAL program listed in Figure 9.6 defines

FIGURE 9.5   PAL Connections for 10L8 Memory Controller

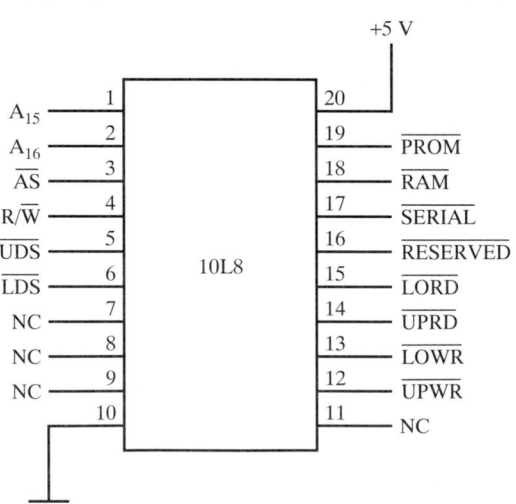

```
PAL10L8                                    PAL DESIGN SPECIFICATION
68K.MEM                                    JAMES L. ANTONAKOS
68000 MEMORY CONTROLLER
BROOME COMMUNITY COLLEGE
A15 A16 /AS RW /UDS /LDS NC NC NC GND
NC /LOWR /UPWR /LORD /UPRD /RES /SER /RAM /PROM VCC

IF (VCC) PROM = /A16 * /A15 * AS          ;PROM --- 000000-007FFF
IF (VCC) RAM = /A16 * A15 * AS            ;RAM ---- 008000-00FFFF
IF (VCC) SER = A16 * /A15 * AS            ;SERIAL - 010000-017FFF
IF (VCC) RES = A16 * A15 * AS             ;RESVD -- 018000-01FFFF

IF (VCC) LOWR = /RW * LDS                 ;LOWER WRITE
IF (VCC) UPWR = /RW * UDS                 ;UPPER WRITE
IF (VCC) LORD = RW * LDS                  ;LOWER READ
IF (VCC) UPRD = RW * UDS                  ;UPPER READ

FUNCTION TABLE
/AS A16 A15 RW /UDS /LDS
/PROM /RAM /SER /RES /LOWR /UPWR /LORD /UPRD

;/AS A16 A15 RW /UDS /LDS /PROM /RAM /SER /RES /LOWR /UPWR /LORD /UPRD
-----------------------------------------------------------------------
 H   X   X  X  X    X    H     H    H    H    X     X     X     X
 L   L   L  X  X    X    L     H    H    H    X     X     X     X
 L   L   L  X  X    X    L     H    H    H    X     X     X     X
 L   H   L  X  X    X    H     H    L    H    X     X     X     X
 L   H   H  X  X    X    H     H    H    L    X     X     X     X
 X   X   X  L  L    L    X     X    X    X    L     L     H     H
 X   X   X  L  L    H    X     X    X    X    H     L     H     H
 X   X   X  L  H    L    X     X    X    X    L     H     H     H
 X   X   X  H  L    L    X     X    X    X    H     H     L     L
 X   X   X  H  L    H    X     X    X    X    H     H     H     L
 X   X   X  H  H    L    X     X    X    X    H     H     L     H
-----------------------------------------------------------------------
```

FIGURE 9.6   PAL Program for 68000 Memory Control

the input/output signal names and the required equations needed to implement the desired functions. A truth table is also included that allows the PAL to be tested and verified. The use of this single PAL eliminates the need for two ICs, the 74LS32 quad OR gate, and the 74LS139 decoder. Even though we have only reduced our chip count by one so far, we have gained a tremendous advantage in the area of flexibility. Changing the addresses of all main sections (EPROM, RAM, SERIAL) is now as easy as reprogramming and replacing this one PAL!

Child's Play Toy

In Figure 9.7 the diagram of a child's play toy is given. A GAL is used to perform all the digital functions. Four rolling wheels are connected to the GAL so that any of four positions can be sensed for each wheel. Other control inputs are used to turn the device on and off (it is never really off, just in a low power mode), play a silly sentence based on the wheel positions, and do other fun things. A digital output (pulse-width modulated) drives an audio amplifier so the GAL output can be heard. With thousands of gates available in the GAL, there is nothing to prevent us from hard coding the digital audio waveforms directly into logic, as ROM data. Other gates are wired as counters, shift registers,

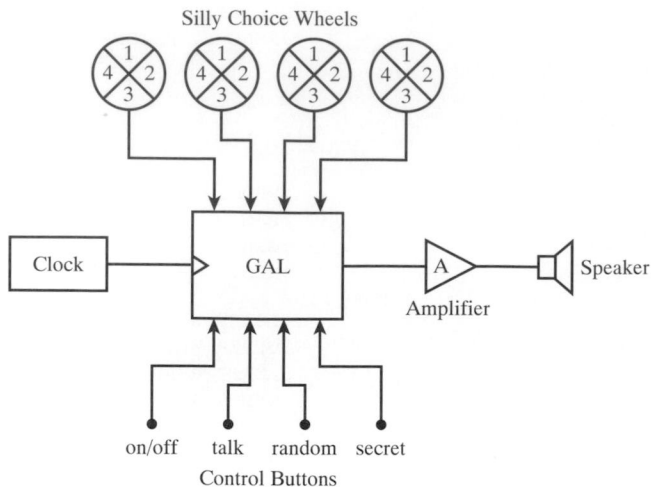

Silly Choice Wheels

FIGURE 9.7 **Using a GAL in a Child's Toy**

and all the other goodies we need to make the circuit work. It would be worthwhile for you to try to draw a detailed block diagram of the logic required, and then estimate how many PALs would be needed instead, or worse, how many TTL packages.

9.7 DESIGN CONSIDERATIONS

Before rushing to convert all your designs over to programmable logic, stop and consider a few things. First, is it really necessary to use a PAL? If there are only three ICs in the design, replacing them with a PAL may not be a good idea. It may be more expensive to buy the single PAL than the three ICs. A circuit with 15 ICs that is reduced to one containing only two PALs is a better choice.

Next, carefully add up the power requirements of the ICs being replaced. The PAL could end up using much more power than the original ICs, due to extra, unconnected gates that use power but perform no function.

Also, once the PAL is in the circuit, there is little you can do but replace it if the equations are wrong. This could be an expensive troubleshooting experience for the beginner. Always check your equations carefully before programming.

9.8 TROUBLESHOOTING TECHNIQUES

Troubleshooting a PAL design is tricky because you cannot examine the output of individual gates in the circuit, only the final PAL output signals. If an equation is incorrect, or the PAL was not programmed correctly, there is little that can be done beyond exam- ining the inputs and outputs with a logic analyzer. If possible, apply the input signals stat- ically (set them low or high and leave them there) and go through the truth table line by line until the faulty output pattern is found. This may help identify an incorrect equation or bad device.

SUMMARY

In this chapter we explored the operation of programmable logic. We saw that there are many types of PALs that provide a large number of internal gates that are connected during programming to be any circuit we wish. The many benefits to using programmable logic will only increase as newer devices hit the market.

STUDY QUESTIONS

1. List two reasons why you should use programmable logic.

2. List two reasons why you should not use programmable logic.

3. What are the steps in a PAL design? Apply them to the design of a small combinational logic circuit with several inputs and outputs (like an adder).

4. What is the logic equation for the PAL in Figure 9.8?

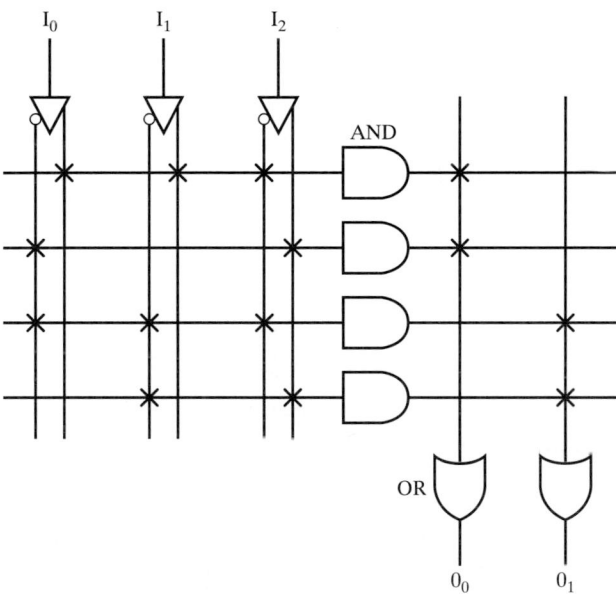

FIGURE 9.8   For Question 9.4

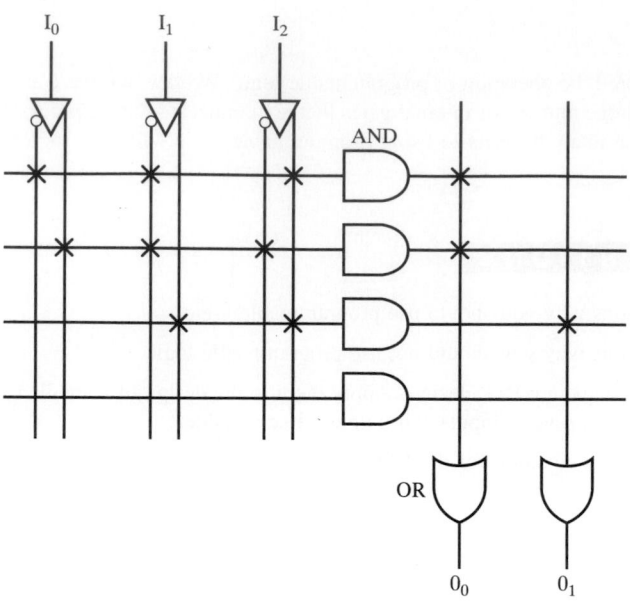

FIGURE 9.9 **For Question 9.5**

5. Repeat Question 4 for the PAL in Figure 9.9.

6. What is the logic associated with this equation:

    ```
    IF (VCC) VAL = /A * B + B * C + /C
    ```

7. What is the first line of the PAL specification used for?

8. What are the fourth and fifth lines of the PAL specification used for?

9. What type of PAL has a flip-flop in its output stage?

10. What is the meaning of the `:=` in a PAL statement?

11. Show a truth table for the full adder in PAL programmer format.

12. How does VHDL differ from PAL programming?

13. How is gate delay specified in a VHDL statement?

14. In the 68000 memory controller, what are all the Boolean equations for the output signals?

15. What type of PAL should be used to make a counter?

16. What software packages for PAL programming are available for free over the web?

COMPLEX LOGIC FUNCTIONS

INSTRUCTIONAL OBJECTIVES

When finished with this chapter you should be able to:

1. Use selectors/multiplexers to switch large quantities of data into smaller, manageable groups.
2. Use decoders/demultiplexers to scan large quantities of data with smaller control words.
3. Design and use comparators.
4. Understand the various ways to make adders and subtractors.
5. Explain the basics of A/D and D/A conversion.

SELF-EVALUATION QUESTIONS

Keep the following questions in mind and try to answer them when you have completed the chapter.

1. When does an enable line become useful?
2. How does a selector work?
3. What is meant by (a) multiplexing, (b) demultiplexing?
4. What is an SAR?
5. How are BCD and binary numbers converted?

10.1 INTRODUCTION

In this chapter we cover more sophisticated logical operations. We will study the data control TTL gates. Data control gates come in two types: selectors/multiplexers and decoders/demultiplexers. Selectors/multiplexers are very useful in parallel to serial conversion, and also in the switching of parallel data buses. Decoders/demultiplexers are very useful for converting coded data back into its original form (e.g., binary-to-decimal conversion). We will study many packages from each group and also consider some possible applications. We will also examine the operation of such common logic devices as the comparator, adder, and D/A and A/D converters.

10.2 SELECTORS/MULTIPLEXERS

A *selector* is a logic element that chooses one signal from a group of signals. The simplest selector is shown in Figure 10.1(a). One of two inputs *A* and *B* is routed to the output *Y* by the select input *S*. Either the upper AND gate is enabled (when *S* is low) and passes its data to the output, or the lower AND gate is enabled (*S* is high). This is reflected in the truth table shown in Figure 10.1(b). The simplified logic symbol in Figure 10.1(c) is commonly used to represent the selector, which we call a *2-line to 1-line* selector, or a 2:1 selector.

Figure 10.2 presents a number of examples of the selector in use. In particular note the waveforms in Figure 10.2(c). The selector is only choosing the *B* waveform in the figure, but could also be made to switch rapidly back and forth from *A* to *B*, thus sending

FIGURE 10.1   Two-Input Selector (2-Line to 1-Line)

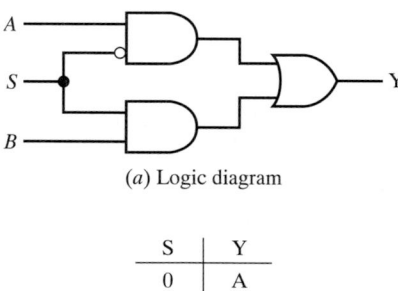

(*a*) Logic diagram

S	Y
0	A
1	B

(*b*) Truth table

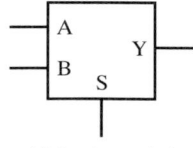

(*c*) Logic symbol

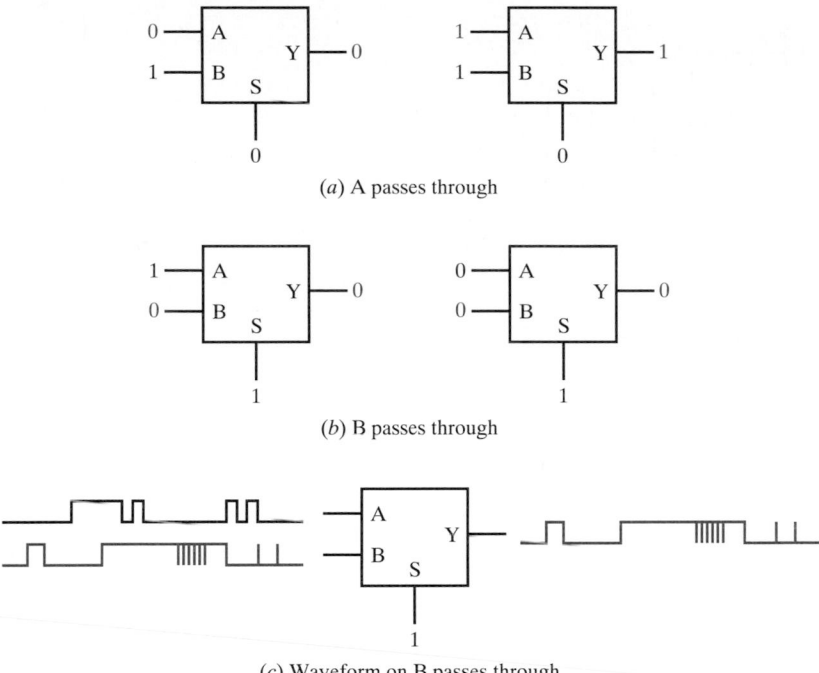

(*a*) A passes through

(*b*) B passes through

(*c*) Waveform on B passes through

FIGURE 10.2 **Some Selection Examples**

both waveforms to the output as a mixture of both. This is called ***multiplexing.*** Using a device called a ***demultiplexer,*** both waveforms are decoded back into their original form. We will examine demultiplexers in the next section.

Making a larger selector can be done a number of ways. In Figure 10.3, three 2:1 selectors are connected to make a 4:1 selector. Two select inputs are required (S_0 and S_1) to specify which of the four input signals to choose.

How many logic gates are used in the 4:1 selector? Each 2:1 selector contains four gates, so the 4:1 selector in Figure 10.3 uses twelve gates. We can save one inverter by using $\overline{S}_0$ twice, but we still require eleven gates. Redesigning the 4:1 selector gives us the circuit in Figure 10.4. Only seven gates are needed to implement the 4:1 selector, a savings of an additional four gates over the initial design.

Prepackaged selectors typically contain an enable signal that only allows the selected output to go high if properly enabled. A high-level enable signal can be added to the 2:1 or 4:1 selectors by connecting an additional input on each AND gate to the enable input. If the enable signal is low, all of the AND gates are disabled.

Let us examine a number of prepackaged selectors/multiplexers.

The 74150 16:1 Data Selector

The 74150 is a 24-pin DIP that contains the logic needed to select one of 16 different inputs and transfer it to an output pin (see Figure 10.5). The desired input source (I_0–I_{15}) is selected by placing the appropriate 4-bit binary word on the data select (ABCD) inputs.

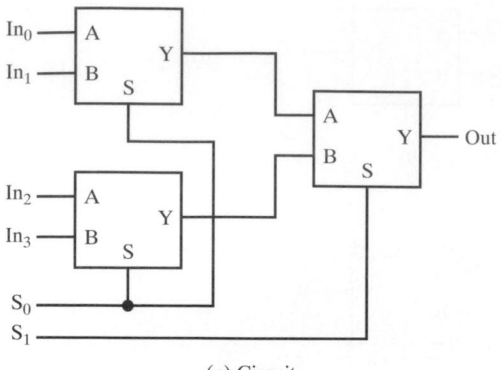

(a) Circuit

S_1	S_0	Out
0	0	In_0
0	1	In_1
1	0	In_2
1	1	In_3

(b) Truth table

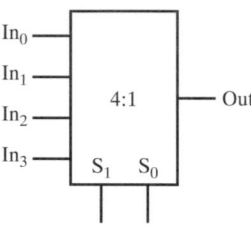

(c) Logic symbol

FIGURE 10.3 Making a 4-to-1 Selector out of 2-to-1 Selectors

FIGURE 10.4 Simplified 4:1 Selector

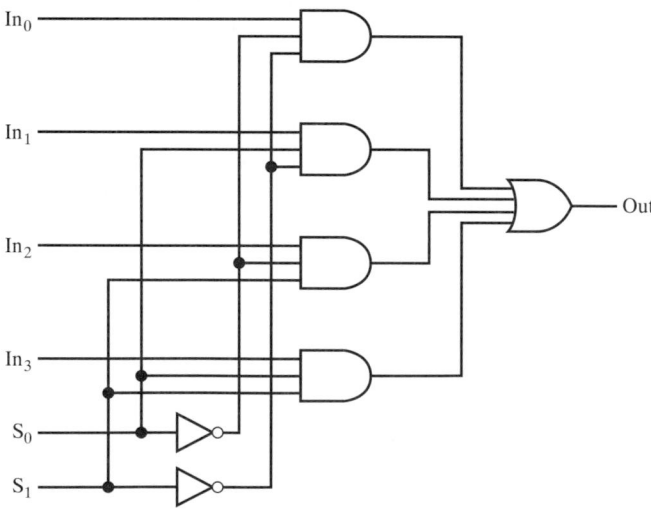

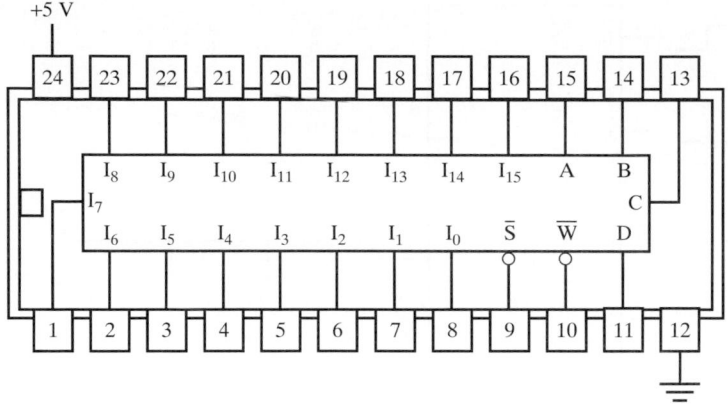

(a) Pinout

D	C	B	A	S	$\overline{W}$
0	0	0	0	0	$\overline{I}_0$
0	0	0	1	0	$\overline{I}_1$
0	0	1	0	0	$\overline{I}_2$
0	0	1	1	0	$\overline{I}_3$
0	1	0	0	0	$\overline{I}_4$
0	1	0	1	0	$\overline{I}_5$
0	1	1	0	0	$\overline{I}_6$
0	1	1	1	0	$\overline{I}_7$
1	0	0	0	0	$\overline{I}_8$
1	0	0	1	0	$\overline{I}_9$
1	0	1	0	0	$\overline{I}_{10}$
1	0	1	1	0	$\overline{I}_{11}$
1	1	0	0	0	$\overline{I}_{12}$
1	1	0	1	0	$\overline{I}_{13}$
1	1	1	0	0	$\overline{I}_{14}$
1	1	1	1	0	$\overline{I}_{15}$
x	x	x	x	1	1

(b) Truth table

FIGURE 10.5   The 74150 16:1 Selector

Once this has been done, the complement of the selected data input will appear on the output pin $\overline{W}$ as soon as the strobe pin $(\overline{S})$ goes low. Whenever the $\overline{S}$ pin is high, the $\overline{W}$ output remains high, regardless of the data present on the data source or select pins. The 74150 could be used to convert 16-bit parallel into serial data with the addition of a 4-bit binary counter on the data select inputs. It could also be used to select serial (TTL only) data from one of 16 sources. Or, it could be used to monitor the status of up to 16 switches, as in a home burglar alarm.

· ·

EXAMPLE 10.1

Figure 10.6 shows the 74150 used to select one of 16 serial data inputs. If the serial data signals are related to each other by integer multiples or other small increments, this could be a building block for an electronic music machine.

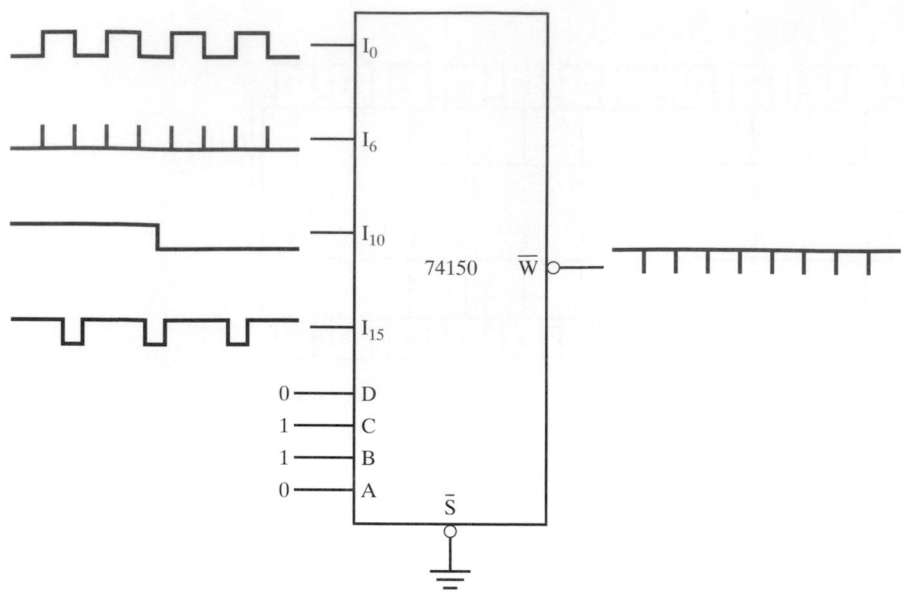

FIGURE 10.6   Choosing a Waveform

EXAMPLE 10.2

In Figure 10.7, the 74150 is used to monitor the status of 16 doors and windows in a simple burglar alarm. The 4-bit pattern on the select inputs is cycled from 0000 to 1111 continuously, identifying all door and window states.

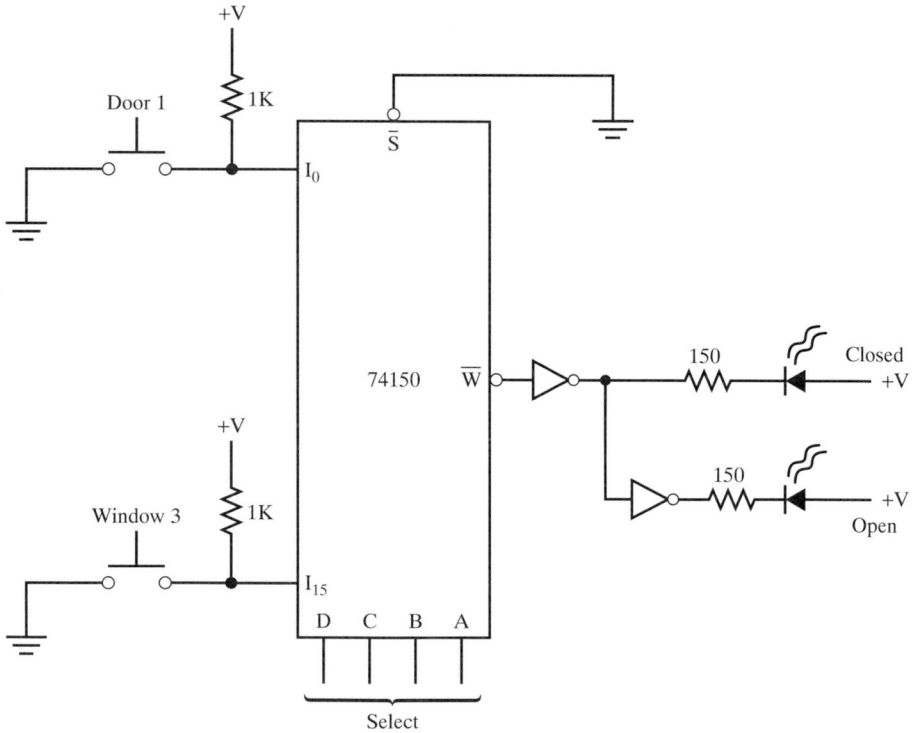

FIGURE 10.7   Simple 16-Input Burglar Alarm

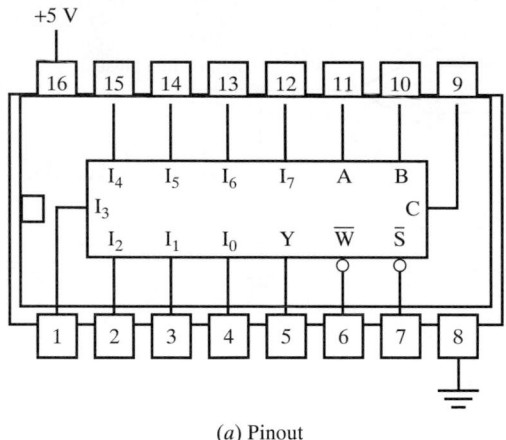

(*a*) Pinout

C	B	A	$\overline{S}$	$\overline{W}$	Y
0	0	0	0	$\overline{I_0}$	I_0
0	0	1	0	$\overline{I_1}$	I_1
0	1	0	0	$\overline{I_2}$	I_2
0	1	1	0	$\overline{I_3}$	I_3
1	0	0	0	$\overline{I_4}$	I_4
1	0	1	0	$\overline{I_5}$	I_5
1	1	0	0	$\overline{I_6}$	I_6
1	1	1	0	$\overline{I_7}$	I_7
x	x	x	1	1	0

(*b*) Truth table

FIGURE 10.8   The 74151 8:1 Selector

The 74151 8:1 Selector

The 74151 (see Figure 10.8) is a 16-pin DIP that is used to select one of eight data inputs. The 74151 is particularly useful for converting 8-bit parallel data to serial data (ideal for microcomputer work) and finds application in video display logic.

To select an input, the 3-bit binary word is placed on the data select (ABC) inputs. Then as soon as the strobe ($\overline{S}$) line is pulled low, the data present on the selected input gets routed to the Y output. The complement of the same data appears on the $\overline{W}$ output. The 74151 could be used in a white noise generator (see Example 10.3). You should realize that a 74150 can be (and should be) used whenever two 74151s are needed to look at 16 inputs. It may be desirable, though, to use two at the same time. That is one advantage of having such a large variety of selectors to choose from.

· ·

EXAMPLE 10.3

In Figure 10.9, 8-bit parallel data is sent to the 74151 (possibly from a computer's memory) and converted into serial data, which is amplified to produce white noise.

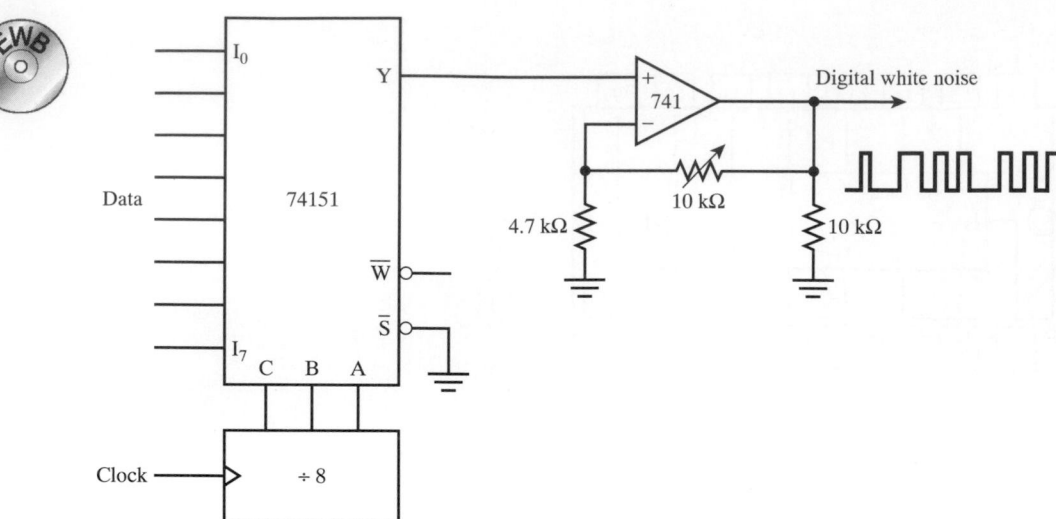

FIGURE 10.9 **Simple White Noise Source**

The 74153 Dual 4:1 Selector

The 74153 (see Figure 10.10) is a 16-pin DIP that contains two 4:1 line selectors. Each selector has its own noninverting output and enable strobe. Both selectors share the same data select (AB) pins.

FIGURE 10.10 **The 74153 Dual 4:1 Selector**

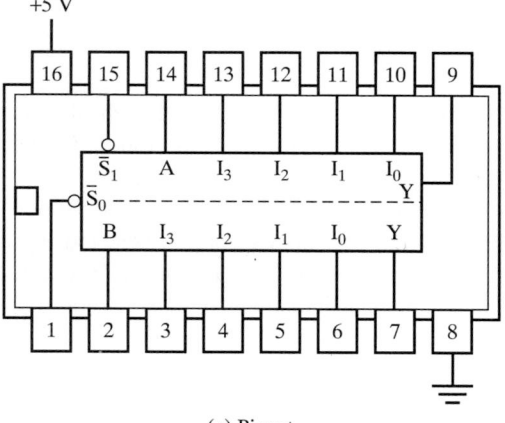

(*a*) Pinout

B	A	$\bar{S}_0$ or $\bar{S}_1$	Y
0	0	0	I_0
0	1	0	I_1
1	0	0	I_2
1	1	0	I_3
x	x	1	0

(*b*) Truth table (each half)

The data is selected by placing a 2-bit word on the data select inputs (AB) and pulling the appropriate strobe line low. Remember that the units operate separately but may still be strobed or enabled at the same time. The 74153 is commonly used in dynamic RAM circuitry to switch the appropriate address lines into the chip.

· ·

EXAMPLE **10.4**

The 74153 is used to select one of four address buses for a dynamic RAM. Each bus (VIDEO and CPU) contains 12 address lines, which must be split into two 6-address pairs for the dynamic RAM. The M_0 and M_1 inputs select one group of six address lines at a time. We will soon study the dynamic RAM operation, but Figure 10.11 shows schematically how a 74153 can be used to select different buses.

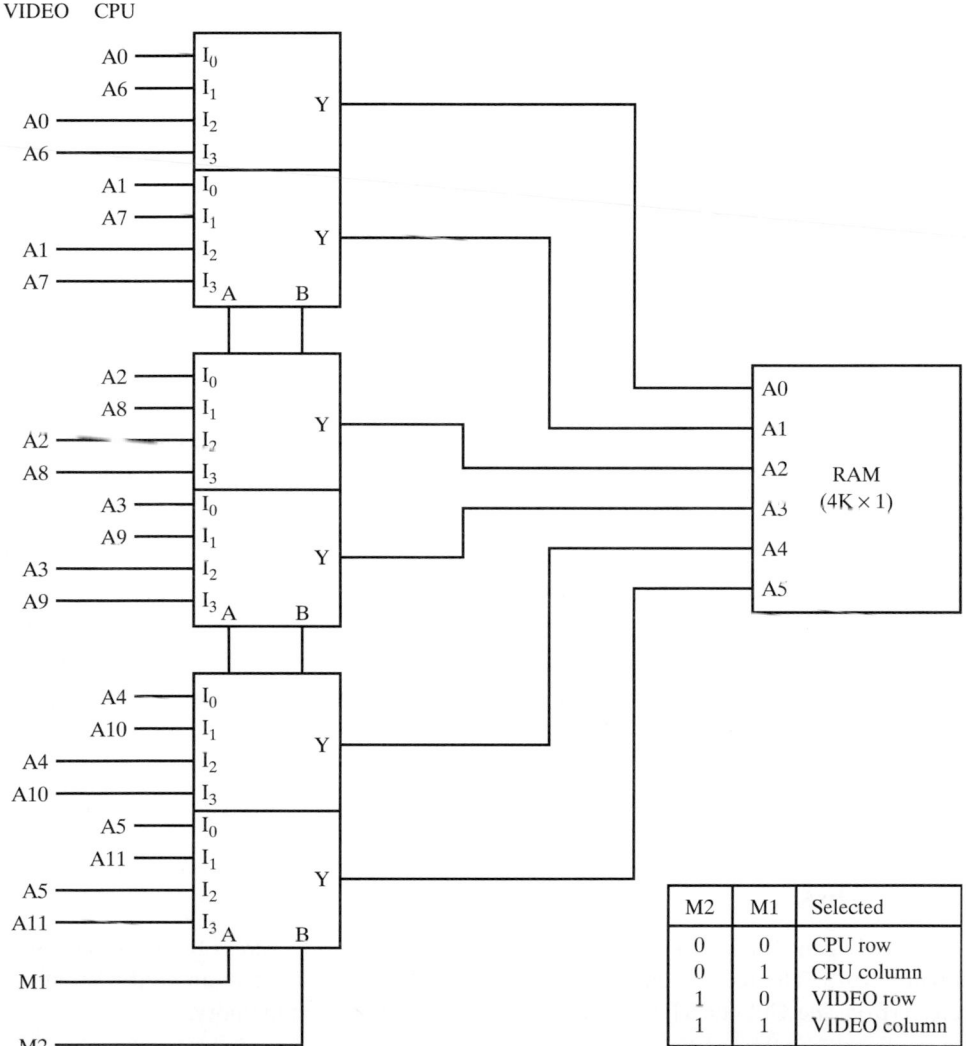

M2	M1	Selected
0	0	CPU row
0	1	CPU column
1	0	VIDEO row
1	1	VIDEO column

FIGURE 10.11 **Selection of Different Buses by the 74153, Which Selects 24 Bits of Address 6 Bits at a Time**

· ·

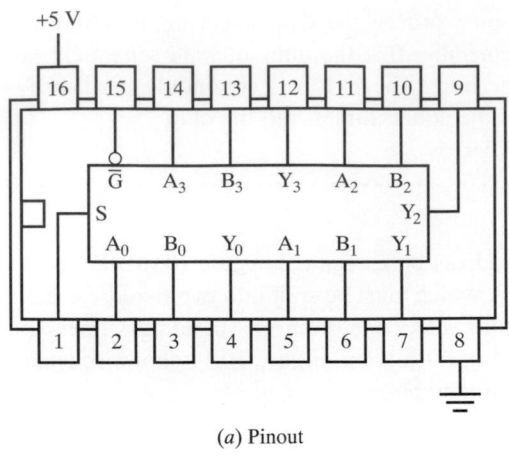

(*a*) Pinout

S	$\overline{G}$	Y
0	0	A
1	0	B
x	1	0

(*b*) Truth table

FIGURE 10.12   The 74157 Quad 2:1 Selector

The 74157 Quad 2:1 Line Selector

Our last selector is the 74157 quad 2:1 data selector (see Figure 10.12). The 74157 is a 16-pin DIP that contains four data selectors, each capable of selecting one of two lines.

The data present on the A inputs appears on the Y outputs when S (select) is low and enable ($\overline{G}$) is low. The data on the B inputs appears when S is high and $\overline{G}$ is low. The idle state exists when $\overline{G}$ is high. In this case, all Y outputs are low (high if the 74158 is used).

The 74157 is very useful to applications in electronic circuits where two control circuits desire access to another circuit (although not at the same time) such as in a PC video card. The central processing unit (CPU) may want to look at the screen memory (to update it) or the video circuitry may want to look at the screen memory (to send some video out to a monitor). The 74157 very easily accomplishes this task.

· ·

EXAMPLE **10.5**

The 74157 can be used to switch between CPU and VIDEO address buses, as shown in Figure 10.13. The three selectors allow the VIDEO signal to select an entire 12-bit address (from the CPU or VIDEO bus) and pass it to the screen memory.

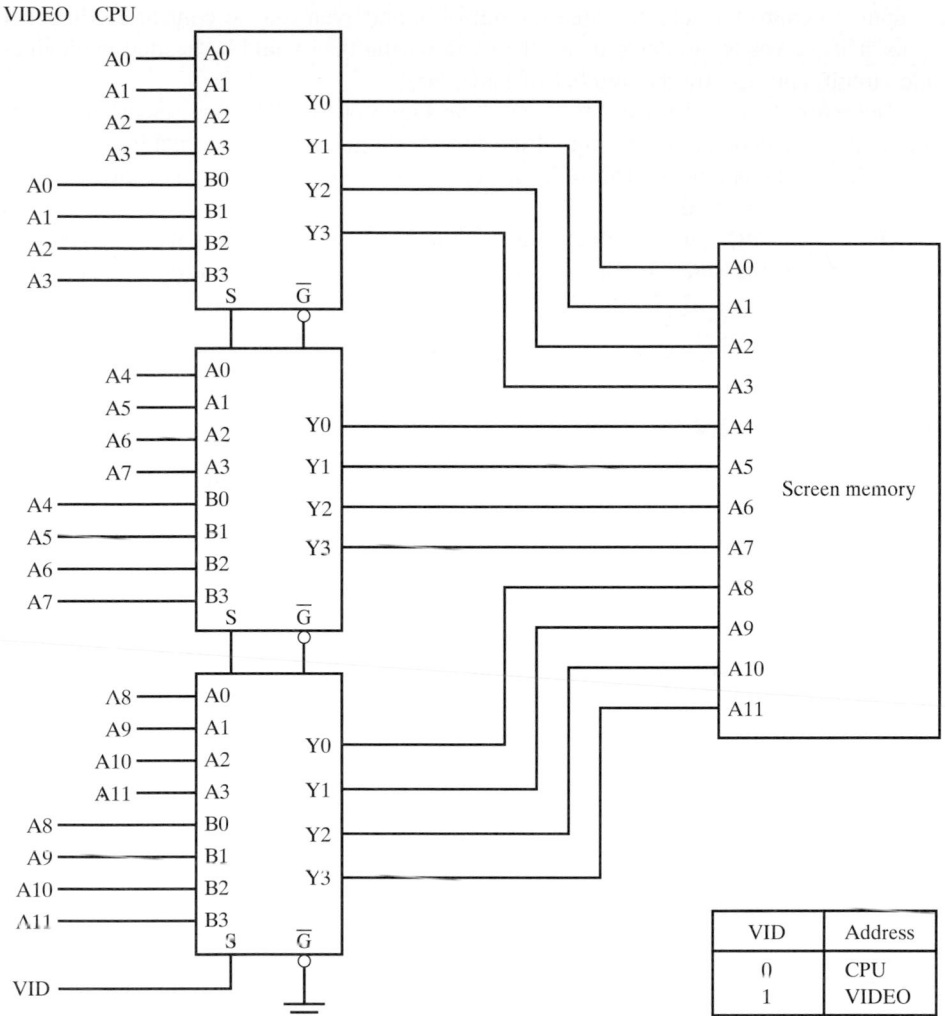

FIGURE 10.13 is referenced by the table:

VID	Address
0	CPU
1	VIDEO

FIGURE 10.13   Circuitry for Simple Address Bus Switches

To summarize, we use a selector whenever we want to narrow large amounts of data to more manageable sizes. We "multiplex" the data by selectively looking at the portions that are important to us. In the next section, we will see how we then "demultiplex" the data whenever we want to control a large number of data paths from a much smaller control word.

10.3 DECODERS/DEMULTIPLEXERS

This section dealing with decoders will become very important when we start our study of computer memory circuits and of computer applications in general. A decoder is used to obtain many outputs from a small source. For instance, with a decoder, we can use two inputs to control 4 outputs, three to control 8, and even four to control 16 different outputs. This serves to cut down on all the extra wiring that would be needed in an electronic circuit, and usually the number of gates, too!

The basic design of a decoder is shown in Figure 10.14. The two select inputs, together with the inverters, choose one of the four AND gates at a time, making its output go high. A NAND gate would cause the selected output to go low. This is called a *2-line to 4-line decoder,* or 2:4 decoder.

Only one AND gate is selected at a time because only one gate receives two 1s depending on the states of the select inputs. Let us look at a number of common decoders.

The 74139 Dual 2:4 Decoder

The 74139 is a 16-pin DIP that enables two inputs to control four outputs. There are two separate 2:4 decoders in the package, and each has its own enable line.

Figure 10.15 shows the pinout and truth table.

FIGURE 10.14   2:4 Decoder (Selected Output Goes High)

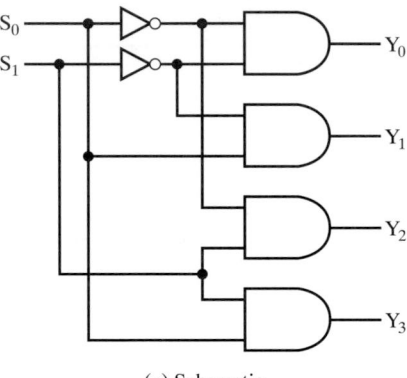

(*a*) Schematic

S_1	S_0	Y_0	Y_1	Y_2	Y_3
0	0	1	0	0	0
0	1	0	1	0	0
1	0	0	0	1	0
1	1	0	0	0	1

(*b*) Truth table

(*c*) Logic symbol

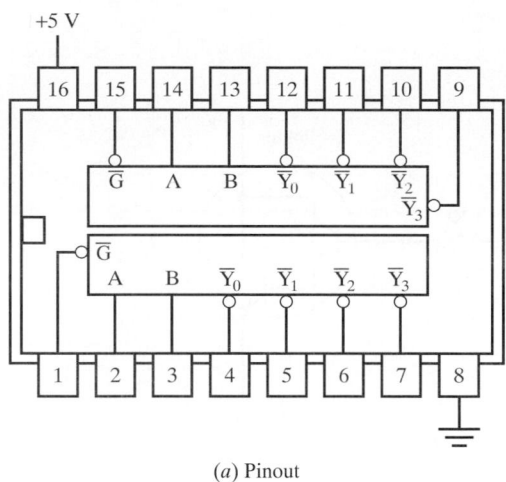

(*a*) Pinout

B	A	$\overline{G}$	$\overline{Y}_0$	$\overline{Y}_1$	$\overline{Y}_2$	$\overline{Y}_3$
0	0	0	0	1	1	1
0	1	0	1	0	1	1
1	0	0	1	1	0	1
1	1	0	1	1	1	0
x	x	1	1	1	1	1

(*b*) Truth table (each half)

FIGURE 10.15   The 74139 Dual 2:4 Decoder

To select one of the four outputs (which are normally high), we simply place the 2-bit binary number corresponding to the output line we wish to decode (0, 1, 2, or 3) on the A, B select inputs, and pull the enable line ($\overline{G}$) low. If $\overline{G}$ is high, the outputs remain high; otherwise, the selected output will go low. Please note that none of the decoders pass data (as the selectors did). They simply enable a certain output, although there are devices that *do* pass data to a selected output. These are called ***demultiplexers,*** which we will get to shortly.

• •

EXAMPLE 10.6

The 74139 is used in two ways in Figure 10.16. One half is used to drive a relay through a Darlington pair, and the other half is used as an indicator. This could be part of a simple control circuit. With A and B low, relay 1 and LED 1 are turned on, and so on. Note that only one relay and LED may be on at any instant.

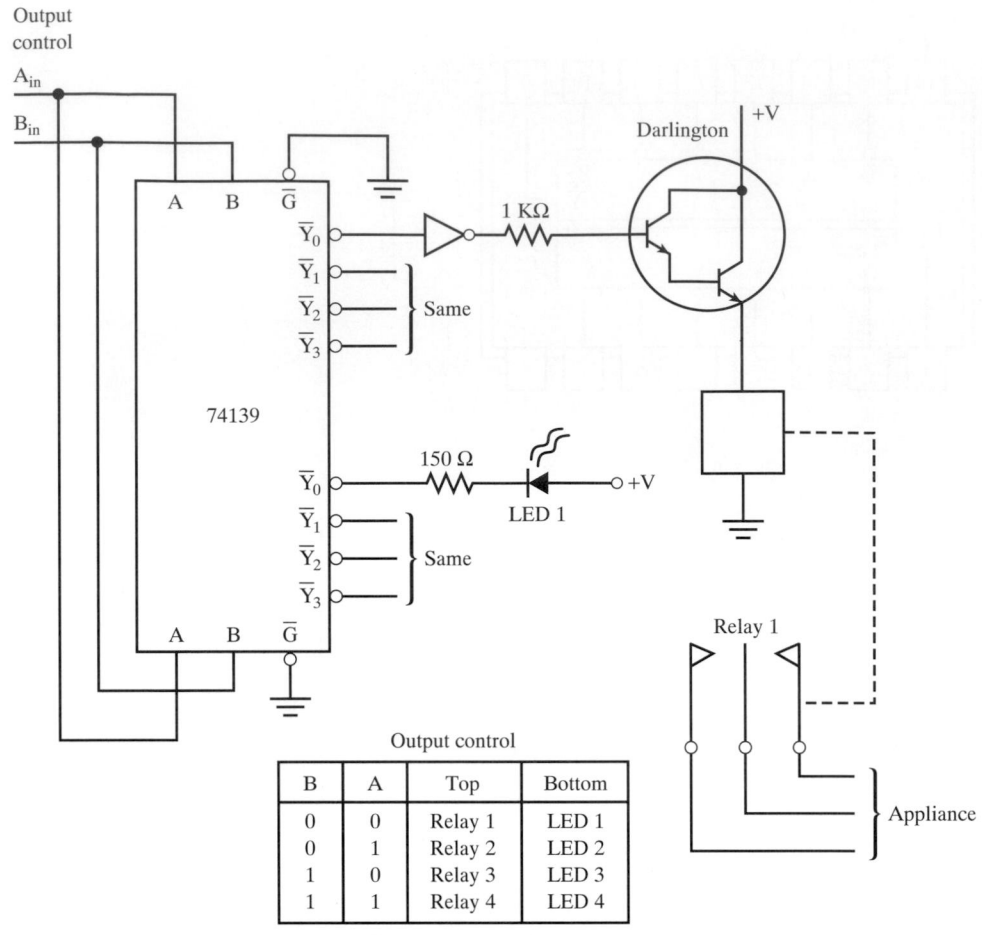

FIGURE 10.16 Simple Control Circuit Using the 74139

B	A	Top	Bottom
0	0	Relay 1	LED 1
0	1	Relay 2	LED 2
1	0	Relay 3	LED 3
1	1	Relay 4	LED 4

Output control

The 74138 3:8 Decoder

If four outputs are not enough, we can turn to the 74138 (see Figure 10.17), a 16-pin DIP that is capable of decoding one of eight output lines from a 3-bit input word. (You should see that the outputs increase by the function 2^N, where N is the number of inputs.) As usual, the binary equivalent of the output we wish to decode is placed on the select inputs A, B, and C. Once the two enable inputs $\overline{G2A}$ and $\overline{G2B}$ have been pulled low, and the third enable input G1 has been pulled high, the selected output will go low.

The enabling of the 74138 is complicated but also very useful. It is often desirable, and sometimes essential, to be able to enable a device from several locations.

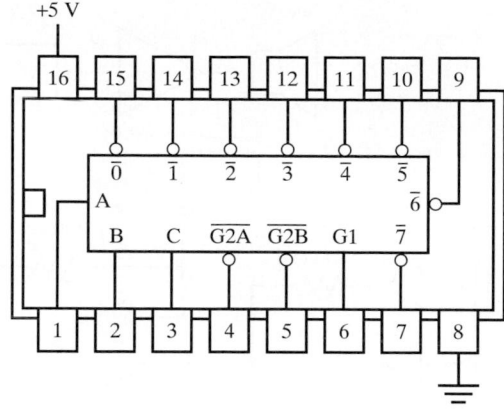

(*a*) Pinout

C	B	A	$\overline{G2A}$	$\overline{G2B}$	G1	$\overline{0}$	$\overline{1}$	$\overline{2}$	$\overline{3}$	$\overline{4}$	$\overline{5}$	$\overline{6}$	$\overline{7}$
0	0	0	0	0	1	0	1	1	1	1	1	1	1
0	0	1	0	0	1	1	0	1	1	1	1	1	1
0	1	0	0	0	1	1	1	0	1	1	1	1	1
0	1	1	0	0	1	1	1	1	0	1	1	1	1
1	0	0	0	0	1	1	1	1	1	0	1	1	1
1	0	1	0	0	1	1	1	1	1	1	0	1	1
1	1	0	0	0	1	1	1	1	1	1	1	0	1
1	1	1	0	0	1	1	1	1	1	1	1	1	0
x	x	x	x	x	0	1	1	1	1	1	1	1	1
x	x	x	1	x	x	1	1	1	1	1	1	1	1
x	x	x	x	1	x	1	1	1	1	1	1	1	1

(*b*) Truth table

FIGURE 10.17   The 74138 3:8 Decoder

....

EXAMPLE 10.7

The 74138 is used to create one of eight tones via a voltage-controlled oscillator (VCO). A voltage-controlled oscillator simply changes frequency as a function of input voltage. Thus lower voltages, in this case, produce lower frequencies. In Figure 10.18, we use the 74138 as an eight-position switch to send one of eight different voltages to the voltage-controlled oscillator.

Since only one of the outputs of the 74138 is low at any one time, seven resistors are pulled to ground, and the last one sources the +5 V. Thus, we have a voltage divider through an op amp with a gain set by the setting of the 100-kΩ potentiometer. The op amp output drives the VCO to produce a tone dependent on the op amp voltage. Changing the binary word on the ABC input changes the output selected, hence the voltage applied to the VCO.

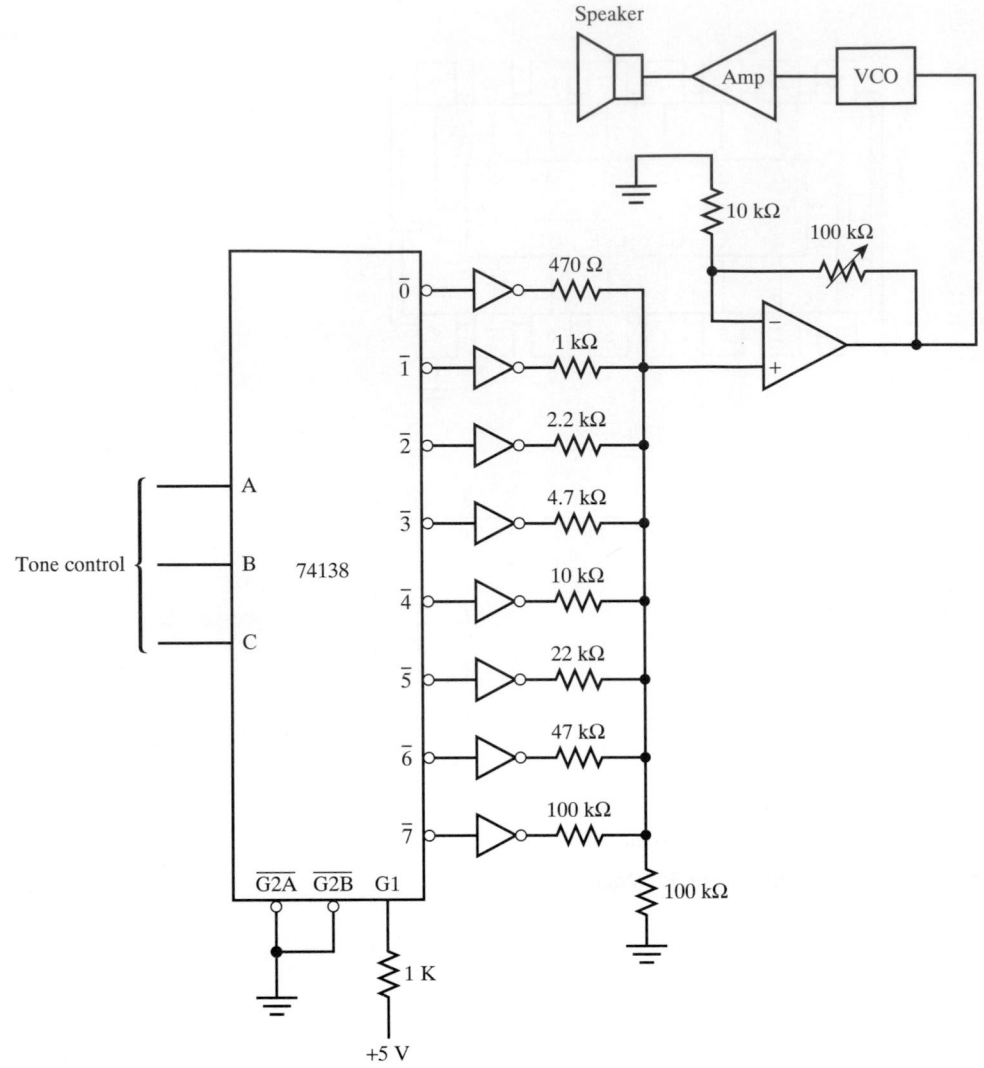

FIGURE 10.18   Simple Tone Generator

The 7442 BCD-to-Decimal Decoder

The 7442 (see Figure 10.19) is a 16-pin DIP that converts a 4-bit input word into one of 10 selected outputs. If the input word is a valid binary representation of one of the output numbers (0–9), the selected output will go low. If the input word is higher than 9 (A–F), all outputs will remain high.

The 7442 was initially used as a decoder for a now obsolete display tube called the nixie tube (see Figure 10.20), which contained a filament and 10 loops of wire shaped to represent the digits 0 through 9. The selected output completed a current path in the nixie tube, causing the proper element (wire number) to glow. The nixie tube has been universally replaced by the seven-segment display.

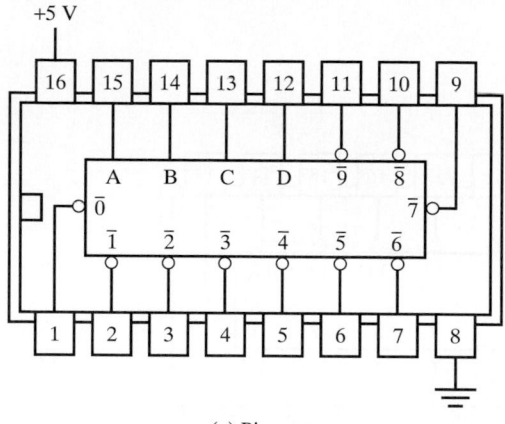

(a) Pinout

D	C	B	A	$\bar{0}$	$\bar{1}$	$\bar{2}$	$\bar{3}$	$\bar{4}$	$\bar{5}$	$\bar{6}$	$\bar{7}$	$\bar{8}$	$\bar{9}$
0	0	0	0	0	1	1	1	1	1	1	1	1	1
0	0	0	1	1	0	1	1	1	1	1	1	1	1
0	0	1	0	1	1	0	1	1	1	1	1	1	1
0	0	1	1	1	1	1	0	1	1	1	1	1	1
0	1	0	0	1	1	1	1	0	1	1	1	1	1
0	1	0	1	1	1	1	1	1	0	1	1	1	1
0	1	1	0	1	1	1	1	1	1	0	1	1	1
0	1	1	1	1	1	1	1	1	1	1	0	1	1
1	0	0	0	1	1	1	1	1	1	1	1	0	1
1	0	0	1	1	1	1	1	1	1	1	1	1	0
1	0	1	0	1	1	1	1	1	1	1	1	1	1
	↓							↓					
1	1	1	1	1	1	1	1	1	1	1	1	1	1

(b) Truth table

FIGURE 10.19   The BCD-to-Decimal Decoder

FIGURE 10.20   A BCD-to-Decimal Nixie Display

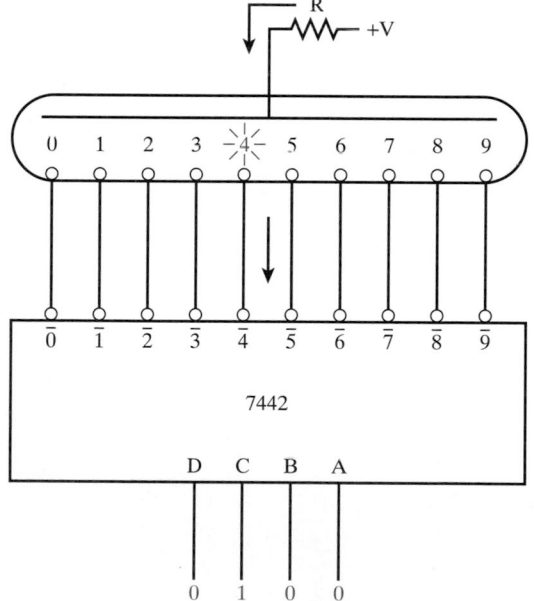

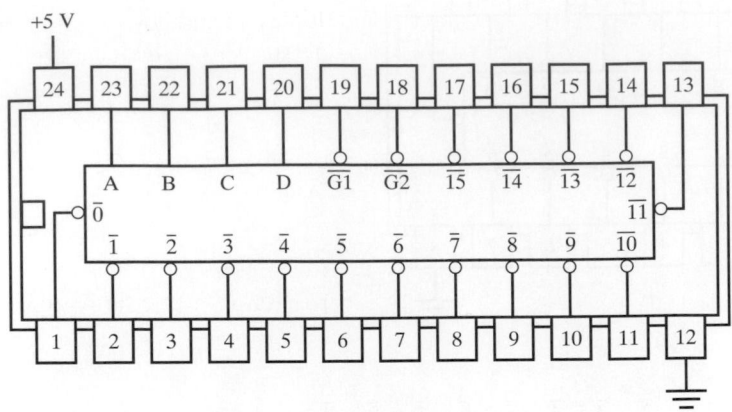

FIGURE 10.21 **The 74154 4:16 Decoder**

The 74154 4:16 Decoder

The 74154 is a 24-pin DIP that decodes a 4-bit word present on its input to one of 16 unique outputs. There are two enable pins ($\overline{G1}$, $\overline{G2}$) to help control the outputs. The desired output is pulled low by placing the correct 4-bit word for that output on the input pins (ABCD) and then grounding both the $\overline{G}$ pins. If either of the $\overline{G}$ pins is high, or if both are, all outputs will remain high. Figure 10.21 illustrates the 74154.

Demultiplexers

The difference between a decoder and a demultiplexer is that a demultiplexer passes data to the selected output, instead of just taking it low or high. Figure 10.22 shows the design and operation of a 1:2 demultiplexer. Both AND gates are supplied with the input signal; only one of them gets a one on the other input controlled by S.

FIGURE 10.22 **1-Line to 2-Line Demultiplexer**

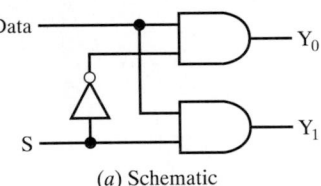

(*a*) Schematic

S	Y_0	Y_1
0	Data	0
1	0	Data

(*b*) Truth table

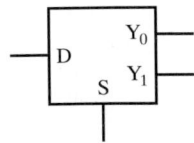

(*c*) Logic symbol

You probably can imagine how we build bigger demultiplexers and how to add an enable input to control all the outputs. Let us look at a different way of making a demultiplexer than you may expect.

· ·

EXAMPLE 10.8

Figure 10.23 shows how a 2:4 decoder is used to make a 2:4 demultiplexer. The selected output of the decoder goes low, allowing the data signal to pass through one of the four OR gates.

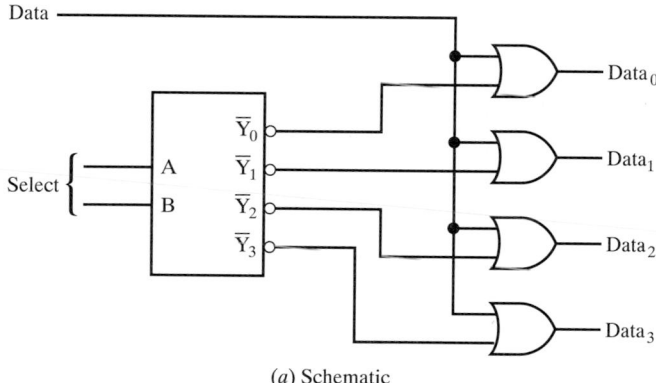

(a) Schematic

B	A	$Data_0$	$Data_1$	$Data_2$	$Data_3$
0	0	Data	1	1	1
0	1	1	Data	1	1
1	0	1	1	Data	1
1	1	1	1	1	Data

(b) Truth table

FIGURE 10.23   Making a 2:4 Demultiplexer

· ·

Multiplexers and demultiplexers are the bread and butter of the telephone company and any business involved in large amounts of data transmission. The telephone company uses multiplexers and demultiplexers to manage special communication lines called *T1 lines,* which support 24 individual data channels over a single pair of wires, rather than 24 pairs. Each data channel provides 64,000 bits/second of information. With special framing bits and other housekeeping data, the total data rate of the T1 line is 1.544 Megabits/second. Figure 10.24 shows the basic operation of a T1 line (one direction is shown; duplicate the hardware for the return signal). The data from all 24 channels is transmitted using a technique called *time division multiplexing (TDM),* where each data channel is given a small slice of time, and the channels are cycled through over and over again. The important thing to do when using TDM is to synchronize both ends of the line to guarantee that the channel data comes out of the right output.

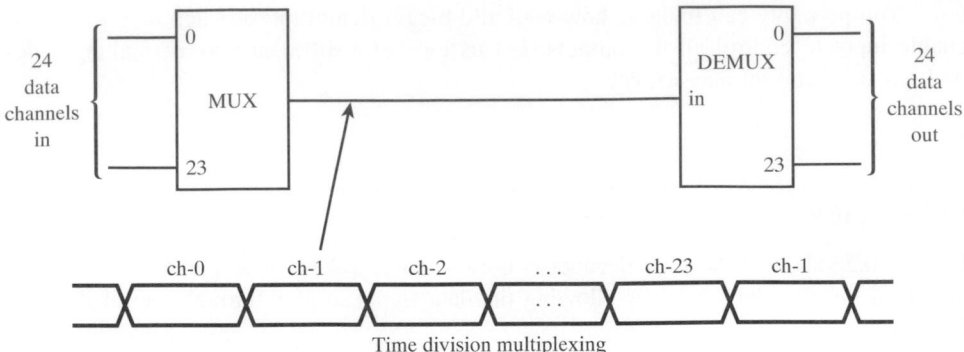

FIGURE 10.24   Basic T1 Operation

10.4   COMPARATORS

A *comparator* is a logic element that compares two binary input numbers *A* and *B* and determines any of the following:

◆ *A* equals *B*

◆ *A* is greater than *B*

◆ *A* is less than *B*

Recall from Chapter 4 that we can use the XNOR gate to compare two *bits,* as indicated in Figure 10.25. The output of the XNOR gate only goes high when the inputs are the same. We use this to our advantage when constructing larger comparators. Figure 10.26 shows four XNOR gates driving the inputs of an AND gate. When all four pairs of input bits are equal there will be four ones on the AND gate inputs. The fifth AND gate input is the cascade $=_{in}$ input. This input is used to connect the $=_{out}$ signal of one comparator to the $=_{in}$ input of the next comparator. If $=_{in}$ is low, the $=_{out}$ signal will also be low, no matter what the XNOR gates are doing. But if $=_{in}$ is high, the $=_{out}$ output will also go high when all four XNOR gates are saying "equal."

Figure 10.27 shows how two 4-bit comparators can be cascaded to make an 8-bit comparator. The upper four bits of each number are the same (1011), but the lower four bits are not. So, even though the $=_{in}$ input of the second comparator is high, its $=_{out}$ output is low, meaning the 8-bit numbers are not equal.

FIGURE 10.25   Comparing Two Bits with an XNOR Gate

bit 1 ──┐
        ⟩ ─○── Equal
bit 2 ──┘

(*a*) Schematic

bit 1	bit 2	Equal
0	0	1
0	1	0
1	0	0
1	1	1

Equal ←→ (rows 0 0 and 1 1)
Not equal ←→ (rows 0 1 and 1 0)

(*b*) Truth table

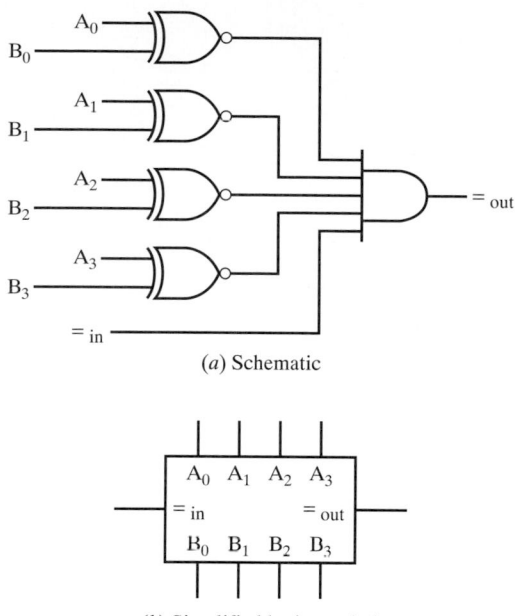

(a) Schematic

(b) Simplified logic symbol

FIGURE 10.26 **4-Bit Comparator**

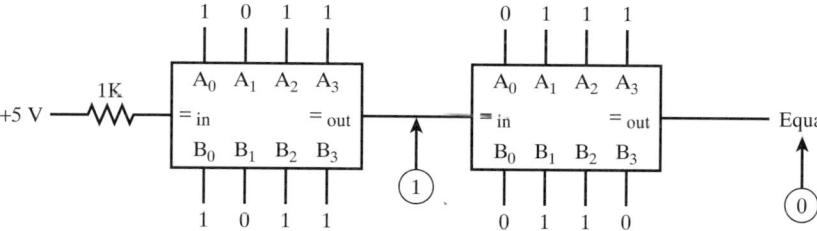

FIGURE 10.27 **An 8-Bit Comparator**

It requires more work to design and build a comparator that determines if two numbers are greater than or less than each other. Figure 10.28 shows the design of a 2-bit greater-than comparator. The $A > B$ output goes high when the 2-bit pattern on A_1A_0 is greater than the pattern on the B inputs. An $A < B$ output can be added easily. Just think of what you need to have an $A < B$ condition. First, A cannot be greater than B, so $A > B$ must be low. Also, A cannot equal B either. So, if we add an $=_{out}$ output to our 2-bit comparator (via some XNOR gates), we can generate the $A < B$ output with the circuit in Figure 10.29. The NOR gate only goes high when both inputs are low, meaning when A does not equal B and A is not greater than B. For practice, try to determine how to make the 2-bit greater-than comparator cascadable.

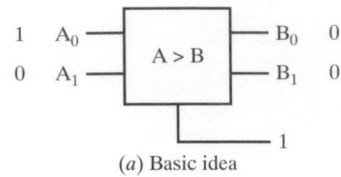

(*a*) Basic idea

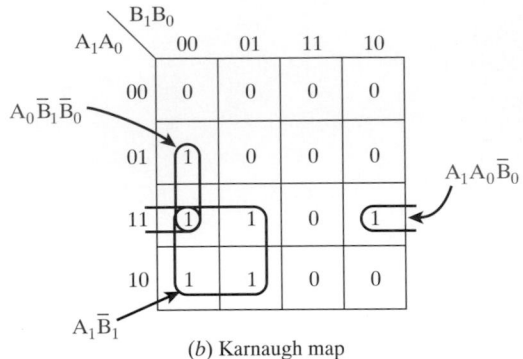

(*b*) Karnaugh map

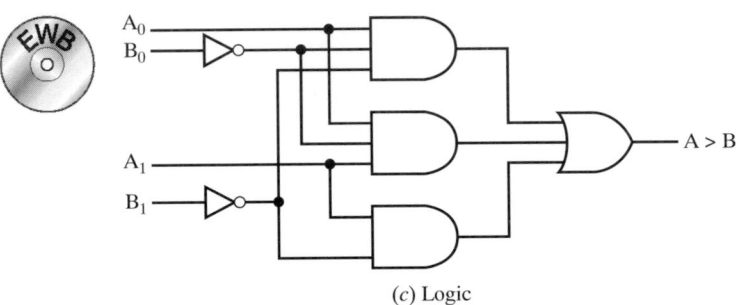

(*c*) Logic

FIGURE 10.28 2-Bit Greater-Than (>) Comparator

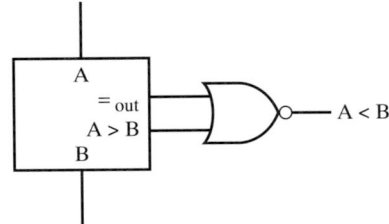

FIGURE 10.29 Simple A < B Testing

Figure 10.30 shows the pinout of the 7485 4-bit magnitude comparator, that performs all the functions we desire, and is cascadable. Comparators are useful in memory and I/O addressing circuitry for allowing user-selectable address ranges. The processor's address line values are compared with the pattern on a DIP switch, and the appropriate action is taken based on the results (enable the memories if equal).

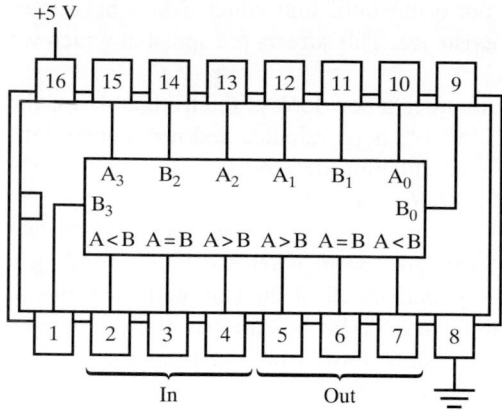

FIGURE 10.30   The 7485 4-Bit Magnitude Comparator

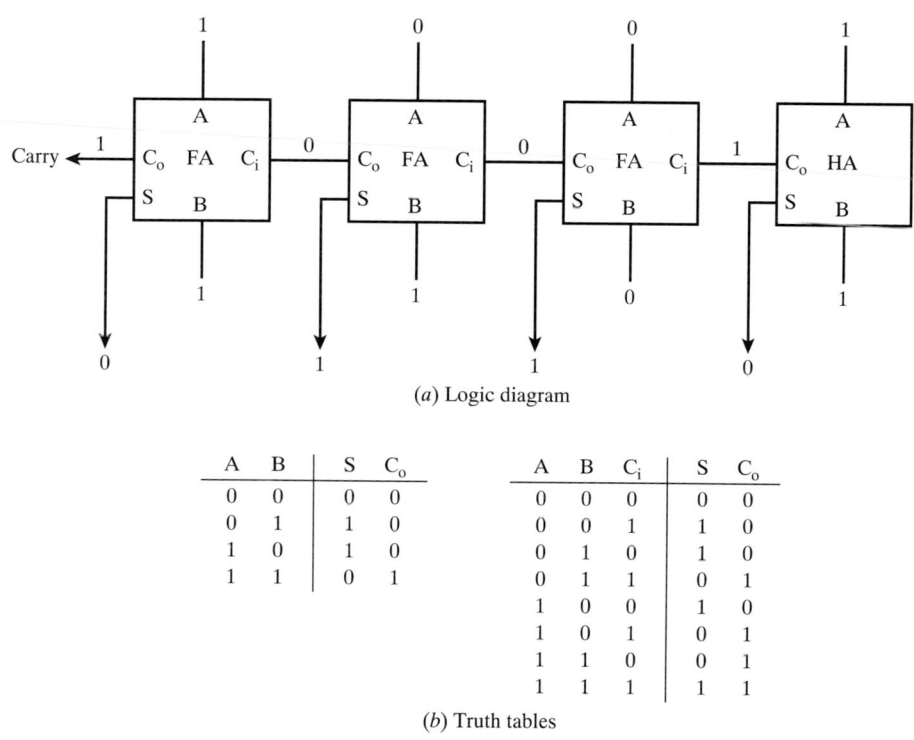

(a) Logic diagram

A	B	S	C_o
0	0	0	0
0	1	1	0
1	0	1	0
1	1	0	1

A	B	C_i	S	C_o
0	0	0	0	0
0	0	1	1	0
0	1	0	1	0
0	1	1	0	1
1	0	0	1	0
1	0	1	0	1
1	1	0	0	1
1	1	1	1	1

(b) Truth tables

FIGURE 10.31   4-Bit Ripple Adder

10.5   ADDERS/SUBTRACTORS

In Chapter 4 we were introduced to the half-adder and full-adder circuits. A 4-bit *ripple adder* can be made by cascading the carry outputs of the half- and full-adders, as indicated in Figure 10.31. Recall that each individual adder has its own propagation delay.

This means that a change in the MSB will not occur until four adder delays have transpired. Thus, the correct output ripples into existence. This affects the speed at which we can change the inputs to the adder.

A different way of designing an adder requires a new look at the truth table for the full adder. Figure 10.32(a) shows how the truth table is reevaluated and new expressions written for the sum and carry outputs. The sum equation developed in Figure 10.32(b) uses a *propagate* term P that is defined as the XOR of *A* and *B*.

The carry equation found in Figure 10.32(c) uses the propagate term and its own *generate* term (*A* ANDed with *B*). These terms are useful when we begin building a **carry lookahead adder (CLA),** an adder that generates all of the sum and carry bits at the same time. It does this by duplicating logic at each bit position so that the outputs are found *independently* of each other. Figure 10.32 shows the equations for the first two bits of a CLA. Notice the large increase in complexity for the second bit's equations. But nowhere in the equations do you see any terms from the first two input bits. If this process is continued, many common terms begin to show up (A_0B_0 in lots of places, etc.). This allows us to reuse some logic and keep the overall gate cost down. Also, the total propagation delay for the CLA is smaller than that of an equally sized ripple adder. For example, an 8-bit ripple adder may have 24 gate delays at the MSB output. An 8-bit CLA may have only six or seven due to all the operations being duplicated (performed in parallel) at each bit.

Figure 10.33 shows the pinout of the 7483 carry lookahead adder. A carry input (C_i) allows us to make larger adders by cascading multiple 7483s.

FIGURE 10.32　Another Look at the Full Adder

A	B	C_i	S		C_o	
0	0	0	0		0	
0	0	1	1	$\overline{A}(B \oplus C_i)$	0	
0	1	0	1		0	
0	1	1	0		1	$\overline{A}BC_i$
1	0	0	1		0	
1	0	1	0	$A(\overline{B \oplus C_i})$	1	$A\overline{B}C_i$
1	1	0	0		1	$AB\overline{C}_i$
1	1	1	1		1	ABC_i

(*a*) Truth table

$$S = \overline{A}(B \oplus C_i) + A(\overline{B \oplus C_i})$$
$$= A \oplus B \oplus C_i$$
$$= P \oplus C_i \text{ where } P = A \oplus B$$

(*b*) Sum equation

$$C = \overline{A}BC_i + A\overline{B}C_i + AB\overline{C}_i + ABC_i$$
$$= (A \oplus B)C_i + AB$$
$$= PC_i + G \text{ where } P = A \oplus B$$
$$\text{and } G = AB$$

(*c*) Carry equation

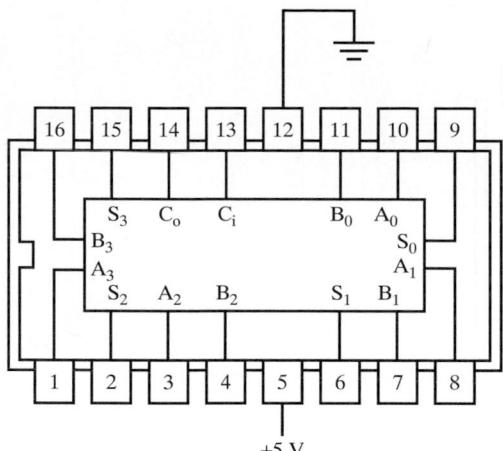

FIGURE 10.33 **4-Bit Carry Lookahead Adder**

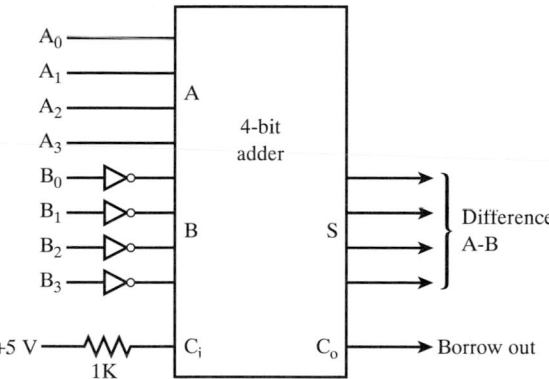

FIGURE 10.34 **Making a 4-Bit Subtractor**

Subtractors

One way to make a binary subtractor is shown in Figure 10.34. The B inputs are inverted before getting to the adder. The carry input is forced high also. Both of these steps are used to generate the 2's complement of the 4-bit B data (invert all bits and add 1). When these inverted bits, plus the carry in, are added to the 4-bit A input, the result is actually the 4-bit difference.

10.6 CODE CONVERTERS

It is sometimes necessary to convert one binary code into another. For example, we might want to convert a 4-bit hexadecimal number into its corresponding BCD equivalent, which means two groups of four bits, because we need to generate the digits 00 through 15.

Figure 10.35 shows the pinouts of the 74184 and 74185 packages. The 74184 is a BCD to 6-bit binary converter, the 74185 is a 6-bit binary-to-BCD converter. Both devices are enabled when the $\overline{G}$ input is low, and are basically hardware lookup tables.

Figure 10.36 shows how BCD and binary numbers are converted.

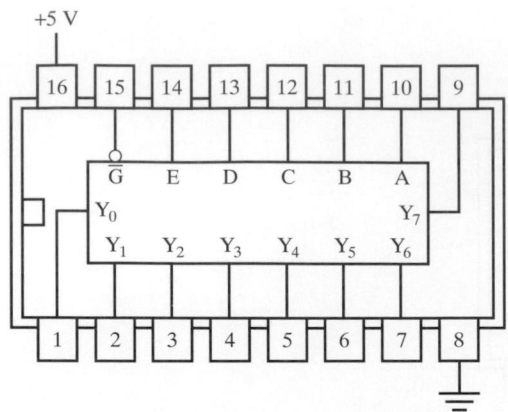

FIGURE 10.35   74184/74185 BCD/Binary Converters

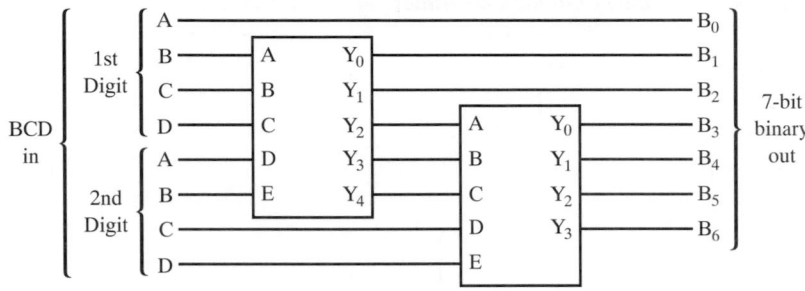

(*a*) BCD to binary conversion using 74184s

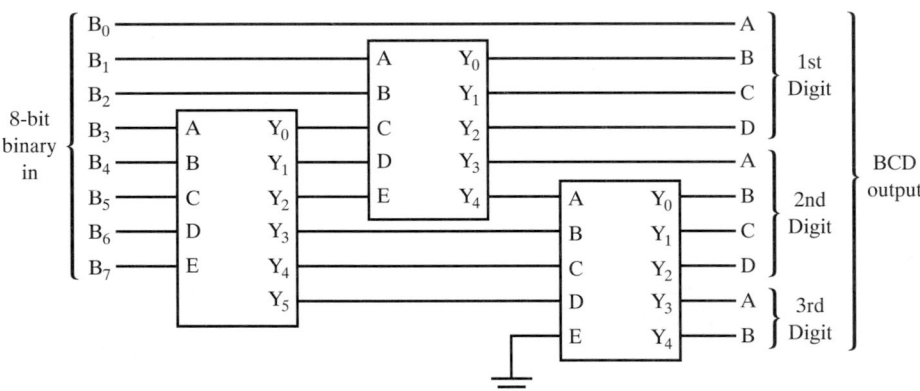

(*b*) Binary to BCD conversion using 74185s

FIGURE 10.36   Conversion Examples

10.7   A/D AND D/A CONVERTERS

When we need to interface a real-world, analog circuit with a digital circuit, two conversions are necessary. ***Analog-to-digital (A/D)*** conversion takes an analog voltage or current and produces an equivalent binary number. ***Digital-to-analog (D/A) conversion*** takes a binary number and makes a corresponding voltage or current out of it. Let us take a brief look at these two processes.

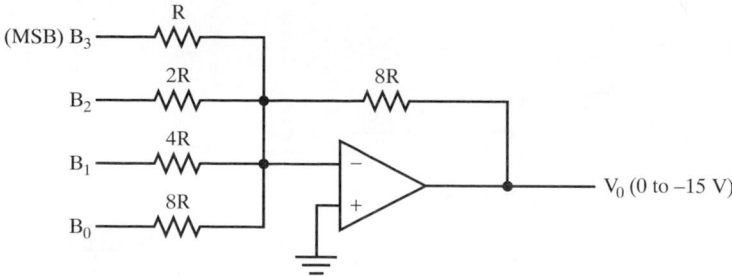

FIGURE 10.37 **4-Bit D/A Converter**

Digital-to-Analog Conversion

Figure 10.37 shows a simple 4-bit D/A converter. The operational amplifier circuit is designed to produce the indicated voltage range by virtue of the ratios between its resistors. The gain of the B_0 input is -1. The other input gains are -2, -4, and -8 (for B_3). So, any binary number on the input, such as 1001, will cause the appropriate weights (-1 and -8) to be added together (giving -9 volts). There are considerations that need to be addressed to make the circuit work with TTL levels (0 to 5 volts). Figure 10.38 shows the Electronics Workbench circuit DTOA that implements the 4-bit D/A. Note the actual values of the resistors. Since the counter is being clocked continuously, we get a *staircase* waveform out of the D/A converter. The output voltage changes by -1 volt with each clock pulse (related to the gain of the LSB input). This means the converter has a *step size* of -1 volts.

Commercial D/A converters, like the 1408 shown in Figure 10.39, offer better control over the output voltage or current. The 1408 provides 256 different output currents (from 0 mA to 1.85 mA in steps of 7.26 μA), which are converted into a range of voltages by the 2.7K ohm resistor. With a 5-volt output swing, the step size for the 1408 in Figure 10.39 is 5 V/255, or 19.6 mV/step. An 8-bit D/A is suitable for synthesizing audio waveforms, such as speech and music.

Analog-to-Digital Conversion

Determining the correct binary number for an analog input is a trickier matter than the simple way a D/A converter works. Figure 10.40 shows one way to build an A/D converter. The unknown analog input is compared with a generated voltage. If higher, the counter counts up. If lower, the counter counts down. The output of the counter is sent to a D/A converter to generate the new analog voltage to compare. Thus, the binary output tracks the analog input, and any changes to the input voltage are discovered instantly and the output is adjusted.

It may take a number of clock cycles to get the counter in Figure 10.40 to the right value. For very fast A/D conversion, a more complex converter called a *successive approximation register (SAR)* is used to control the conversion. The counter in Figure 10.40 is replaced with an SAR that determines each bit of the output one clock cycle at a time. Figure 10.41 shows how a 4-bit conversion is made using an SAR. Assume that the step size for the D/A is 1 volt. On the first clock pulse, the output of the SAR becomes 1000 (the MSB is set). This produces an output of 8 volts, which is less than the 11-volt input voltage. So the MSB is kept high. During the second clock cycle the next bit is set high, producing an output of 12 volts. This is greater than the input voltage, so the SAR clears the bit and moves on.

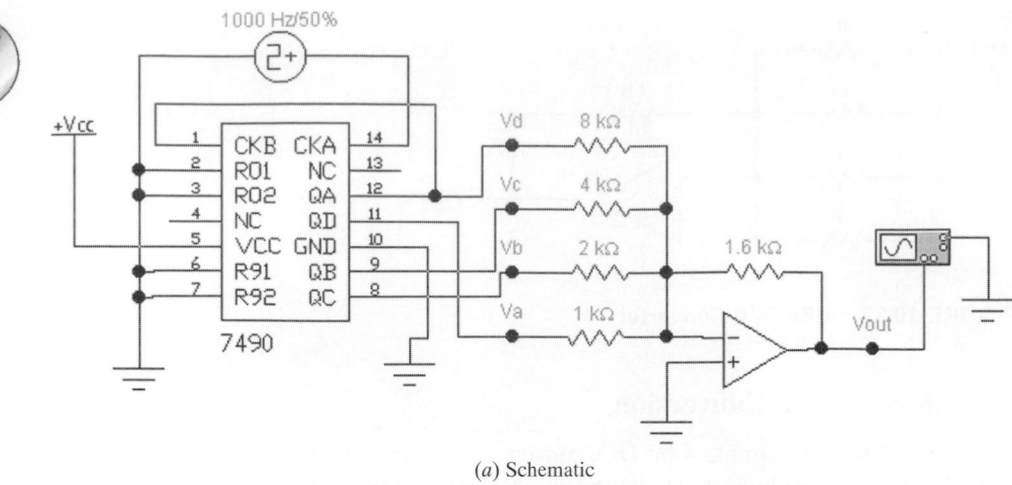

(*a*) Schematic

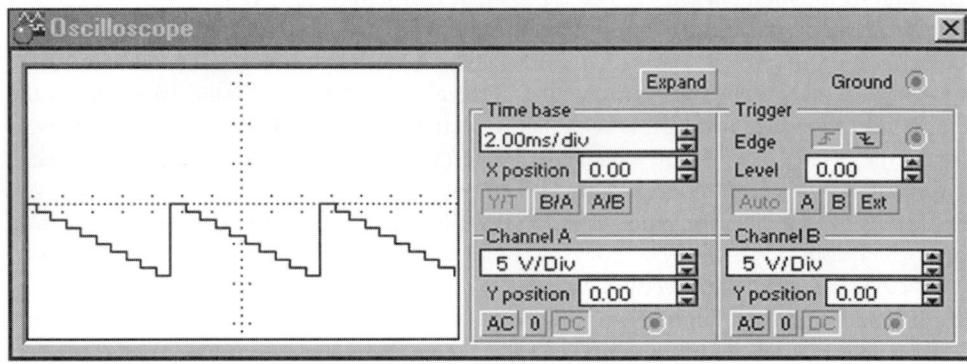

(*b*) Oscilloscope display of output

FIGURE 10.38   Electronics Workbench File DTOA

FIGURE 10.39   1408 8-Bit D/A Converter

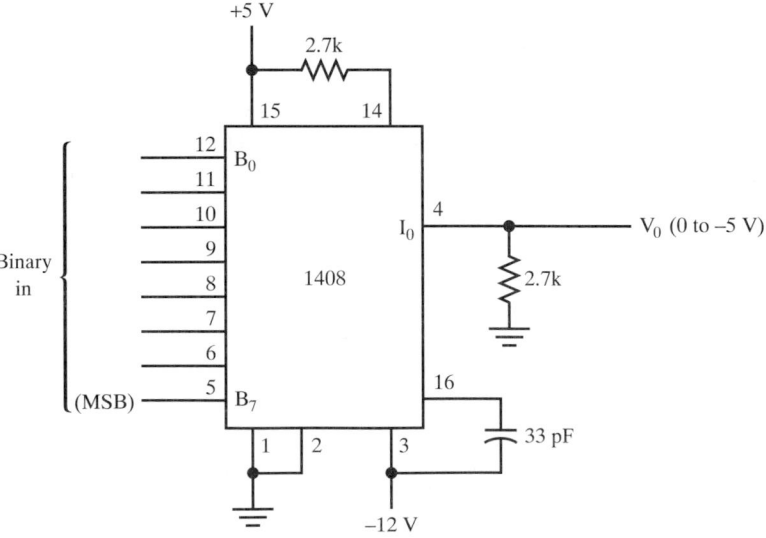

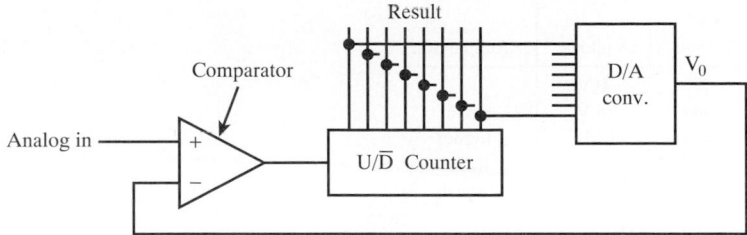

FIGURE 10.40   Analog-to-Digital Converter Using a D/A

⋯⋯C/C++ HELPER⋯⋯⋯⋯⋯⋯⋯⋯⋯⋯⋯⋯⋯⋯⋯

The DAC program on the companion CD-ROM is used to determine the voltage controlled by each bit of a digital-to-analog converter. The number of bits in the D/A converter and the output voltage range are the only input values required by the program. Examine the following sample execution:

```
C> DAC
Enter the number of bits ---> 8
Enter the voltage range ----> 5
Number of steps: 255
   Bit           Voltage
10000000      2.5098 Volts
01000000      1.2549 Volts
00100000      0.6275 Volts
00010000      0.3137 Volts
00001000      0.1569 Volts
00000100      0.0784 Volts
00000010      0.0392 Volts
00000001      0.0196 Volts
```

The DAC program determines the number of steps (255) based on the number of bits (8) entered by the user. The weight of each bit (1, 2, 4, 8, etc.) is divided by the number of steps and multiplied by the voltage range.

Use the DAC program to find the similarities between 8-, 10-, and 16-bit D/A converters with the same voltage range.

Clock cycle	SAR Output	Voltage	Comparison
Reset	0000	0 V	–
1	1000	8 V	lower
2	1100	12 V	higher
3	1010	10 V	lower
4	1011	11 V	equal

(*a*) Conversion steps

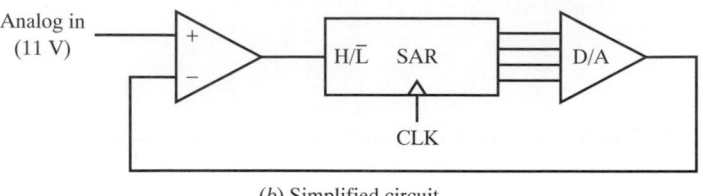

(*b*) Simplified circuit

FIGURE 10.41   SAR Method for A/D Conversion

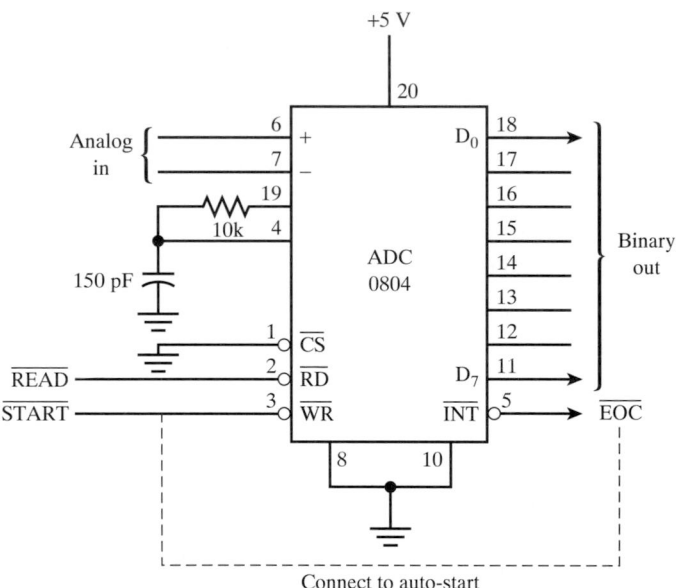

FIGURE 10.42   8-Bit A/D Converter

The third and fourth clock pulses cause the next two bits to generate 10 V and finally 11 V, the actual input voltage. One nice feature of the SAR A/D converter is that all conversions take the same time (in this case, always four clock pulses).

Figure 10.42 shows a commercial A/D converter, the ADC0804. The 0804 is ideal for interfacing with a microprocessor due to its various control inputs. An R-C combination is used to control the speed of the internal conversion clock. The 0804 digitizes over 8000 samples/second with good results.

10.8 PRACTICAL APPLICATIONS

Let us now take the time to study a number of practical applications for all of the devices encountered in this chapter.

Digital Clock Alarm Circuit

In Chapter 7 we were presented with the design of a simple digital clock. In this example, we add an alarm to the clock. Just as several 7490 decade and modulo-6 counters were used to keep track of the time, we now add more counters to store the alarm time. To determine if the alarm should go off, we must compare the clock time with the alarm time. This is done on a digit-by-digit basis, using circuitry like that shown in Figure 10.43. A 4-bit comparator is used to compare the two BCD digits in the stage. Cascade signals are used to extend the comparator to all stages.

Digital Clock Display Logic

Once we add an alarm to the digital clock, the display logic must be changed, since we must now be able to view two sets of digits on the displays. Figure 10.44 shows the display logic for one stage of the clock. A quad 2:1 selector is used to send one group of BCD bits or another to the display decoder. It would be worthwhile to estimate the total

FIGURE 10.43 Digital Clock Alarm Circuitry (1^s Stage)

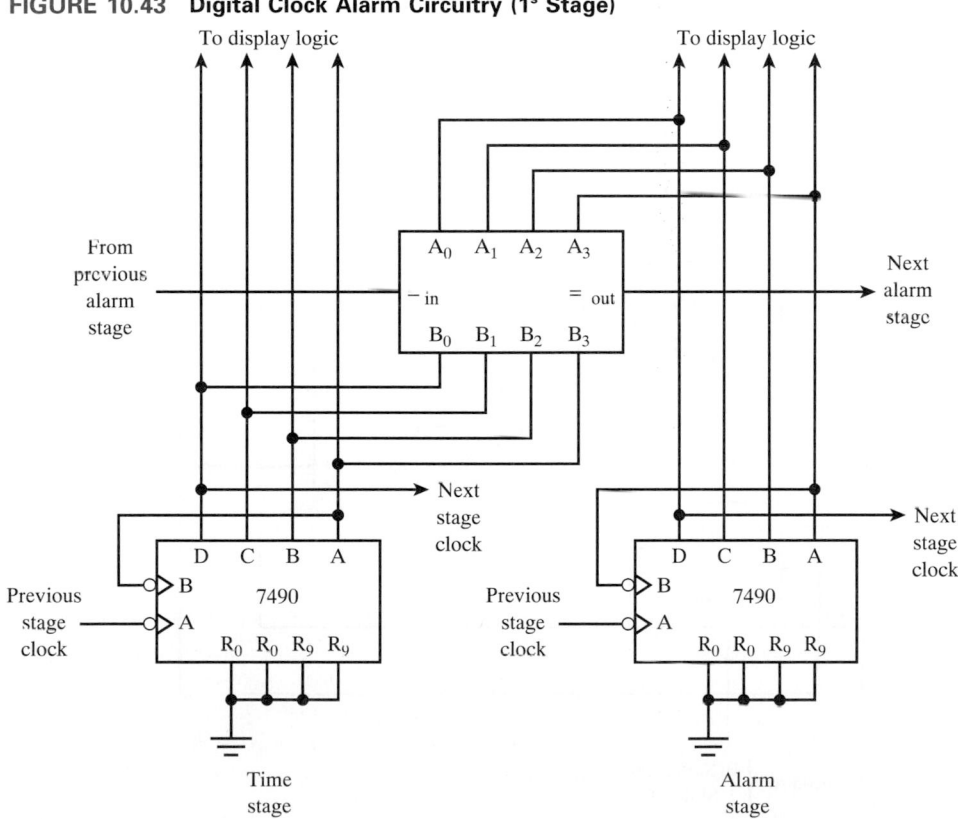

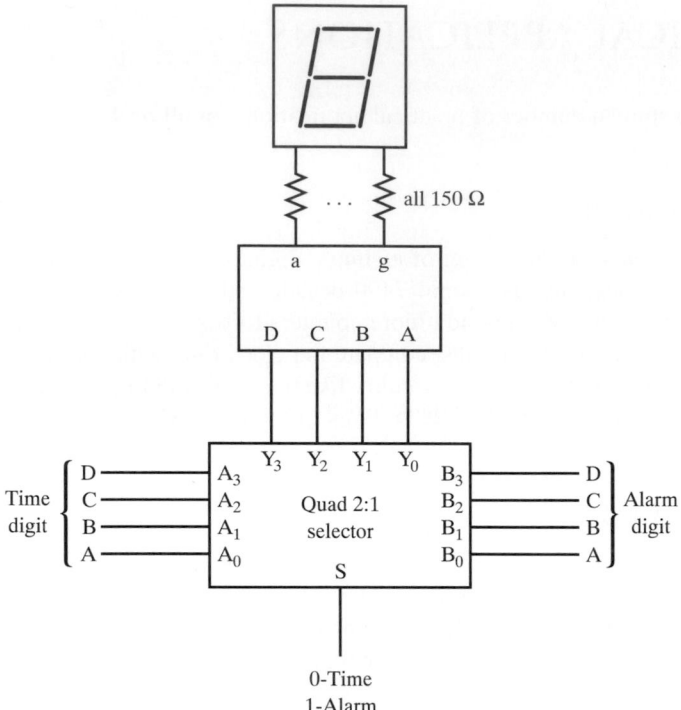

FIGURE 10.44   Digital Clock Display Logic

FIGURE 10.45   One Stage of a Right/Left Shift Register

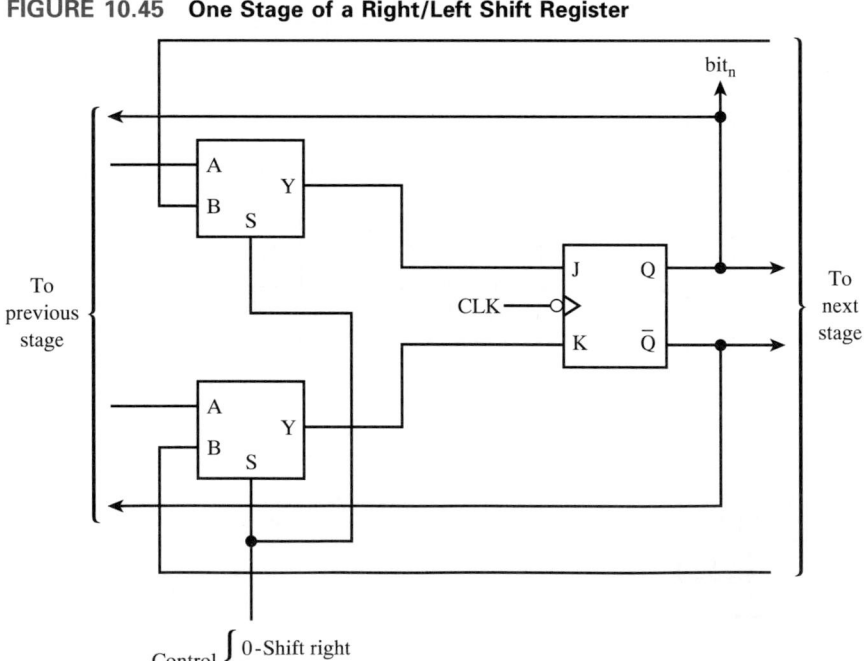

number of devices required in the digital clock, now that alarm functions have been added. Also, with the same methods, we can add a calendar to the clock as well. The only difficulty in the calendar logic will be getting the months to roll over correctly.

Right/Left Shift Register

The shift registers we covered in Chapter 7 included one that was able to shift left and right. In Figure 10.45 we see that data selectors are required to implement the shift register. The right/left control input causes the 2:1 selectors to pass the appropriate stage's output to the J-K inputs.

Multiplexed Display

Let us look at one final application that uses a handful of gates to drive, or multiplex, an 8-digit display. The display can be used for computer output, as in the status of the address and data buses, or as output for an electronic clock or calendar.

In Figure 10.46 we see how the display is generated. A 500-Hz oscillator drives a divide-by-8-counter. For every count of 0 to 7, one seven-segment display is enabled by

FIGURE 10.46   Circuitry to Drive an 8-Digit Multiplexed Display

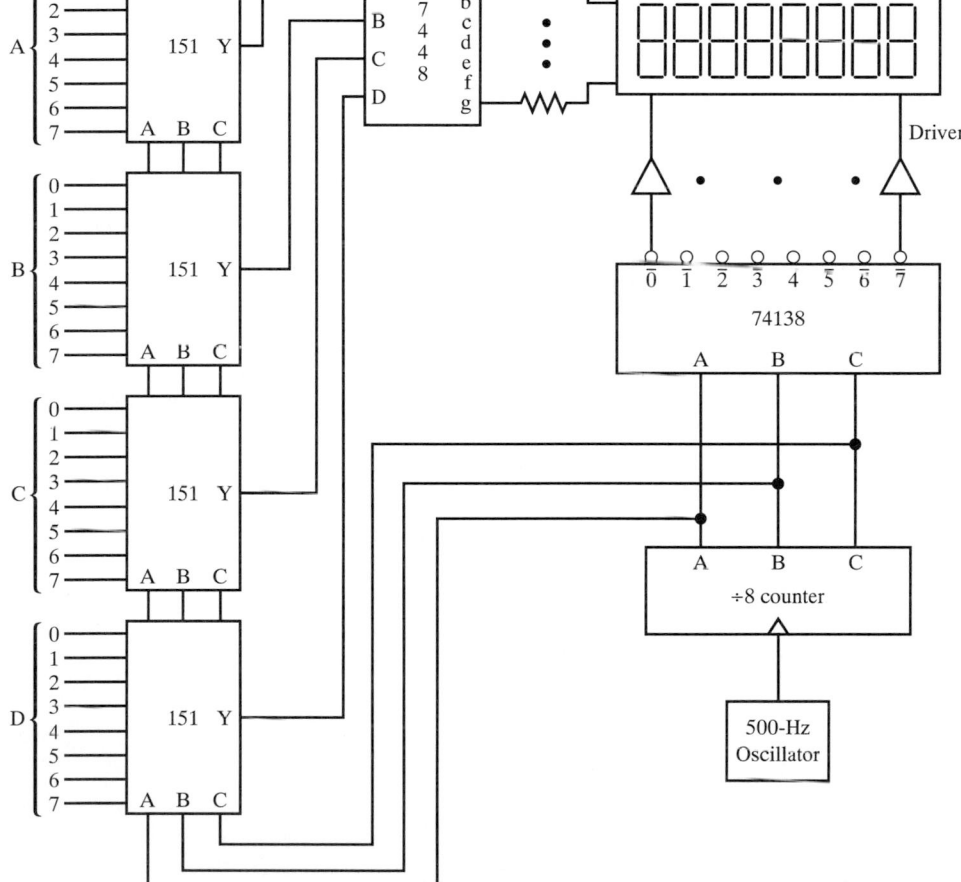

the 74138. In addition, each of the four 74151s looks at one of eight inputs. The inputs are separated so that each 74151 gets one bit of the input word for each digit. The outputs of the 74151s are fed to a 7448 BCD/seven-segment decoder, which drives the display. At 500 Hz, our eyes cannot tell that the displays are flashing in sequence, and we get the appearance of a solid 8-digit display.

10.9 TROUBLESHOOTING TECHNIQUES

We were only able to touch on a handful of the hundreds of logic elements available today. There are plenty more types of counters, shift registers, buffers, comparators, converters, and selectors/decoders on the market. Settling down with a TTL data book for a few hours to survey what it contains would be time well spent. In the future, you might be in the middle of a complicated design, wringing your hands because you require a very complex logic element. You have spent hours designing it and it still does not work. Then a colleague more familiar with the TTL data book than you walks in and says, "Oh, all that stuff is in the 74181 Arithmetic Logic Unit."

You will learn a lot that day.

SUMMARY

In this chapter we began with the integrated circuits that perform very special functions: selectors/multiplexers and decoders/demultiplexers. Through the use of selectors, we are able to handle large quantities of data by looking at smaller, controlled portions of it as we desire. Decoders allow us to control large amounts of outputs with a small number of inputs. As we saw, the number of inputs or outputs that we can control or look at is a function of the form 2^N, where N is the number of control bits. We also covered comparators, adders, subtractors, A/D and D/A conversion, and other practical circuits.

STUDY QUESTIONS

General

1. How many inputs can be multiplexed with a 5-bit control word? How many outputs can be multiplexed with an 8-bit control word?

2. Show that the circuit in Figure 10.47 is a 2:1 multiplexer.

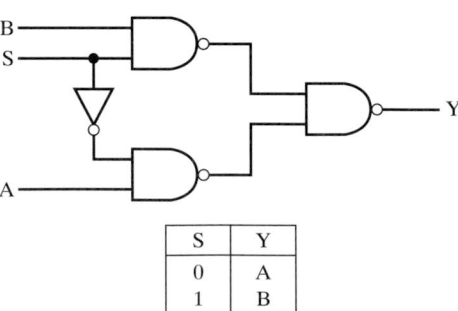

S	Y
0	A
1	B

FIGURE 10.47   Circuit for Question 10.2

3. Which circuit is more practical, Figure 10.48(a) or Figure 10.48(b). Why?

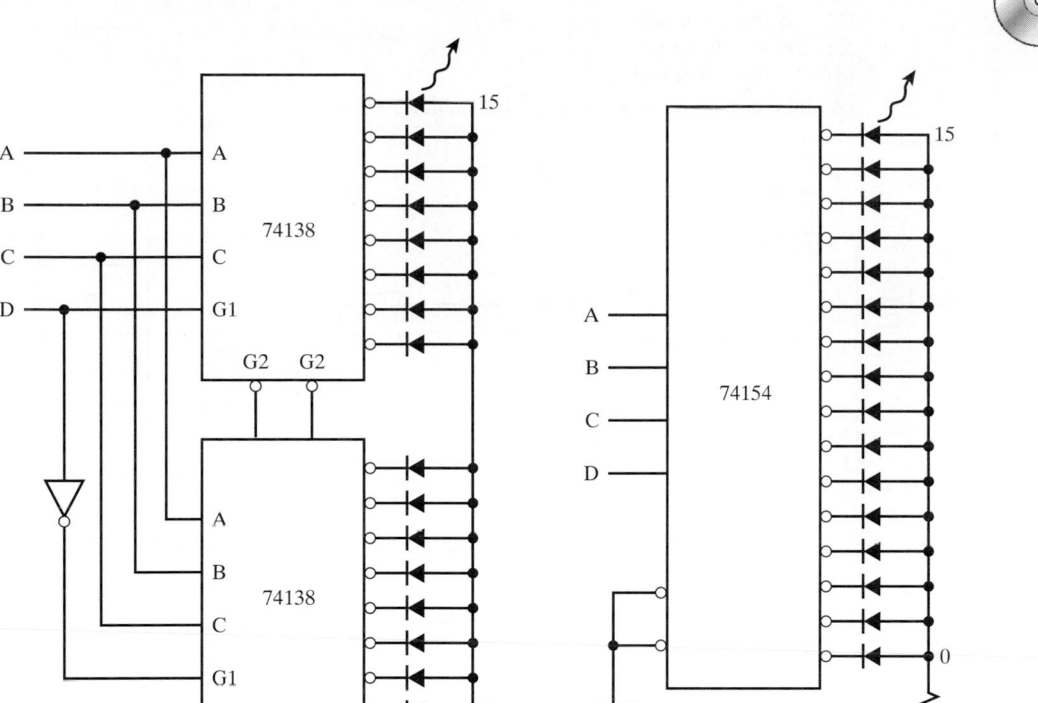

(a)   (b)

FIGURE 10.48 Circuits for Question 10.3 (all LEDs): (a) One of 16 for 74138 (b) One of 16 for 74154

4. Which LEDs in Figure 10.49 will light when:
   (a) *A* is low, *B* is low
   (b) *A* is high, *B* is low
   (c) *A* is low, *B* is high
   (d) *A* is high, *B* is high
   (Assume that all devices are properly enabled.)

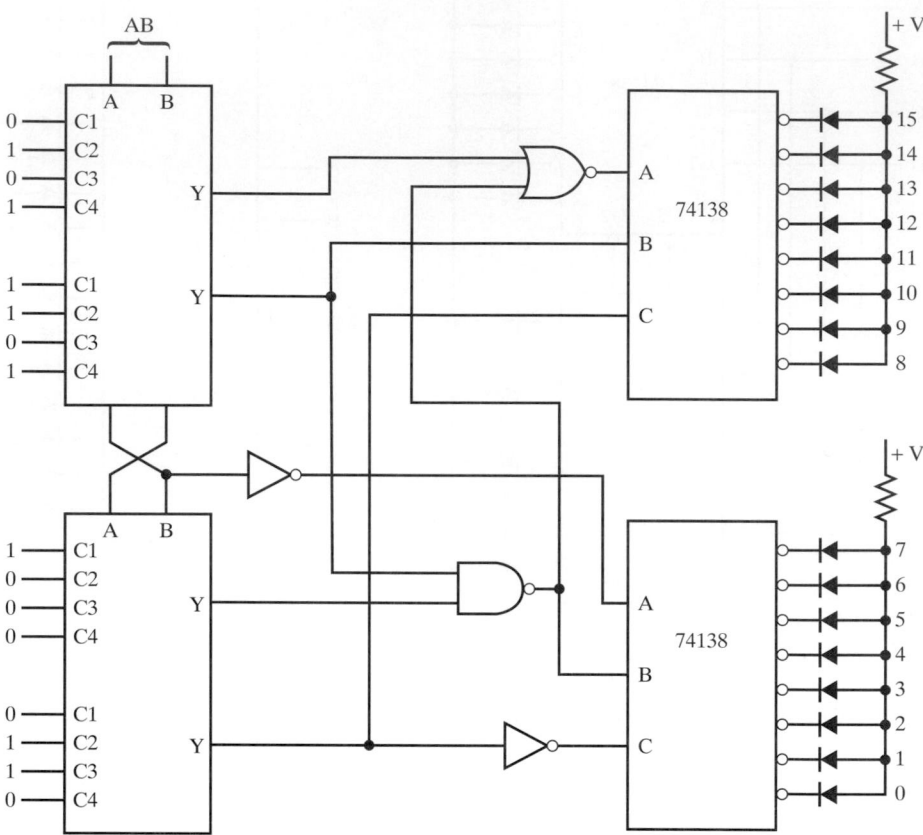

FIGURE 10.49   Circuit for Question 10.4

5. Draw at least four cycles of the output waveforms ∅A and ∅B in Figure 10.50. [Assume counter at 0 and both latches cleared (Q = 0) to start.] *Hint:* Study the R-S flip-flop of Chapter 6 and first draw the output waveforms of the 7442 only.

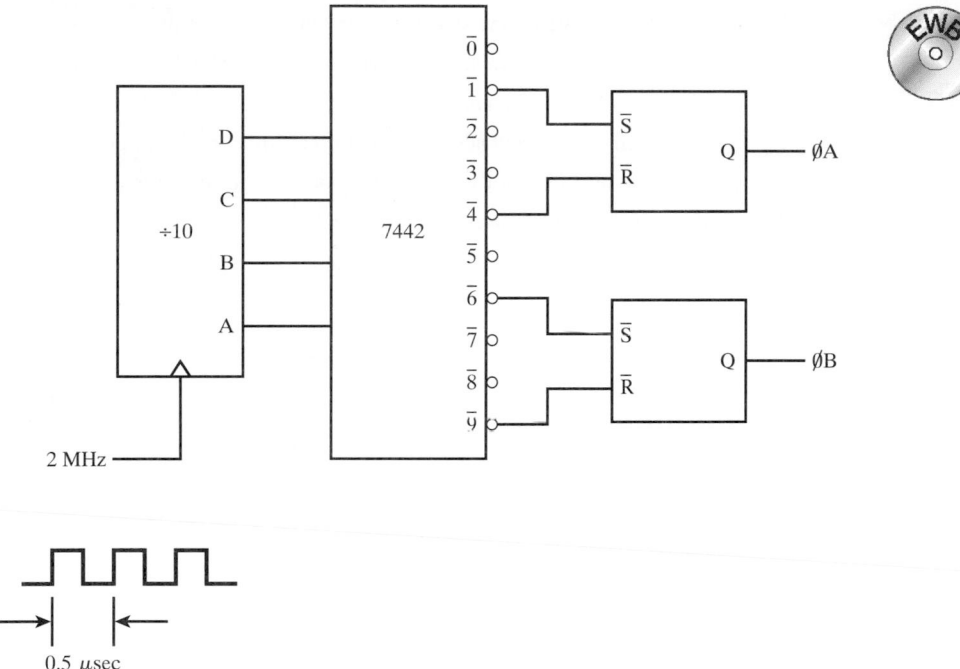

FIGURE 10.50   Circuit for Question 10.5

6. Design a circuit to pass a digital signal to one of eight outputs. The outputs are selected by three binary inputs A, B, and C.

7. Add an enable input to the circuit of Question 6 so that the circuit decodes when $\overline{EN}$ is low and the outputs are all high when $\overline{EN}$ is high.

8. Find the smallest number of TTL chips that will select one of 256 outputs. The input select word is 8 bits long and the circuit must contain some kind of enable.

9. Design a circuit that will look at input bits 1 and 5 when X_1 is low and bits 3 and 7 when X_1 is high. Bits 1 and 3 are selected when X_2 is low and bits 5 and 7 are selected when X_2 is high. (*Hint:* Use an 8-to-1 device.)

10. A control word consists of 9 bits A0 through A8. Design a decoding circuit that will pull $\overline{IN}$ low if the control word equals 0BC hex and will pull $\overline{OUT}$ low when the control word equals 0BD hex. (*Hint:* Use an 8-input NAND gate, some inverters, and some OR gates.)

11. Modify the circuit of Question 10 so that $\overline{IN}$ and $\overline{OUT}$ will go low only if the control word is correct AND an additional input $\overline{STROBE}$ is also low.

12. Design a circuit that will pull $\overline{RD}$ low when an 8-bit control word is any of the following: 00, 10, 20, 30, ... , E0, F0. All numbers are in hex.

13. Modify the circuit of Question 12 so that an additional output $\overline{WR}$ is pulled low when the control word is any of the following: 08, 18, 28, 38, ... , E8, F8. Make sure that $\overline{RD}$ and $\overline{WR}$ are never low at the same time.

14. Pick any decoder/demultiplexer from the TTL data book and trace one signal from input to output.

15. Repeat Question 14 for any selector/multiplexer.

16. What is time division multiplexing?

17. Show how 16 data channels are multiplexed so that one bit from each channel is output at a time over a single wire. It takes 16 clock cycles to send one bit from each channel.

18. What is the data rate on the single wire in Question 17 if the bit time is 10 ns?

19. Show how the 2-bit greater-than comparator can be extended to 4 bits.

20. Design a 2-bit less-than comparator.

21. What is a carry lookahead adder? How does it differ from a ripple adder?

22. Design a one-bit subtractor with borrow input and borrow output.

23. What is a successive approximation register?

24. Design a state machine that will implement a 4-bit SAR.

Electronics Workbench

25. Use Electronics Workbench to determine the resistor values for a 5-bit D/A converter based on the 4-bit op amp circuit.

26. Verify your solutions to questions 6, 9, 11, and 20 using Electronics Workbench.

Programmable Logic

27. Write the PAL equations for
    (a) a 2:4 decoder
    (b) an 8:1 multiplexer
    (c) a 4-bit comparator ($A = B$)
    (d) a 2-bit greater-than comparator

28. Where could programmable logic be used in the digital clock example?

MEMORIES

When finished with this chapter, you should be able to:

1. Identify the various attributes of the different memory types.
2. Tell the difference between RAM and ROM.
3. Understand the uses of static versus dynamic memories.

4. Show how a byte-oriented memory is constructed from a single-bit RAM.
5. Describe the benefits of using an EPROM.
6. Describe the operation of a decoder or selector in selecting a particular memory.
7. Summarize the different types of memories used in the personal computer.

Keep the following questions in mind and try to answer them when you have completed the chapter:

1. Why are shift register memories inherently slow compared to random-access memories?
2. What are the advantages and disadvantages of ROM, PROM, EPROM, and EEPROM?

3. After an address has been sent to a RAM or ROM, why must there be a delay before reading data?
4. Which lines of a memory (RAM, ROM) are tri-stated and why?
5. What is the purpose of a character generator?
6. How can a memory system be tested?

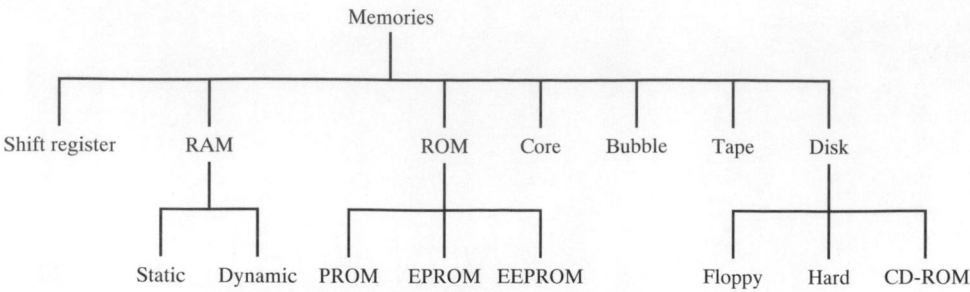

FIGURE 11.1 Breakdown of Solid State Memory Families

11.1 INTRODUCTION

We have already covered a few of the simpler memory devices, namely, the flip-flop and the shift register. Although very useful in digital circuits for storing temporary results, these devices are not practical for use in more sophisticated circuitry, such as microcomputers. For instance, in a digital circuit, we may use flip-flops to keep track of an 8-bit binary count, or some other word of information. In a microcomputer, where we frequently deal with thousands of bits of information, it becomes impractical and uneconomical to use single flip-flop packages for memory.

In this chapter we will see how a single integrated circuit is used to store these thousands of bits, and how they are used to implement a microcomputer memory.

Figure 11.1 shows a breakdown of the various types of memory. We will study many of them in detail and also learn about their function in the microcomputer. We will see how power requirements, speed, and density all become important when it is necessary to decide which type of memory is to be used.

11.2 RANDOM ACCESS VS. SEQUENTIAL ACCESS

Memories are classified by access method, along with other parameters. A ***random-access memory,*** or ***RAM,*** is a device that allows any location to be accessed at any time. This is done by supplying an address to the device during a read or write operation.

A sequential-access memory, such as a shift register, may require several clock cycles before the desired data is available.

Both types of memories have their use in the digital world.

Shift Register Memories

In Chapter 6 we saw how shift registers are used to delay signals and aid in the hardware multiplication process. Although not particularly useful as storage devices for microcomputers, shift registers do have some applications. Their relatively slow speed and low density are not suited for computers, but these properties are ideal for applications that call for repetitive data. Consider the operation of a television typewriter in which ASCII* data from an ordinary electronic keyboard is displayed on a television set. To maintain a constant image on the television screen, the data required to fill the screen must be sent to the television set 60 times a second. By using a shift register, we are able to *recirculate* the data to give a fresh image every 60th of a second. The term "recirculate" refers to the method of feeding the output of a shift register back into the input.

In Figure 11.2, we see how a simple shift register memory is made, and how, through a control switch, we are able to recirculate the data in the shift register. With the switch in the Load position, data present on the Data in line is loaded or shifted into the shift register. Once we have completed loading the data into the shift register, we flip the switch into the Recirculate position. Now when we clock data out of the shift register, we load or recirculate the same data back into the shift register. By combining seven shift registers and the necessary recirculate logic, we can very easily make a memory able to store the ASCII data needed to make our TV typewriter function.

A TV typewriter may be capable of showing data on a television screen in a variety of formats. Let's assume that a 16-line display of 32 characters per line is desired. This means that $32 \times 16 = 512$ ASCII characters can be displayed. Since a television screen must be continuously updated to maintain a constant display (30 frames/sec, or 60 fields/sec), the ASCII data must be repeatedly available. Storage in shift registers is ideal because the data can be recirculated and is continuously available for display.

Seven 512-bit shift registers can be used to form the memory as shown in Figure 11.3. The clock line of each register is connected to a common clock, to rotate all data at the same rate. This principle is used in many data terminals and shows an important application of a shift register memory.

TV typewriters are among the few devices that use shift registers. For another example, we can design a circuit that employs special recirculate logic to produce a random pattern of ones and zeros.

---

*ASCII is a standard communications code that is covered in Chapter 12. For now, think of it as a 7-bit code.

FIGURE 11.2 Simple Shift Register Memory

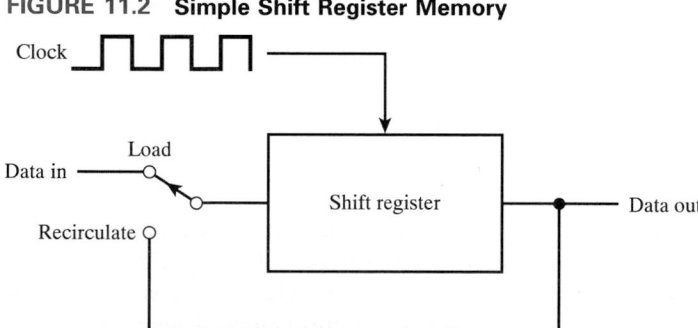

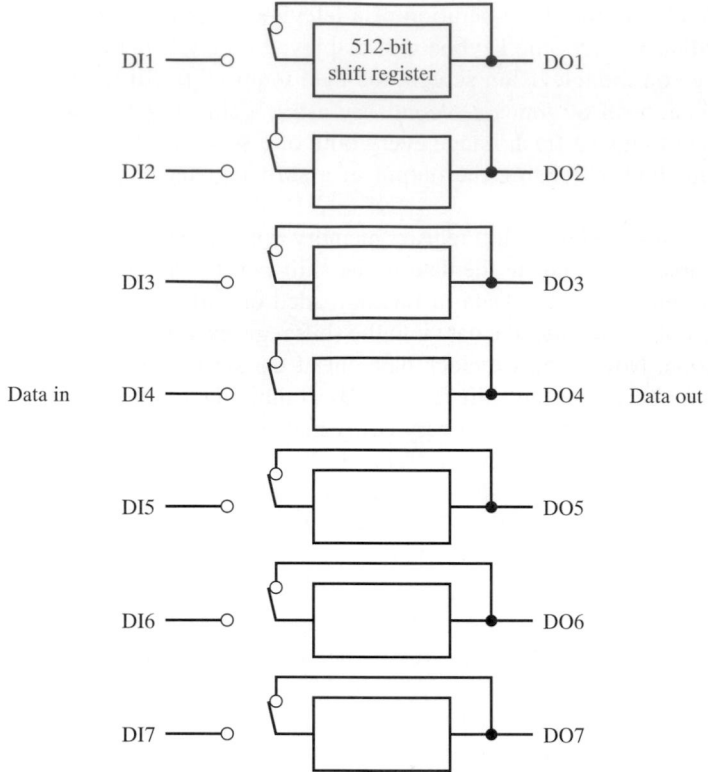

FIGURE 11.3 **Seven 512-Bit Shift Registers**

11.3 STATIC RAM VS. DYNAMIC RAM

Before we continue with our discussion of memories, we take time out to define some important terms. When dealing with integrated memory devices, we frequently come across the terms *static* and *dynamic*. **Static memory** devices are composed of actual flip-flop packages, whereas **dynamic memory** devices contain capacitive elements for storage. The advantages of a dynamic device include its higher bit density, lower cost, and low power requirement. The static device has the advantage of increased speed. The speed of a memory is measured in terms of the time required to get data from a particular memory location. This is known as the **access time**. The access time for a static memory is as low as 10 ns; the dynamic memory has an access time of 60–70 ns. The designer is faced with making certain trade-offs when building a microcomputer system: high speed, but more power and space required; or lower speed, lower power, and less space. Larger systems employ dynamic memories favoring high density and lower power requirements.

Figure 11.4 shows a static memory cell (flip-flop) and a dynamic RAM cell (capacitor). The very small capacitors that make up the dynamic memory cell may be charged to represent a logic one and discharged to represent a logic zero. The capacitor cannot remain charged very long and must be recharged every 2 ms. The process is called a **refresh** operation, and dynamic RAM requires a separate refresh clock. The act of reading or writing can refresh the dynamic RAM, but separate refresh circuitry is usually used.

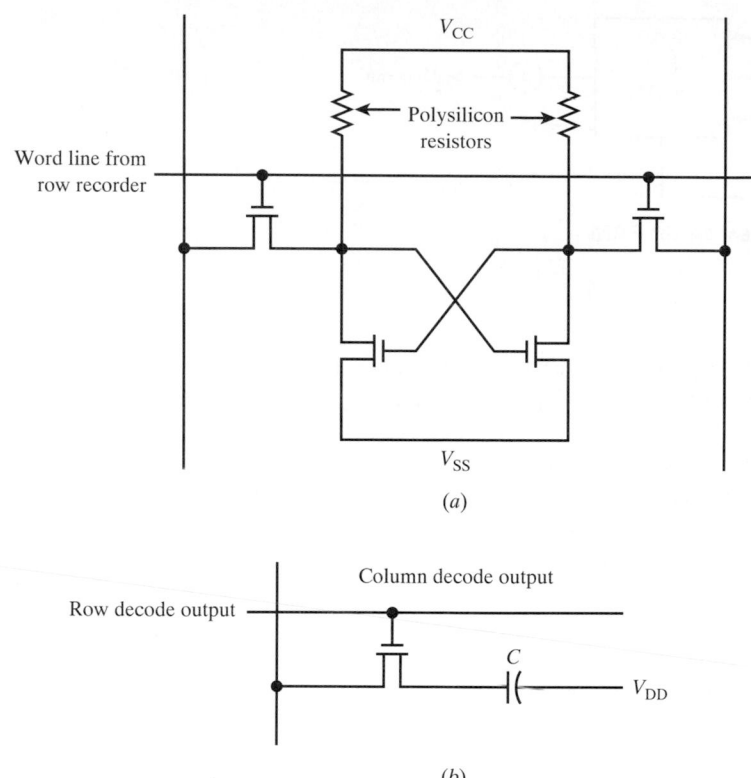

FIGURE 11.4 (*a*) Static Memory Cell (*b*) Dynamic Memory Cell

11.4 STATIC RAM

Our study in memories now takes us to the world of *random-access memories* and *read-only memories*. Unlike the shift register, which obliges us to clock the devices to get at the desired data, the random-access and read-only memories are organized as storage elements in sequential order. To get a piece of data out of a RAM or a ROM, we simply send the device a binary word that is actually an address of an internal storage area, and the device gives us our data.

RAMs and ROMs come in many sizes (number of bits or bytes that you can store) and are usually referred to as 256 × 4 bit, 1K by 1 bit (where K = 1024), or 1K × 8 bits. In the case of 256 × 4 bits, we have a memory consisting of 256 words of 4 bits each. To get at a particular word of 4 bits, we send the device one of 256 possible addresses, and the 4 bits stored at that location appear at the output.

In the 1K × 8 case, we can choose any of 1024 addresses to retrieve an 8-bit word (or byte). This kind of memory is usually referred to as a 1K byte RAM or ROM.

Now that we know a few basics, let's study the RAM in detail and see how we use it in a microcomputer memory.

RAM

For our study of the RAM, we will first look at an ideal RAM, then at a few of the most common RAMs used in microcomputer memory circuits. Notice the three RAM control functions shown in Figure 11.5: Address, read or write, and chip enable. The ***chip enable***

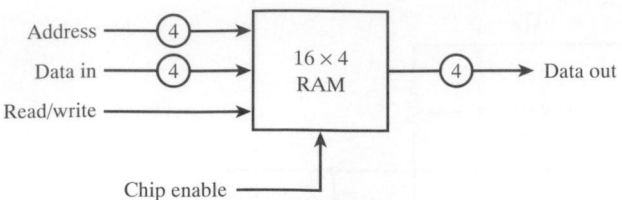

FIGURE 11.5 An Ideal 64-Bit RAM

function becomes *very* important when we combine many RAMs to make a large micro-computer memory. In most cases, this input to the RAM is called $\overline{\text{CS}}$, which means "not-chip select" or, essentially, that a logic zero must be present on this line for the chip to operate. (*Note:* $\overline{\text{CS}}$ is a digital signal and should not be confused with the usual power input pins to the device.) If this signal is not present, the outputs of the RAM usually tri-state (go to a high impedance) and are not seen any more, as far as other logic devices are concerned.

With the $\overline{\text{CS}}$ line low, the chip is prepared to accept address and data information and a ***read/write*** signal. In operation, the address is decoded inside the chip, and the particular storage element desired is selected. Once this has been done, the read/write line comes into play. Here we use only one line to carry out two functions. If we aren't *reading,* then we must be *writing.* In practical applications, this line is abbreviated $R/\overline{W}$, which implies that a logic one on this line will read RAM data and a logic zero will write RAM data. If we are reading, the data from the selected storage element or elements is sent to the output lines. If we are writing, the data present at the Data in lines is sent to the selected storage area, where it replaces the data stored there previously. Because the RAM is a static–volatile device, it will retain this data only as long as power is applied. Usually, upon power-up (the first application of power to a digital circuit), random patterns of ones and zeros appear in all RAMs, and data must be loaded into them before a useful memory is available to the user.

Thus far we have been looking at an "ideal RAM" where everything happens instantly. In other words, the desired data location was selected and either read or written into instantly after the chip was selected. Well, we know that *nothing* happens in zero time, and RAMs are no exception. We will now look at an actual memory chip. Figure 11.6 shows a block diagram of the internal circuitry of a RAM.

FIGURE 11.6 Internal Diagram of a RAM

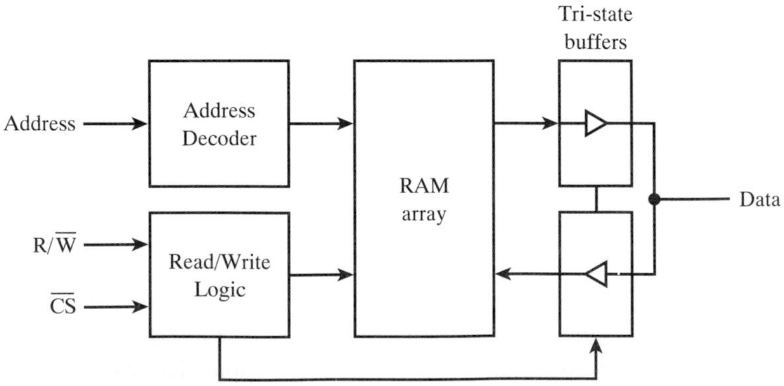

The delays associated with reading or writing data in a real RAM are due to:

1. The delay of the address decoder.
2. The delay of the RAM array.
3. The delay of the input/output buffers.

Together, these individual delays contribute to the overall access time of the device. It is necessary to account for the access time of a memory when designing a memory system.

The 6116 2KB RAM

The 6116 is a 2KB RAM containing a *bidirectional data bus.* This means that data comes and goes on the same set of data lines. This is possible through the use of tri-state buffers (as previously indicated in Figure 11.6). The 6116 requires 11 address lines (recall that 2^{11} equals 2048) to access one of 2048 locations, as indicated in Figure 11.7(a). The 6116 is ideal for small memory applications that do not require access times shorter than 100 ns.

FIGURE 11.7   6116 2KB RAM

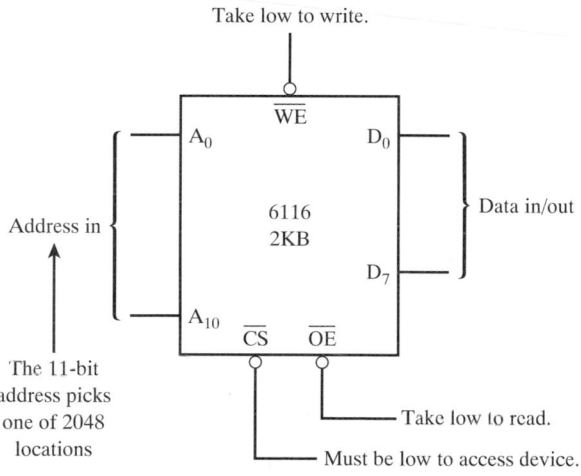

(a) Schematic symbol

1	A_7	+5 V	24
2	A_6	A_8	23
3	A_5	A_9	22
4	A_4	$\overline{WE}$	21
5	A_3	$\overline{OE}$	20
6	A_2	A_{10}	19
7	A_1	$\overline{CS}$	18
8	A_0	D_7	17
9	D_0	D_6	16
10	D_1	D_5	15
11	D_2	D_4	14
12	GND	D_3	13

6116

(b) Pinout

Figure 11.7(b) shows the pinout of the 6116. Note that every pin on the 24-pin DIP is used for some purpose. This indicates that a larger memory will not fit into a 24-pin package.

The 6264 8KB RAM

Containing four times the storage capability of the 6116, the 6264 stores 8192 bytes of information. This requires the addition of two new address lines, as indicated in Figure 11.8.

FIGURE 11.8   The 6264 8KB RAM

1	N/C	+5 V	28
2	A_{12}	$\overline{WE}$	27
3	A_7	CS_2	26
4	A_6	A_8	25
5	A_5	A_9	24
6	A_4	A_{11}	23
7	A_3	$\overline{OE}$	22
8	A_2	A_{10}	21
9	A_1	$\overline{CS_1}$	20
10	A_0	D_7	19
11	D_0	D_6	18
12	D_1	D_5	17
13	D_2	D_4	16
14	GND	D_3	15

6264

FIGURE 11.9   32KB Memory System

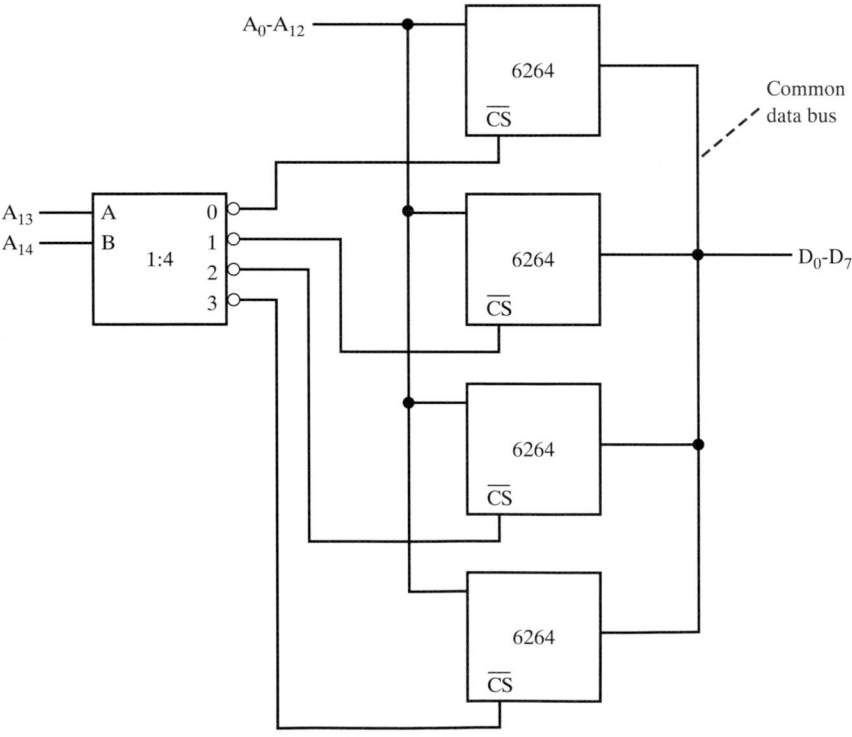

When larger memory systems are needed, it is possible to connect multiple 6116s or 6264s together, enabling one at a time via the $\overline{CS}$ inputs. Figure 11.9 shows one way to make a 32KB memory. Here, four 6264s are connected in parallel, with their individual CS inputs coming from a 1:4 decoder. Recall that when a memory is not enabled, its data bus is tri-stated. This allows us to wire the eight data outputs together in parallel without the individual chips interfering with one another. The additional address lines control which of the four decoder outputs goes low to enable its specific RAM.

It would be a good idea to spend some time searching the Web for newer RAMs to see how access time has decreased and capacity has increased with each new device.

11.5 DYNAMIC RAM

What Is Dynamic RAM?

Dynamic RAM is a special type of RAM memory that is currently the most popular form of memory used in large memory systems for microprocessors. Let us discuss a few of the specific differences between static RAMs and dynamic RAMs. Static RAMs use digital flip-flops to store the required binary information, whereas dynamic RAMs use MOS capacitors. Because of the capacitive nature of the storage element, dynamic RAMs require less space per chip, per bit, and thus have larger densities.

In addition, static RAMs draw more power per bit. Dynamic RAMs employ MOS capacitors that retain their charges (stored information) for short periods of time, whereas static RAMs must saturate transistors within the flip-flop to retain the stored binary information, and saturated transistors dissipate maximum power.

A disadvantage of the dynamic RAM stems from the usage of the MOS capacitor as the storage element. Left alone, the capacitor will eventually discharge, thus losing the stored binary information. For this reason the dynamic RAM must be constantly refreshed to avoid data loss. During a refresh operation, all of the capacitors within the dynamic RAM (called DRAM from now on) are recharged.

This leads to a second disadvantage. The refresh operation takes time to complete, and the DRAM is unavailable for use by the processor during this time.

Older DRAMs required that all storage elements inside the chip be refreshed every 2 ms. Newer DRAMs have an extended 4-ms refresh time, but the overall refresh operation ties up an average of 3 percent of the total available DRAM time, which implies that the CPU has access to the DRAM only 97 percent of the time. Because static RAMs require no refresh, they are available to the CPU 100 percent of the time, a slight improvement over DRAMs.

In summary, we have static RAMs that are fast, require no refresh, and have low bit densities. DRAMs are slower and require extra logic for refresh and other timing controls, but are cheaper, consume less power, and have very large bit densities.

Accessing Dynamic RAM

A major difference in the usage of DRAMs lies in the way in which the DRAM is addressed. A 64K bit DRAM requires 16 address bits to select one of 65,536 possible bit locations, but its circuitry contains only 8 address lines. A study of Figure 11.10 will show how these 8 address lines are expanded into 16 address lines with the help of two additional control lines: $\overline{RAS}$ and $\overline{CAS}$.

The 8 address lines are presented to row and column address buffers, and latched accordingly by the application of the $\overline{RAS}$ and $\overline{CAS}$ signals. To load a 16-bit address into

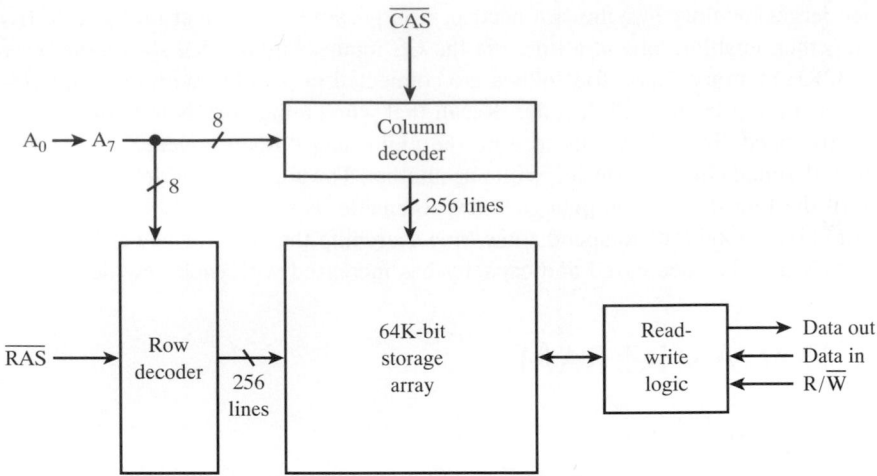

FIGURE 11.10 Internal Block Diagram of a 64K-Bit Dynamic RAM

the DRAM, 8 bits of the address are first latched by pulling $\overline{\text{RAS}}$ low. Then the other 8 address bits are presented to A_0 through A_7, and $\overline{\text{CAS}}$ is pulled low. By adding just one more address line to the DRAM, the addressing capability is increased by a factor of 4, because one extra address line signifies an extra row and column address bit. This explains why DRAMs tend to quadruple in size with each new release.

The actual method for addressing the DRAM is presented in Figure 11.11. First, the 8 row address bits are applied to A_0 through A_7, and $\overline{\text{RAS}}$ is pulled low. Then A_0 through A_7 receive column address information, and $\overline{\text{CAS}}$ is pulled low. After a short delay, the circuitry inside the DRAM will have decoded the full 16-bit address, and reading or writing may commence.

The row address strobe and column address strobe signals must be generated within 100 ns of each other to avoid data loss. The specific timing requirements for the DRAM depend on the manufacturer.

External logic is needed to generate the $\overline{\text{RAS}}$ and $\overline{\text{CAS}}$ signals, and also to take care of presenting the right address bits to the DRAMs. The circuit of Figure 11.12 shows an example of the required logic.

FIGURE 11.11 DRAM Cycle Timing

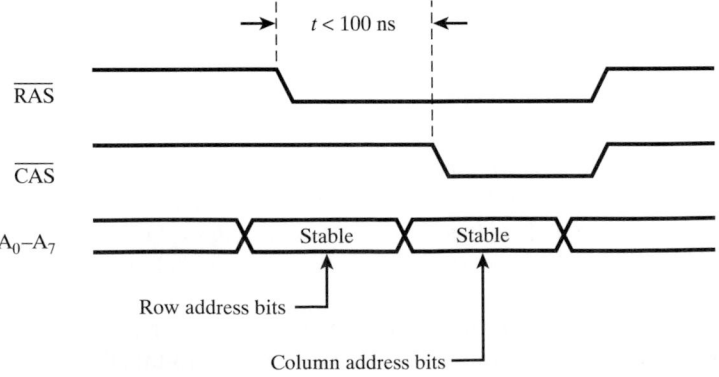

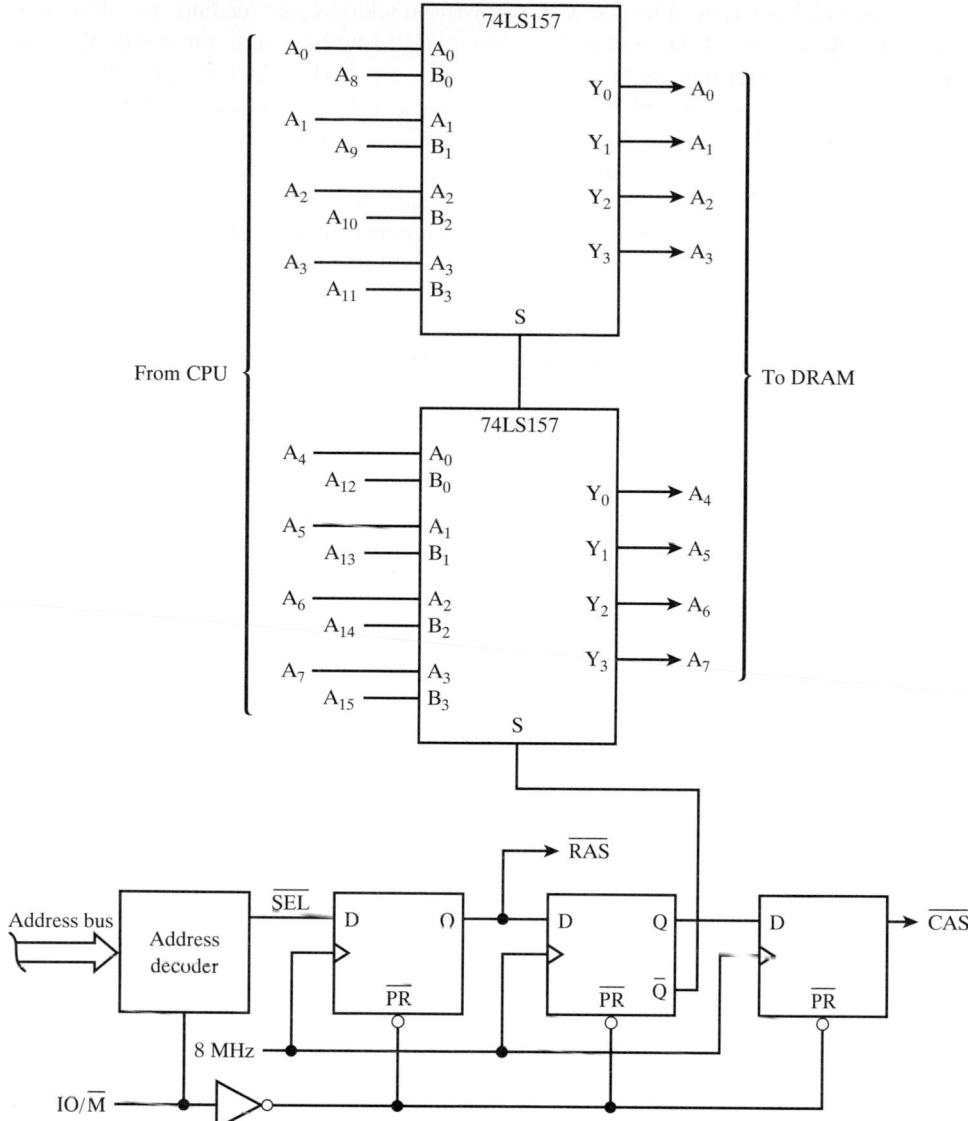

FIGURE 11.12 Address Bus Selector for DRAM

The operation of this circuit is as follows. The address decoder monitors the address bus for an address in the desired DRAM range, and outputs a logic 0 when it sees one. Normally the three Q outputs of the shift register are all high. The first clock pulse will shift the logic 0 from the address decoder to the output of the first flip-flop, causing $\overline{RAS}$ to go low. Because the output of the second flip-flop is still high, the 74LS157s (quad 2-line to 1-line multiplexers) are told to pass processor address lines A_0 through A_7. This is how we load the ROW address bits into the DRAM.

The second clock pulse will shift the logic 0 to the second Q output (the first is still low also), which causes the 74LS157s to select the processor address lines A_8 through A_{15}. These address bits are recognized and latched by the DRAM when the third clock pulse occurs, because the logic 0 has been shifted to the third Q output, which causes $\overline{CAS}$ to go low. The DRAM has been loaded with a full 16-bit address, and reading or writing may commence. At the end of the read or write cycle, $IO/\overline{M}$ will go high, presetting all three flip-flops via the preset line, and the shift register reverts back to its original state.

This sequence will repeat every time the address decoder detects a valid address.

Figure 11.13 shows a complete DRAM addressing circuit with read-write logic. When the 74LS138 detects a valid memory address, one of its eight outputs will go low, removing the 74LS175 quad D flip-flop from its forced-clear state. All four Q outputs are high at this time. As a logic 1 shifts through the 74LS175 (connected as a 4-bit shift register), the

FIGURE 11.13 Complete DRAM Addressing Circuit

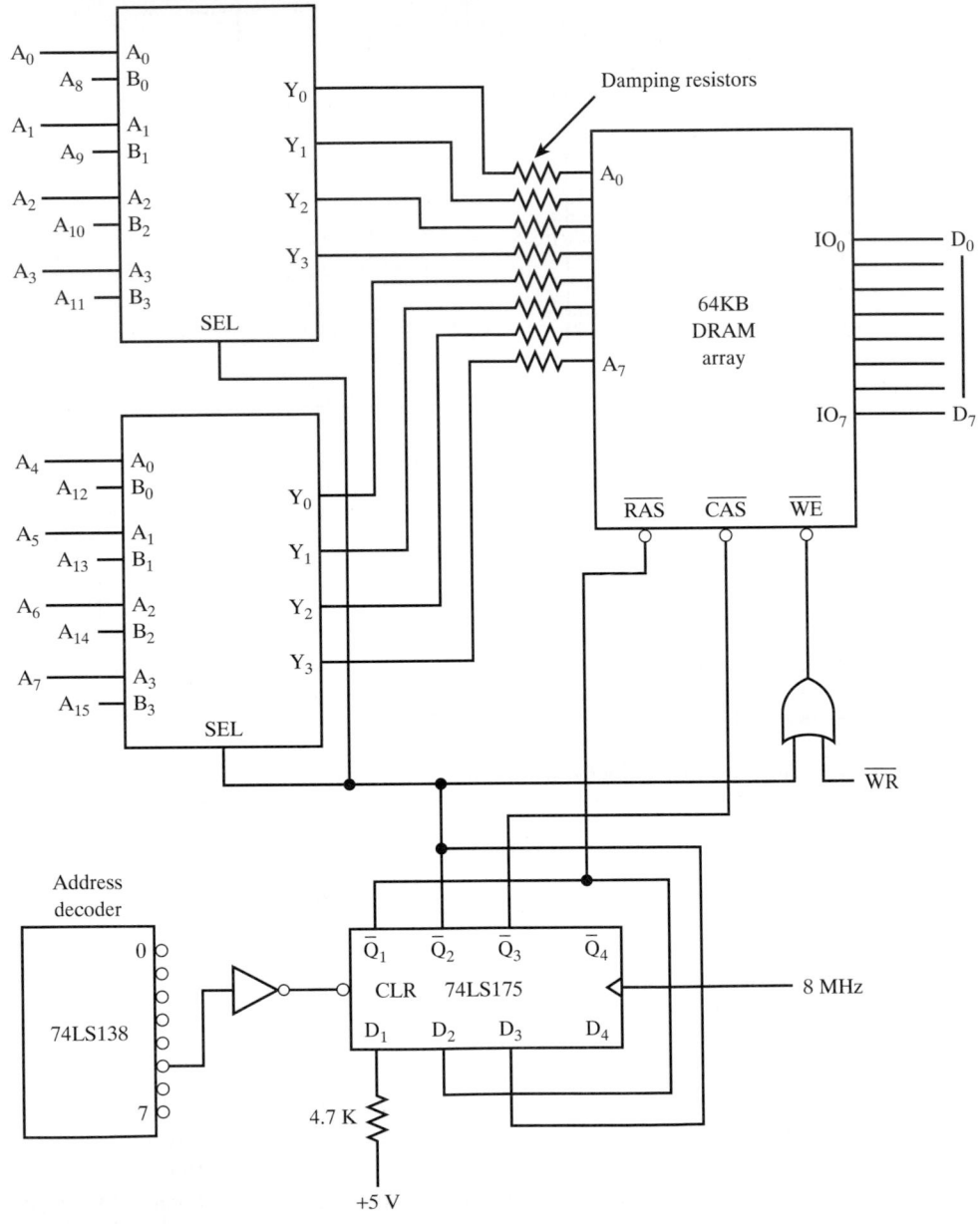

$\overline{RAS}$, $\overline{WE}$, and $\overline{CAS}$ signals will be generated. The resistors in the address and control lines are called damping resistors, and are used to control the waveshape of the digital signals to the DRAMs. The damping resistors reduce ringing and other noise that would normally occur in a high-speed digital system. The only circuitry missing from Figure 11.13 is the required refresh logic, which we will study in the next section.

Refreshing Dynamic RAM

Previously we learned that DRAMs need to be refreshed, or the MOS capacitors that retain the binary information will discharge and data will be lost. Older DRAMs required that all cells (storage elements) be refreshed within 2 ms. Although the process of reading or writing a DRAM cell is a form of refresh, it is possible that entire banks of DRAM remain inactive while the CPU addresses other memories or I/O devices, so a safe designer will include a refresh circuit in the new DRAM system.

Newer DRAMs (such as the MCM6664) contain a single control line called $\overline{REF}$ that automatically refreshes the DRAM whenever it is pulled low. We will instead look at the process that is used to refresh a DRAM and the circuitry needed to control the process.

DRAMs are internally designed as a grid of memory cells arranged as a matrix, with an equal number of rows and columns (hence the $\overline{RAS}$ and $\overline{CAS}$ control signals). A 4K-bit DRAM would need 12 address lines: 6 for the row decoder and 6 for the column decoder. Each decoder would pick one row and column out of a possible 64. During a refresh operation, all 64 column-cells would be refreshed by the application of a single $\overline{RAS}$ signal. This is called RAS-only refresh. To refresh all 4096 bits, it is necessary only to $\overline{RAS}$ select all 64 rows. A larger DRAM, a 64K bit one for example, would require $\overline{RAS}$ selecting more rows (256 in this case). The easiest way to ensure that all rows get selected during a refresh operation is to use a binary counter and connect the output of the counter to the DRAM address lines during a refresh. To ensure that the DRAMs get refreshed periodically, a timer is needed to generate a REFRESH signal. The REFRESH signal will suspend processor activity while the DRAM is refreshed. Figure 11.14 shows how a 555

FIGURE 11.14   555 Timer Generates Refresh Signals Every 100 μs

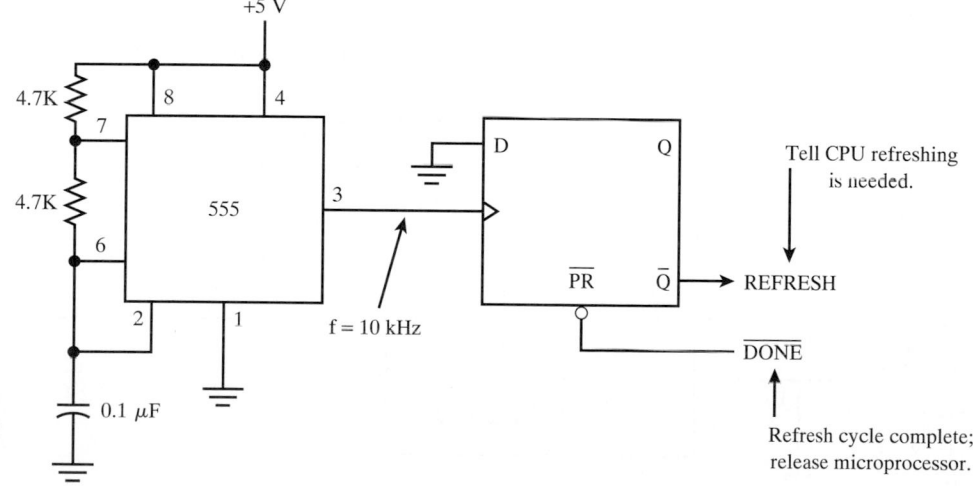

timer can be used to generate a REFRESH signal every 100 μs. The 555 timer clocks a D-type flip-flop, whose output is REFRESH. When the refresh cycle is completed, $\overline{\text{DONE}}$ is used to preset the flip-flop and remove the REFRESH request, until the 555 times out again. Figure 11.15 shows how the refresh timer, together with the $\overline{\text{RAS}}$ refresh circuitry, is used to refresh the DRAMs. When the 555 timer initiates a refresh cycle, REFRESH will go high, issuing a HOLD request to the CPU. The processor will respond by asserting HLDA, which allows a 0 to be clocked into the 2-bit shift register used to control $\overline{\text{RAS}}$ and $\overline{\text{DONE}}$. When $\overline{\text{RAS}}$ is active, bits A_0 through A_7 of the address bus will contain the 8-bit counter value (the current state of the 74LS393). When $\overline{\text{DONE}}$ is active, the refresh flip-flop is preset, which removes the HOLD signal. This causes the processor to release HLDA, which in turn causes the 2-bit shift register to be loaded with 1s. At this point, the bus request is over, and the processor resumes execution. Because the 555 timer also clocks the 8-bit counter, a unique row address is generated each refresh cycle.

A Dynamic RAM Controller

We may conclude that the circuitry required to address, control, and refresh DRAMs is both complicated and extensive (which may translate into *expensive*). There must be a simpler way.

There is!

Various companies make DRAM controller devices that take care of all refreshing and timing requirements needed by the dynamic RAMs. All circuitry is contained in a single package in most cases. The DRAM controller does its work independently of the

FIGURE 11.15   DRAM Refresh Generator

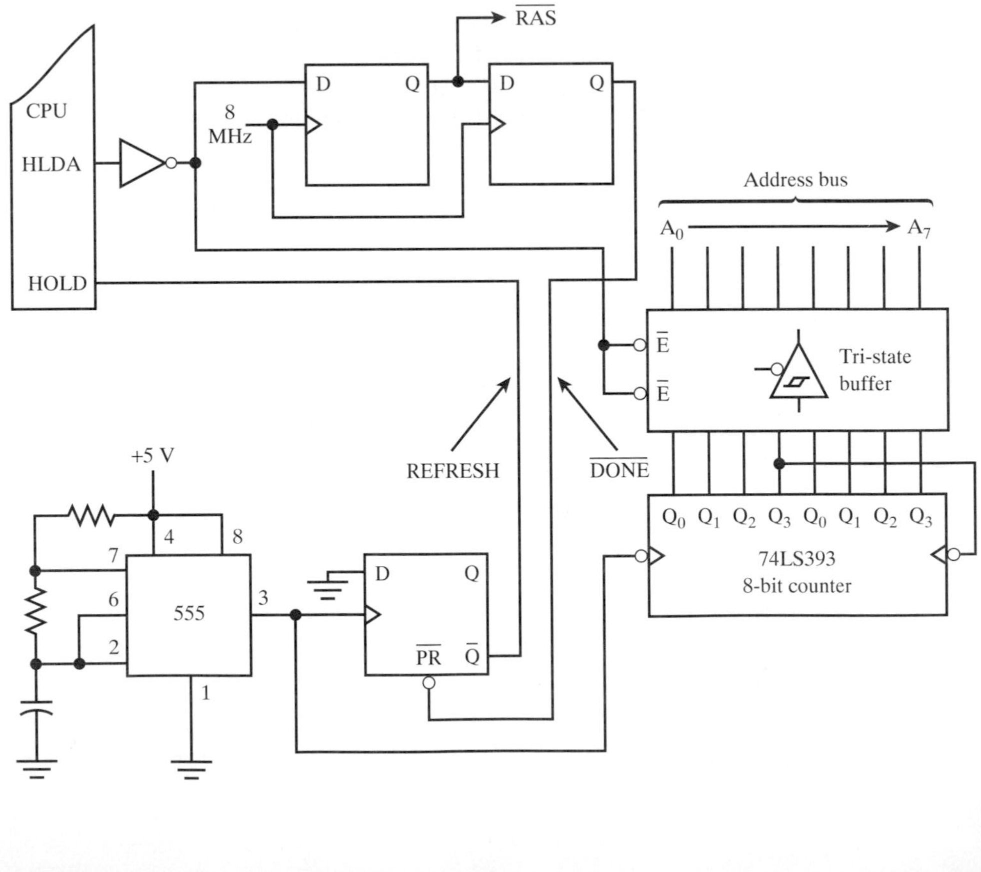

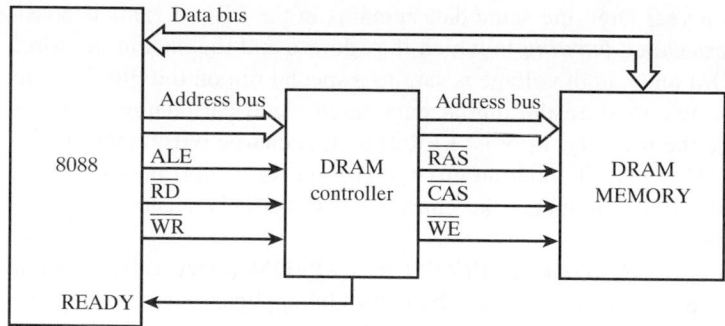

FIGURE 11.16   DRAM Controller Interfacing

processor. This means that the DRAM controller will issue a wait to the processor when the processor tries to access memory during a refresh cycle. Figure 11.16 shows how a DRAM controller is used in an 8088-based system. Using a dedicated DRAM controller minimizes the time required to design, debug, and eventually troubleshoot DRAM memory systems. It may also be more cost-effective in the long run.

Dynamic RAM Summary

Our study of DRAMs has shown that they are slow and require complicated circuitry to get them to work (unless a DRAM controller is used). On the other hand, DRAMs are cheaper, per bit, than static RAM, they consume less power, and have much larger bit densities. With the advance of the microcomputer into the word processing arena, where very large memories are needed to store and manipulate text, dynamic RAM becomes a very economical solution. Image processing, large informational databases, and virtually any large storage system make the use of dynamic RAMs an ideal choice. Furthermore, interfacing dynamic RAMs is made easier with the use of a DRAM controller.

11.6   READ-ONLY MEMORY

Read-only memory (ROM), unlike RAM, retains its data when power is turned off. The data is typically programmed into the ROM once, and then read as many times as necessary. Many digital circuits require ROM to control their operation, guaranteeing that good data is present in memory when power is turned on. The personal computer uses a system ROM to store critical instructions capable of booting the computer when it is turned on.

There are many types of ROM. Their functions are as follows:

ROM   Read-only memory. Programmed once at the factory.

PROM   *Programmable ROM.* Programmed once by the user.

EPROM   Erasable PROM. Can be programmed and erased (via ultraviolet light) by the user.

EEPROM *Electrically Erasable PROM.* Can be erased with a signal instead of ultraviolet light.

Every time we read information from a PROM after it has been programmed, we will get the same data back. Even if power is turned off, and then back on again two

seconds, one day, even a year later, the same data remains in the PROM. This is possible by virtue of a process called ***burning*** in which the address and the data to be stored are supplied to the PROM and a high voltage is sent to a special pin on the PROM. This causes the internal circuitry to store the digital data as an electrical charge, which is held permanently inside the memory. In most EPROMs, this charge will remain indefinitely unless the EPROM is placed under an ultraviolet light for a short period of time (20 min). The ultraviolet light dissipates the charge and essentially clears the ROM; hence the term "erasable."

The ***EEPROM (electrically erasable PROM)*** or EAPROM (electrically alterable PROM, E^2PROM) is a type of memory that can be erased by application of a voltage instead of ultraviolet light. Either a specific location can be changed or the entire memory cleared. The 2815 EEPROM is intended to be pin-for-pin compatible with the 2716 EPROM as indicated in Figure 11.17. With it comes the convenience of a RAM-type memory, except that the data stored in the EEPROM is retained when power is withdrawn from the device.

Table 11.1 lists several standard EPROM types and their capacities. It is not uncommon to find several EPROMs on a circuit board in microprocessor-based equipment. The relationship between the part number and the capacity is found by determining the number of *bits* stored in the EPROM. For example, the 2716 stores 16K bits, and is thus a 2KB EPROM.

FIGURE 11.17 Programmable ROMs

1	A_7		+5 V	24
2	A_6		A_8	23
3	A_5		A_9	22
4	A_4		V_{PP}	21
5	A_3		$\overline{OE}$	20
6	A_2	2716	A_{10}	19
7	A_1		$\overline{CE}$	18
8	A_0		O_7	17
9	O_0		O_6	16
10	O_1		O_5	15
11	O_2		O_4	14
12	GND		O_3	13

(*a*) 2KB EPROM

1	A_7		+5 V	24
2	A_6		A_8	23
3	A_5		A_9	22
4	A_4		V_{PP}	21
5	A_3		$\overline{OE}$	20
6	A_2	2516	A_{10}	19
7	A_1		$\overline{CE}$	18
8	A_0		IO_7	17
9	IO_0		IO_6	16
10	IO_1		IO_5	15
11	IO_2		IO_4	14
12	GND		IO_3	13

(*b*) 2KB EEPROM

TABLE 11.1 Standard EPROMs

EPROM	CAPACITY
2708	1KB
2716	2KB
2732	4KB
2764	8KB
27128	16KB
27256	32KB
27512	64KB

There are also ROMs that are programmable but not erasable. These ROMs, simply called PROMs, may be burned only once. They are burned in the same way as an EPROM except that instead of storing a voltage, the burn signal pops a small fuse at the desired storage location inside the PROM. Once popped, this fuse can never be fixed or replaced. So, if you make a mistake while burning a PROM, unless you can make a holiday ornament out of it, you have to throw it out and start over with a new one. This is an advantage of an EPROM over a PROM, but EPROMs do cost more because they are erasable.

Least expensive is simply the ROM. ROMs are custom programmed at the factory by altering a mask in the design process. Like EPROMs and PROMs, the ROM retains its data with power off. Factory-programmed ROMs have a variety of functions and are very useful. They are used as character generators for TV typewriters (see Figure 11.18), sine–cosine lookup tables for electronic trig function generators, code converters, and in many other applications. An EPROM or PROM may be used for specific applications that are desired by the computer hobbyist or industrial factory technician, as opposed to general uses.

We will now look at one of the most common ROMs available, and one of those best suited for microcomputers.

· ·

EXAMPLE 11.1

Figure 11.18 shows a 2513 character generator ROM in a typical TV typewriter application. The 2513 contains all the letters of the alphabet, plus the numbers 0–9 and some special symbols, organized in a 5×7 dot matrix format. To display a character on a video display, each character is represented using a dot matrix. The 5×7 dot matrix of the 2513 offers the lowest resolution of available character generators, but it is simple to use. Other character generators offer better resolution of characters with 7×9, 9×11, and 11×13 dot matrices.

Figure 11.19 shows the 5×7 dot matrix (five columns and seven rows) and the 2513 pinout. The character generator contains the information to produce a character when

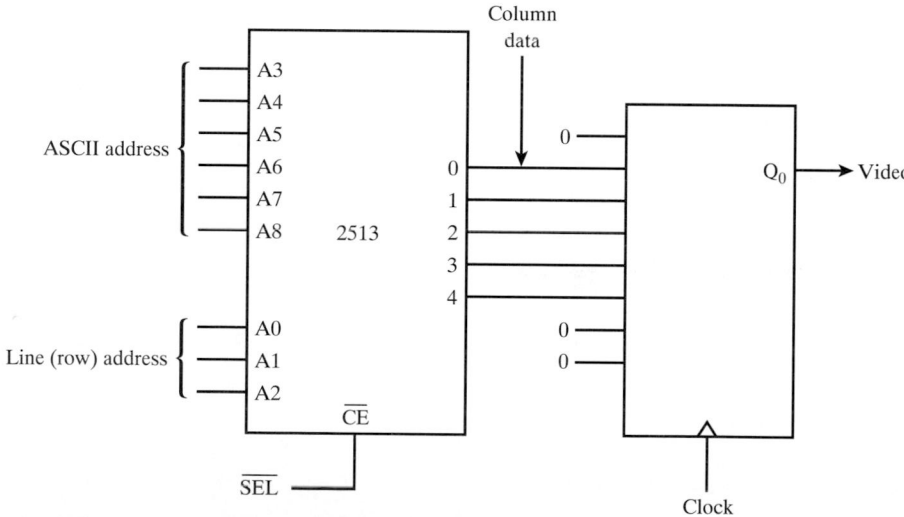

FIGURE 11.18   A Character Generator ROM

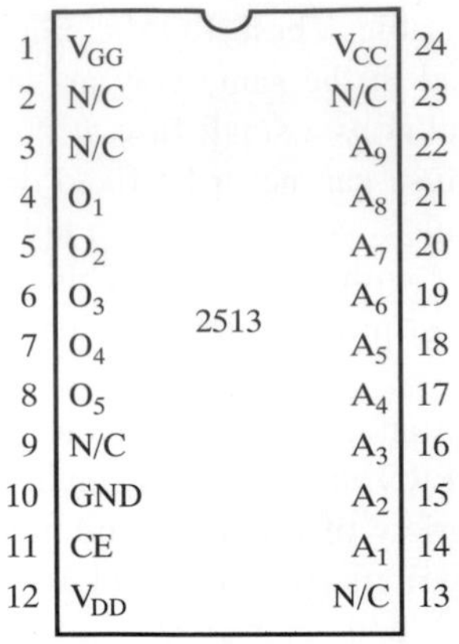

(*a*) Pinout

	O_5	O_4	O_3	O_2	O_1
Selected by $A_1\,A_2\,A_3$	0	0	0	0	0
	1	1	1	1	1
	1	0	0	0	0
	1	0	0	0	0
	1	1	1	1	0
	1	0	0	0	0
	1	0	0	0	0
	1	1	1	1	1

(*b*) Sample output data

FIGURE 11.19 2513 Details

an ASCII code is presented to the ROM. The inputs are the six address lines (ASCII data) and three row inputs. The circuitry that obtains information from the ROM must take it a row at a time. The ASCII code for a letter or symbol doubles as the letter (or symbols) address in the ROM (A4–A8). Output data is sent to an 8-bit shift register with the other three inputs set to zero to provide for blank spacing between characters.

EXAMPLE 11.2

In Figure 11.20, a 74154 4-line to 16-line decoder is used to make a 16 × 8 bit ROM. The output data is shown in Table 11.2.

TABLE 11.2 Diode ROM Data

A3	A2	A1	A0	DATA OUT
0	0	0	0	7E
0	0	0	1	BD
0	0	1	0	DB
0	0	1	1	E7
0	1	0	0	BB
0	1	0	1	D5
0	1	1	0	B6
0	1	1	1	BF
1	0	0	0	BB
1	0	0	1	ED
1	0	1	0	AB
1	0	1	1	FD
1	1	0	0	FE
1	1	0	1	F5
1	1	1	0	D9
1	1	1	1	F7

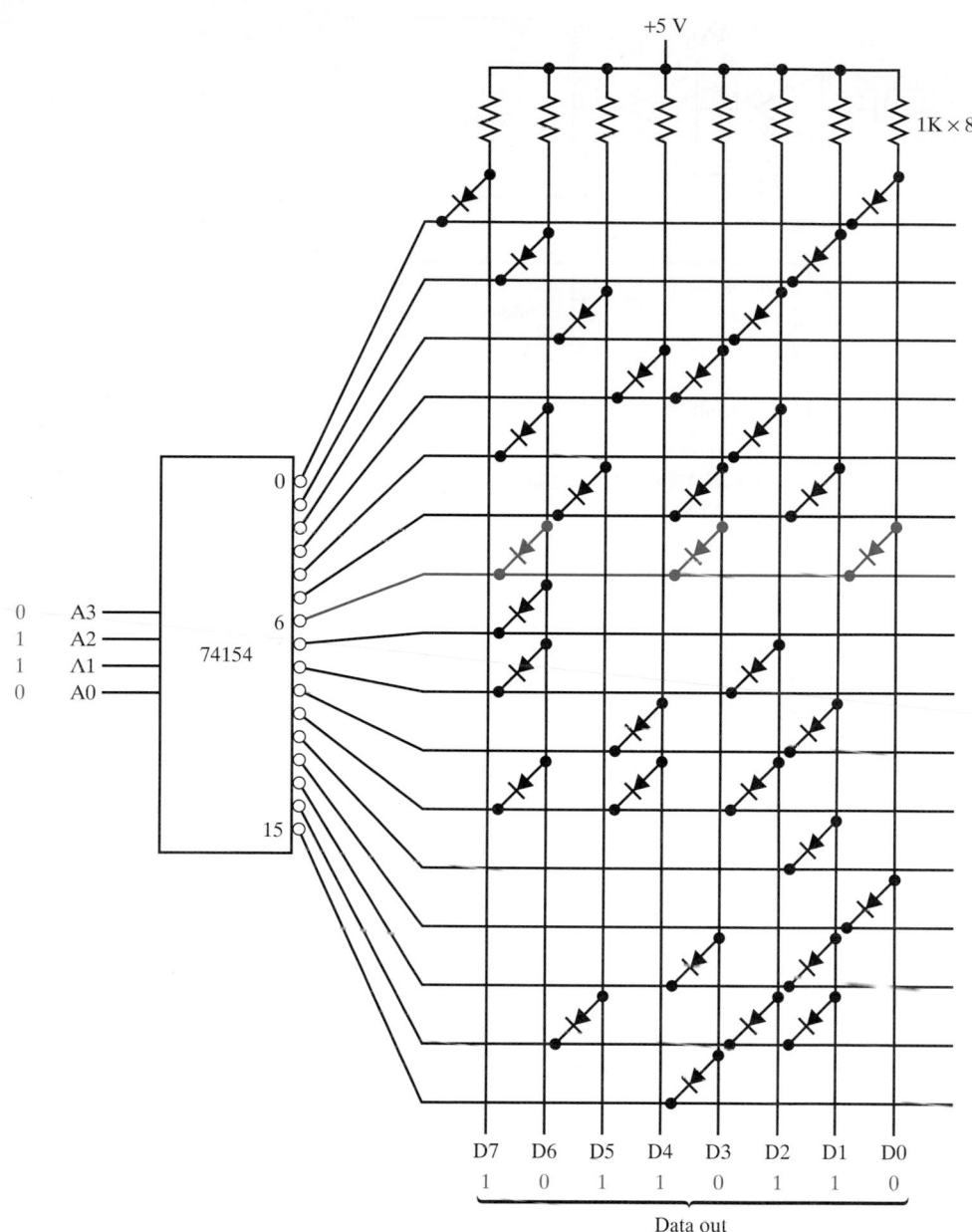

FIGURE 11.20 Diode Matrix ROM for Example 11.2

When a selected output of the 74154 goes low, it also pulls combinations of output lines low through the diodes connecting them. This type of memory is called a *diode matrix ROM* and is very useful when small ROMs are needed.

11.7 PRACTICAL APPLICATIONS

In this section we explore a few examples of how memory circuitry is used in everyday applications.

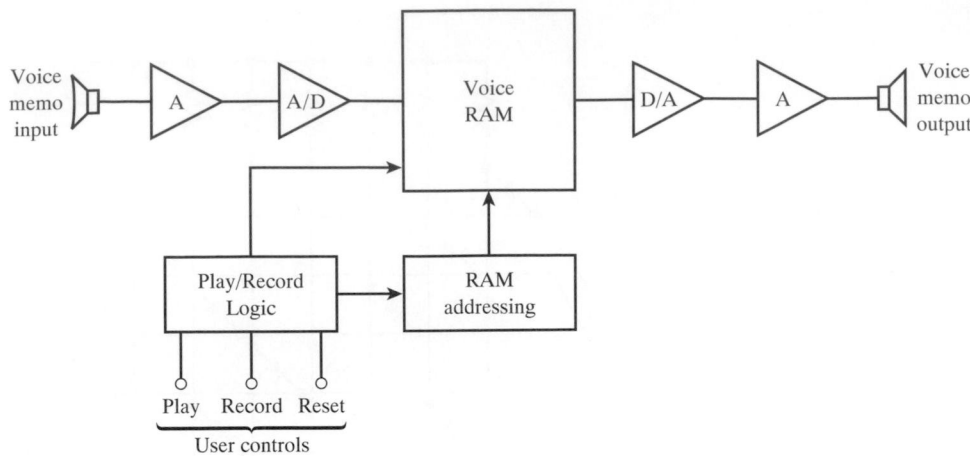

FIGURE 11.21   Personal Memo Machine

Personal Memo Machine

People on the go are increasingly using personal memo machines to record their thoughts. These machines contain a small microphone and speaker, and all circuitry required to digitize, record, and playback speech. Figure 11.21 shows a simplified block diagram of a memo machine.

Typically, human speech has frequency components that are at or below 4 KHz. So, to accurately sample a person's voice, a sample rate of 8000 samples/second is used. Imagine that the A/D converter in Figure 11.21 generates 8-bit samples. At the rate of 8000 samples/second, a 256KB voice RAM would be capable of storing over 32 seconds of speech. Think of how many seconds (or minutes) would be possible if a 4MB voice RAM were used.

Network Switch

The rapid expansion of computer networking into practically every institution we come in contact with (education, business, personal computing) has pushed the need for faster communication speeds into the 100s and even 1000s of megabits/second. All of these bits need to be stored, temporarily, as they are moved from one network to another, and even within a single network. One important component of a computer network is the *switch,* a device that connects 8, 16, or more computers together, and acts like a high-speed traffic cop, directing data packets from one point to another. When many packets arrive at the switch at the same time, several may need to be buffered before they are forwarded to their destination. An 8-port switch may employ a one-MB RAM for buffering packets. Assuming a minimum packet size of 64 bytes, a one-MB RAM would be capable of storing over 16,000 packets if required. Without the internal memory, the effectiveness of the switch would be greatly reduced.

Memory Usage in the Personal Computer

Several different types of memories are used in the personal computer. Let us briefly examine their individual types and features.

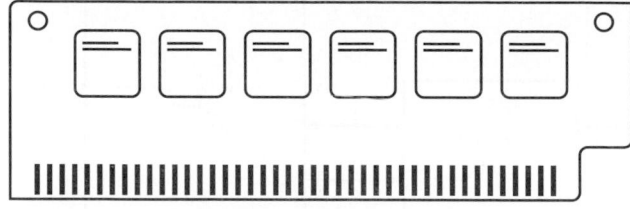

FIGURE 11.22 Single-In-Line Memory Module (SIMM)

Single-in-Line Memory Module (SIMM)

The *single-in-line memory module,* or *SIMM,* is one way of physically organizing memory. It is a small "boardlet" with several memory chips soldered to it. This boardlet is inserted into a system slot. Figure 11.22 shows a SIMM.

SIMMs came about in an attempt to solve two problems. The first problem was "chip creep." Chip creep occurs when a chip works its way out of a socket as a result of thermal expansion and contraction. The old solution to this problem—soldering memory chips into the board—wasn't a good solution, because it made them harder to replace. So the SIMM was created. The only problem with the SIMM is that if only 1 bit in any of its chips goes bad, the whole SIMM must be replaced. This is more expensive than replacing only one chip. SIMMs come in 256KB, 1MB, 4MB, and 16MB sizes. A similar type of memory module, called a SIPP, contains metal pins that allow the SIPP to be soldered directly onto the motherboard.

Regarding parity bits in a SIMM, a 32-bit SIMM is non-parity, and a 36-pin SIMM stores one parity bit for each byte of data. Pentium processors incorporate parity in their address and data buses.

Dual in-Line Memory Module (DIMM)

The DIMM was created to fill the need of Pentium-class processors containing 64-bit data buses. A DIMM is like having two SIMMs side by side, and come in 168-pin packages (more than twice that of a 72-pin SIMM). Ordinarily, SIMMs must be added in pairs on a Pentium motherboard to get the 64-bit bus width required by the Pentium.

Synchronous DRAM (SDRAM)

This type of RAM is very fast (up to 100-MHz operation) and is designed to synchronize with the system clock to provide high-speed data transfers.

Extended Data Out DRAM (EDO DRAM)

This type of DRAM is used with bus speeds at or below 66 MHz and is capable of starting a new access while the previous one is being completed. This ties in nicely with the bus architecture of the Pentium, which is capable of back-to-back pipelined bus cycles. Burst EDO (BEDO RAM) contains pipelining hardware to support pipelined burst transfers.

Video RAM (VRAM)

Video RAM is a special *dual-ported* RAM that allows two accesses at the same time. In a display adapter, the video electronics needs access to the VRAM (to display the Windows desktop, for example) and so does the processor (to open a new window on the desktop). This type of memory is typically local to the display adapter card.

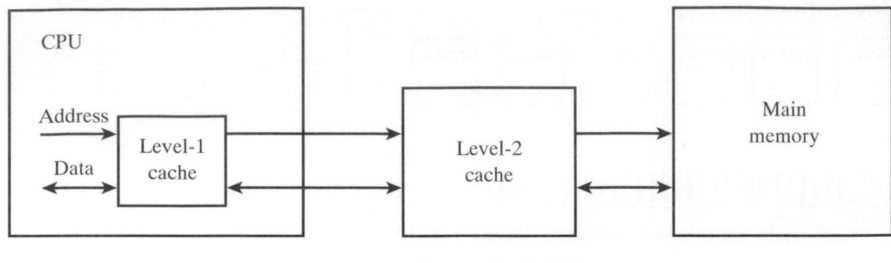

Contained in the same package
in the Pentium Pro

FIGURE 11.23   Using Cache in a Memory System

Level-2 Cache

Cache is a special, high-speed memory capable of providing data within one clock cycle, and is typically ten times faster than regular DRAM. Although the processor itself contains a small amount of internal cache (8KB for instructions and 8KB for data in the original Pentium), you can add additional *level-2 cache* on the motherboard, between the CPU and main memory, as indicated in Figure 11.23. Level-2 cache adds an additional 64KB to 2MB of external cache to complement the small internal cache of the processor. The basic operation of the cache is to speed up the average access time by storing copies of frequently accessed data.

11.8 TROUBLESHOOTING TECHNIQUES

As we have seen before, a good knowledge of binary numbers is beneficial when working with digital circuitry, and necessary to the efficient design of address decoders (and other interfacing circuitry). Once the decoders are designed, however, they must be tested to see if they perform as required. Testing the operation of a memory address decoder can be accomplished any number of ways. The circuit can be set up on a breadboard, simulated via software, or just plain stared at on paper until it seems correct.

In situations like the troubleshooting phase of the single-board computer project in Chapter 14, testing the memory address decoder is often necessary. In addition to checking the wiring connections visually (or via a continuity tester or DMM), a logic analyzer is connected so that the waveforms can be examined. Even an old 8-channel logic analyzer can be used to diagnose difficult problems in a microprocessor-based system. Oscilloscopes typically fall short in showing the associated high-speed timing relationships (unless they are storage scopes).

The logic analyzer is connected so that all of the inputs to the address decoder and as many of the outputs as necessary are sampled. In addition, the logic analyzer is set up so that it triggers (and begins capturing data) at RESET, so that the initial activity of the processor's address bus can be observed. If the single-board's EPROM is not enabled, the system will not function at all. The logic analyzer will show how the address decoder responds at power on.

The logic analyzer can also be used to examine the data coming out of the EPROM or RAM as well. Sampling the data and a few of the address lines should be enough to verify whether the data are correct. It is sometimes possible to spot switched data or address lines this way.

SUMMARY

Memories are an important part of many digital systems. RAM, ROM, EPROM, and high-speed cache all have their place in the many different digital circuits that employ them. Whether they are used to store phone numbers, instructions for booting up a computer, or the digitized image of a photograph, semiconductor memories play a significant role in the digital world. The single-board computer project in Chapter 14 shows how RAM and EPROM are used in the design of a simple microcomputer system.

STUDY QUESTIONS

General

1. A 32K memory is desired for use with an 8-bit microcomputer. If there are 32K memory locations:
   (a) How many locations are there actually?
   (b) How many *bits* are stored?

2. What is the purpose of a chip enable line?

3. (a) Does RAM lose data if power is turned off?
   (b) Does ROM lose data if power is turned off?

4. Show how a single wire is used for read/write operations ($R/\overline{W}$).

5. How many bits are stored in (a) a 2716 type memory? (b) A 27256?

6. How many bytes may be stored on a $3\frac{1}{2}$-in. double-density, double-sided disk? Both sides of the disk store 80 tracks, with eighteen 512-byte sectors per track. What is the size in KB?

7. How many address lines are required for a 4K × 8 memory (4096 × 8)?

8. How many address lines are required for a 2764-type memory?

9. If a memory has 13 address lines, how many memory locations are there?

10. If a memory has 14 address lines, how many memory locations does it have?

11. If an 8KB memory starts at address 2000 H, what is the address of its last location?

12. If a 16KB (16,384) memory starts at address 5000 H, what is the address of its last location?

13. (a) How many address lines are needed for a 12KB memory?
    (b) If this 12KB memory starts at 0, what is the address of the last location?
    (c) If the memory starts at 5000 H, what is the address of the last location?

14. How many memory locations can be addressed by a microcomputer with (a) 16 address lines? (b) 20 address lines? (c) 24 address lines? (d) 28 address lines? (e) 32 address lines?

15. Can you think of applications for a computer that might require as many as 2^{24} memory locations?

16. Design a pseudo-random generator using a 5-bit shift register and an exclusive NOR gate. What output bits must be used to get a full 31-word sequence?

17. Repeat Question 16 for an 8-bit shift register.

18. If a 1K-bit shift register is clocked at 5 MHz, what is the time required to fully recirculate its data?

19. How many 6264s are required for a 32K by 16 memory? How many 6116s are required?

Electronics Workbench

20. What types of memory devices are included in Electronics Workbench? Make a list of the devices and their capacities.

Programmable Logic

21. Explain how a PAL can be used as a small RAM or ROM.

DIGITAL DATA TRANSMISSION

12

When finished with this chapter you should be able to:

1. Describe the ASCII code and its use.
2. Convert 7-bit ASCII to 11-bit transmission code.
3. Explain the importance of "handshaking."
4. Show how modems and telephones are used in data transmission.
5. Describe the start, stop, and parity bits used in the 11-bit transmission code.

Keep the following questions in mind and try to answer them when you have completed the chapter:

1. Why is the "start" bit used in the 11-bit transmission code?
2. Why are CTS and RTS important?
3. What is the difference between a TTL and an RS232C transmitted code?
4. How are bits 6 and 7 used to separate the ASCII code into four groups of characters?
5. What is baud rate?
6. What are the advantages and disadvantages of serial and parallel transmission.

277

12.1 INTRODUCTION

To relay commands and data, it is essential to be able to communicate with a computer or digital circuit. Often this is done from a keyboard arranged in a manner similar to a typewriter. Each time a key is pressed, a binary code unique to that key is generated. The code is then transmitted in either parallel or serial form to the computer. Usually the code for the character is *echoed* back to the terminal by the computer so that it may be displayed on the screen of a computer terminal. This involves the transmission of digital character data. Pure binary data may be sent back and forth from a CPU to its memory or from an image processing ASIC to a DMA controller. In this chapter we examine the various ways serial and parallel data transmission may take place.

12.2 PARALLEL VS. SERIAL COMMUNICATION

There are two ways to transmit digital data from one location to another: parallel and serial. Let us imagine that we want to transmit 8-bits of data, the value 10001100 B.

The 8-bit data may be transmitted in two ways.

1. All 8 bits may be transmitted at once on eight separate lines. Such *parallel* transmission is used only for high-speed data transfer. Parallel transfer of data is very common *inside* computers. In this case the parallel paths are referred to as *buses*. Parallel communication outside the computer is typically used on the printer port. Figure 12.1(a) shows the sample 8-bit data being sent from the computer to the printer.

FIGURE 12.1 **Data Transmission**

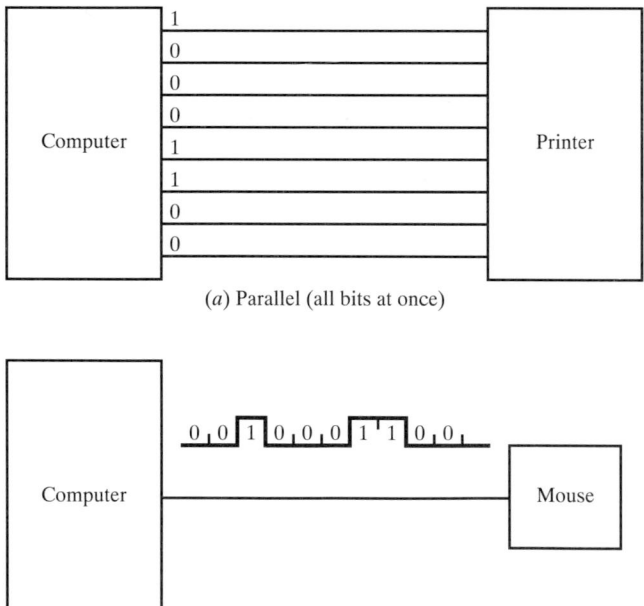

(*a*) Parallel (all bits at once)

(*b*) Serial (one bit at a time)

2. The 8 bits may be transmitted one at a time over a single wire. Serial transmission is used for lower data rates and has the advantage of taking only one line. Keyboard and mouse data generally are transmitted serially to avoid having to run eight wires to the computer. This would be a waste of wire for the low-speed data character information usually represented. Figure 12.1(b) illustrates serial transmission. We will take a closer look at each transmission method in the next few sections.

12.3 ASCII

The *ASCII* code (pronounced as-kee) is a 7-bit code for upper- and lowercase letters, digits, and special characters (!, $, %, etc.) and the control characters. Table 12.1 shows the 128-character ASCII code in binary format. The same code is given in Table 12.2 in a hexadecimal format.

Let us look at the ASCII code for the capital letter A.

$$\begin{array}{ccc} & b_6 & b_0 \\ \text{ASCII "A"} = & 1 \ \ 00000 \ 1 & = 41 \text{ H} = 65 \text{ D} \\ & \uparrow \qquad\qquad \uparrow & \\ & \text{MSB} \qquad \text{LSB} & \end{array}$$

The code can be expressed in binary, hex, or decimal. The right-most bits represent a count with A = 41 H, B = 42 H, C = 43 H, and so on. Knowing that A is 41 H, the other capital letters can be determined without having to consult a reference. For instance, L is twelfth letter of the alphabet.

$$12 \text{ D} = 0\text{CH} = 1100 \text{ B}$$

Therefore,

$$L = 64 \text{ D} + 12 \text{ D} = 76 \text{ D} \quad \text{and} \quad L = 40 \text{ H} + 0\text{CH} = 4\text{CH}$$

and

$$L = 1000000 \text{ B} + 1100 \text{ B} = 1001100 \text{ B}$$

The two most significant bits of the ASCII code (b_5, b_6) determine the type of characters being transmitted. You can verify this by consulting Table 12.3. The control characters are generated by the keyboard when the user simultaneously depresses a control key and a letter key. The control characters ($b_5 = 0, b_6 = 0$) and the capital letters ($b_5 = 0, b_6 = 1$) are very similar, the only difference being b_6.

Depressing the control key forces b_5 and b_6 low. We can see from Table 12.1 that the letter A is in the same row as SOH (Start of Heading). The letter next to A is Q, which is in line with DC1 (Device Control 1). This juxtaposition points out that you can overlay the capital letter columns on the control columns. Table 12.4 lists some of the more important control codes. The symbol ↑ (up arrow) is used to indicate a control character. Most keyboards have a backspace key and a "new line" (carriage return) key. However, these functions may also be generated by depressing ↑ H and ↑ M, respectively.

The ASCII code for a character may be transmitted in parallel (to a printer) or serial (over a modem).

TABLE 12.1 ASCII Code System and Character Set

b_3	b_2	b_1	b_0	COLUMN→ ROW↓	0	1	2	3	4	5	6	7
0	0	0	0	0	NUL	DLE	SP	0	@	P	`	p
0	0	0	1	1	SOH	DC1	!	1	A	Q	a	q
0	0	1	0	2	STX	DC2	"	2	B	R	b	r
0	0	1	1	3	ETX	DC3	#	3	C	S	c	s
0	1	0	0	4	EOT	DC4	$	4	D	T	d	t
0	1	0	1	5	ENQ	NAK	%	5	E	U	e	u
0	1	1	0	6	ACK	SYN	&	6	F	V	f	v
0	1	1	1	7	BEL	ETB	'	7	G	W	g	w
1	0	0	0	8	BS	CAN	(	8	H	X	h	x
1	0	0	1	9	HT	EM	)	9	I	Y	i	y
1	0	1	0	10	LF	SUB	*	:	J	Z	j	z
1	0	1	1	11	VT	ESC	+	;	K	[	k	{
1	1	0	0	12	FF	FS	,	<	L	\	l	|
1	1	0	1	13	CR	GS	–	=	M	]	m	}
1	1	1	0	14	SO	RS	.	>	N	∧	n	~
1	1	1	1	15	SI	US	/	?	O	—	o	DEL

□ PRINTABLE CHARACTERS ▨ PRINTER CONTROL CHARACTERS

▨ CODES GENERATED AND TRANSMITTED BY THE TERMINAL. BUT NO ACTION IS TAKEN.

USASCII CONTROL CHARACTERS
(From USA Standards Institute Publication X3.4–1968)

ACK	acknowledge		ETX	end of text
BEL	bell		FF	form feed
BS	backspace		FS	file separator
CAN	cancel		GS	group separator
CR	carriage return		HT	horizontal tabulation
DC1	device control 1		LF	line feed
DC2	device control 2		NAK	negative acknowledge
DC3	device control 3		NUL	null
DC4	device control 4 (stop)		RS	record separator
*DEL	delete		SI	shift in
DLE	data link escape		SO	shift out
EM	end of medium		SOH	start of heading
ENQ	enquiry		STX	start of text
EOT	end of transmission		SUB	substitute
ESC	escape		SYN	synchronous idle
ETB	end of transmission block		US	unit separator
			VT	vertical tabulation

*not strictly a control character

TABLE 12.2 Hex-ASCII Table

00	NUL	1A	SUB	34	4	4E	N	68	h
01	SOH	1B	ESC	35	5	4F	O	69	i
02	STX	1C	FS	36	6	50	P	6A	j
03	ETX	1D	GS	37	7	51	Q	6B	k
04	EOT	1E	RS	38	8	52	R	6C	l
05	ENQ	1F	US	39	9	53	S	6D	m
06	ACK	20	SP	3A	:	54	T	6E	n
07	BEL	21	!	3B	;	55	U	6F	o
08	BS	22	"	3C	<	56	V	70	p
09	HT	23	#	3D	=	57	W	71	q
0A	LF	24	$	3E	>	58	X	72	r
0B	VT	25	%	3F	?	59	Y	73	s
0C	FF	26	&	40	@	5A	Z	74	t
0D	CR	27	'	41	A	5B	[	75	u
0E	SO	28	(	42	B	5C	\	76	v
0F	SI	29	)	43	C	5D	]	77	w
10	DLE	2A	*	44	D	5E	^ (↑)	78	x
11	DC1 (X-ON)	2B	+	45	E	5F	– (←)	79	y
12	DC2 (TAPE)	2C	,	46	F	60	`	7A	z
13	DC3 (X-OFF)	2D	-	47	G	61	a	7B	{
14	DC4	2E	.	48	H	62	b	7C	|
15	NAK	2F	/	49	I	63	c	7D	}
16	SYN	30	0	4A	J	64	d		(ALT MODE)
17	ETB	31	1	4B	K	65	e	7E	~
18	CAN	32	2	4C	L	66	f	7F	DEL
19	EM	33	3	4D	M	67	g		(RUB OUT)

TABLE 12.3 Effect of b_5, b_6

b_6	b_5	CHARACTER
0	0	Control characters
0	1	Digits, symbols
1	0	Capital letters
1	1	Lowercase letters

TABLE 12.4 Important Control Codes

FUNCTION	CONTROL CHARACTER	
BEL	↑G	(Bell)
BS	↑H	(Backspace)
LF	↑J	(Line feed)
CR	↑M	(Carriage return)
ETX	↑C	(End of text)

The Parity Bit

Often in the transmission of data, errors occur. For example, startup of large motors or other electrical equipment generates noise that may interfere with the code being transmitted. The addition of an eighth bit to the ASCII code attempts to determine whether such an error has been produced. Figure 12.2 shows the addition of an unwanted 1 to the code for the letter "A." In this case, bit 3 is received as a one. The code 1 0 0 1 0 0 1 is received. An "A" is sent, but a letter "I" is received. The *parity* bit would catch this type of error. The idea is to use the eighth bit (parity) to indicate that the transmitted character will contain either an even number or an odd number of 1s. If we assume that a circuit is designed for even-parity checking, the parity bit is established as a zero or a one, as necessary, to make the number of 1s an even number. Since an "A" already has an even number of ones, the even-parity bit is zero. Table 12.5 lists a few characters and their parity bits.

Recall that the XOR function can be used to determine parity. This is shown in Figure 12.3.

FIGURE 12.2   Noise During Data Transmission of an "A"

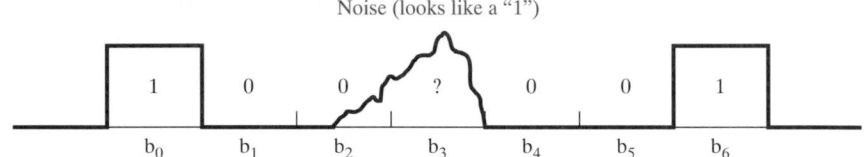

TABLE 12.5   Parity Bits

CHARACTER		P_{even}	P_{odd}
A	1 0 0 0 0 0 1	0	1
B	1 0 0 0 0 1 0	0	1
G	1 0 0 0 1 1 1	0	1
L	1 0 0 1 1 0 0	1	0
W	1 0 1 0 1 1 1	1	0

FIGURE 12.3   Generating an Even Parity Bit

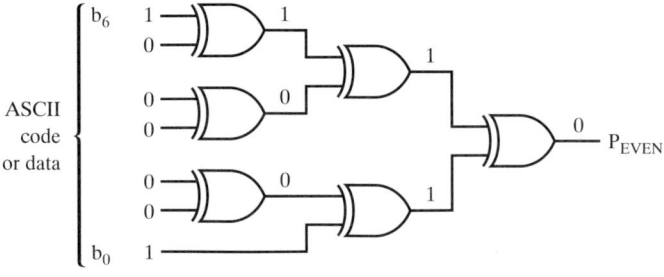

If the transmitted parity bit is supposed to be even, and an odd number of ones is received (as in the "A" transmission example), it is assumed that an error has occurred and an error message will be issued. In noncritical applications, it is not necessary to check the parity of received data. Some computers, however, generate either an even or an odd parity bit for memory data, to help ensure correctness.

To sum up, an eighth bit is added to the 7-bit ASCII code. This bit is called the parity bit, and the new code is sometimes referred to as 8-bit ASCII.

The Extended Binary Coded Decimal Interchange Code (EBCDIC)

Although ASCII is an extremely popular code that is used widely in data communication and with microcomputers, another code exists and is used primarily in IBM (International Business Machines Corporation) equipment. This is an 8-bit code (in contrast with ASCII's 7 bits). The EBCDIC code uses 8 data bits and encodes data characters using zeros and ones, just as ASCII does. However, the codes are *not* the same. For instance, an ASCII space is 20 H while an EBCDIC space is 40 H. Some of the codes are compared in Table 12.6.

Clearly, the codes are quite different. Moreover, with 8 bits, 256 character codes are possible (as opposed to 128 with 7 bits). Table 12.7 shows the EBCDIC codes for uppercase letters.

It may be necessary to convert from EBCDIC to ASCII or vice versa. This is easily accomplished with a lookup table (in software or hardware).

TABLE 12.6 Comparison of Some Codes

CHARACTER	ASCII	EBCDIC
Space	20 H	40 H
A	41 H	C1 H
B	42 H	C2 H
J	4A H	D1 H
S	53 H	E2 H
Carriage return	0D H	15 H

TABLE 12.7 EBCDIC Uppercase Letters

CHARACTER	CODE	CHARACTER	CODE	CHARACTER	CODE	CHARACTER	CODE
Space	40					0	F0
A	C1	J	D1			1	F1
B	C2	K	D2	S	E2	2	F2
C	C3	L	D3	T	E3	3	F3
D	C4	M	D4	U	E4	4	F4
E	C5	N	D5	V	E5	5	F5
F	C6	O	D6	W	E6	6	F6
G	C7	P	D7	X	E7	7	F7
H	C8	Q	D8	Y	E8	8	F8
I	C9	R	D9	Z	E9	9	F9

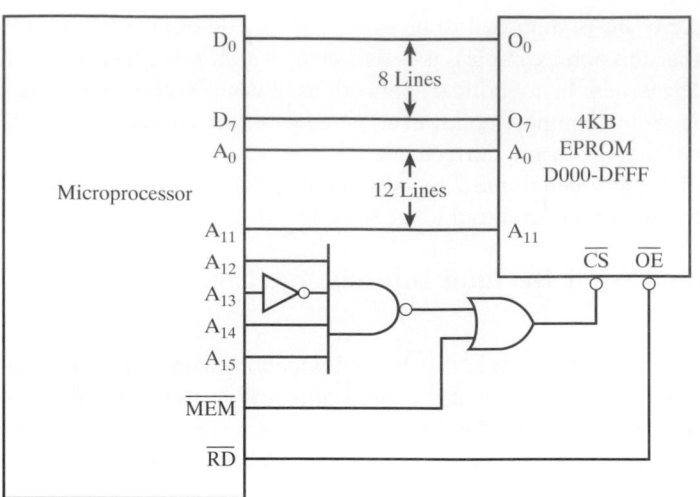

FIGURE 12.4 Interfacing an EPROM to a Microprocessor

12.4 PARALLEL COMMUNICATION

Parallel communication is the fastest way to send digital data from one place to another. Figure 12.4 shows how a parallel connection is used to connect the data bus of a microprocessor to an EPROM. The data on the parallel bus is transferred to the microprocessor when the control signals $\overline{\text{MEM}}$ and $\overline{\text{RD}}$ are active (low), and there is a valid address on the upper address lines.

Applications that only require seven data bits (such as parallel ASCII transmission) may choose to use the eighth bit as a *strobe* signal. A strobe signal is a signal that goes low or high at a particular time to signify that data is available.

Data is also transferred between integrated circuits in parallel as well. Connecting a decade counter to a BCD-to-seven-segment decoder, performing a parallel load on a counter or shift register, and adding or comparing two binary numbers all may be done in parallel for high speed.

12.5 SERIAL COMMUNICATION

Transmitting ASCII codes in serial requires that a pattern be established that describes the number and meaning of each of the bits in the serial waveform. An 11-bit transmission code may be used to transmit serial ASCII data. The line starts at a logical high level. If no character is being transmitted, the line is a one. The transmission begins with the line dropping low for a **start bit**. See Table 12.8 and Figure 12.5. The data bits follow the start bit beginning with the LSB (b_0), then the parity bit, and finally the **stop bits**.

There are two ways to transmit serial data: synchronously and asynchronously (see Figure 12.6). The 11-bit transmission code is an **asynchronous transmission** method, since the receiver uses the start bit to synchronize itself with the incoming data. **Synchronous transmission** requires both a data line and a clock line. The separate clock line is used for synchronization and allows for higher data rates.

TABLE 12.8 Eleven-Bit Transmission Code

BIT	CODE
1	Start bit (always = 0)
2	Data bit b_0 (LSB)
3	Data bit b_1
4	Data bit b_2
5	Data bit b_3
6	Data bit b_4
7	Data bit b_5
8	Data bit b_6 (MSB)
9	Parity bit b_7
10	Stop bit (always = 1)*
11	Stop bit (always = 1)

*May be 1, $1\frac{1}{2}$, or 2 stop bits.

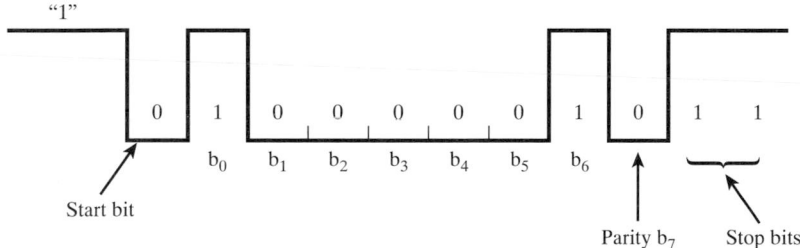

FIGURE 12.5 **Eleven-Bit Transmission Waveform**

FIGURE 12.6 **Data Transmission**

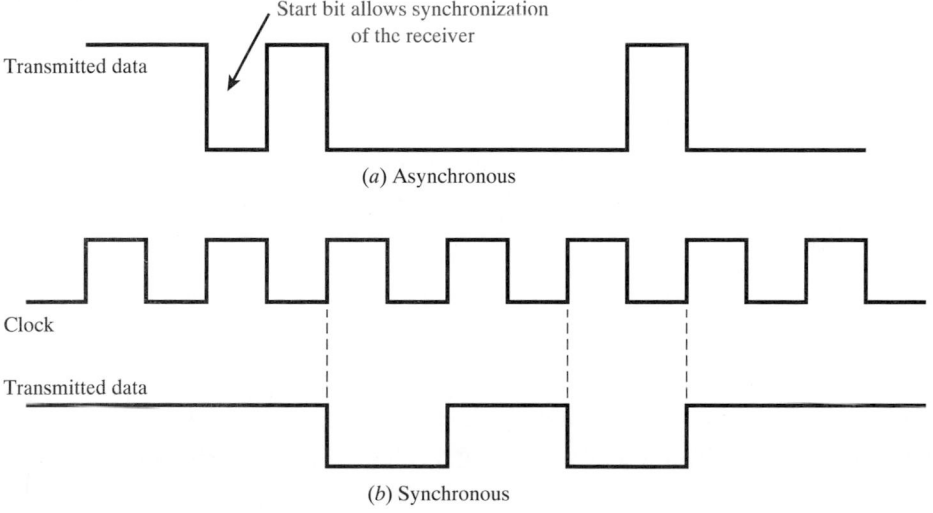

The 11-bit transmission code is sometimes modified in the following ways:

1. Stop bits: may be 1, $1\frac{1}{2}$ or 2.
2. Parity: may be odd or even; may be mark (1) or space (0); or it may be eliminated.
3. Data bits: for non-ASCII data may be set from 5 to 8.

A common method of transmission is 8 data bits, no parity, and one stop bit. Several examples are shown in Table 12.9.

The next subject is the *rate* at which we transmit these bits. This is referred to as the **baud** rate, named for J. E. (Emile) Baudot, who did much work on an earlier code. Baud rate actually refers to the number of transitions, not bits, per second. But baud rate is generally thought of as bits/sec. Standard rates for transferring serial data are 110, 300, 600, 1200, 2400, 4800, 9600, 19,200, 28,800, and 33,600 baud and higher. If data is transmitted at 300 baud (loosely 300 bits/sec), the time per bit is 1/300 sec or 3.333 ms. Since there are 11 bits used to define a transmitted character, the time to transmit a character is $11 \times 1/300 = 11/300$ or 36.6666 ms. If one character can be transmitted in 11/300 of a second, then at top speed a maximum of 300/11 or 27.2727 characters can be sent per second using the 11-bit transmission code. Table 12.10 lists these characteristics for the different baud rates.

The rate at which data is transmitted is frequently a matter of who is using the data. If a person is working at a data terminal, 1200 or 2400 baud generates data on the screen at a comfortable rate—not too fast, not too slow. High-speed data transfer (e.g., 4800 baud and above) is reserved for applications that do not involve display terminals read by human beings.

TABLE 12.9   Sample ASCII Characters

ASCII LETTER	EVEN PARITY	11-BIT WAVEFORM
R = 1010010	1	0 0 1 0 0 1 0 1 1 1 1
↑G = BEL = 0000111	1	0 1 1 1 0 0 0 0 1 1 1
5 = 0110101	0	0 1 0 1 0 1 1 0 0 1 1

TABLE 12.10   Characteristics of Different Baud Rates

BAUD	TIME/BIT (ms)	TIME/11 BITS (ms)	MAXIMUM NUMBER OF CHARACTERS/SECOND
110	9.09	100.0	10
300	3.333	36.6666	27.2727
600	1.6666	18.333	54.5454
1200	0.8333	9.1666	109.09
2400	0.41666	4.5833	218.18
4800	0.20833	2.2916	436.36

12.6 THE RS232C STANDARD

Standards for data transmission are set by the Electronic Industries Association (EIA). Standard **RS232C** establishes a means of transmitting ASCII data between a terminal and a computer. This includes the actual pin connections on a 25-pin connector common to most data terminals.

The first consideration is the problem of transmitting a TTL waveform down a long cable or line. The line looks like a capacitor, and a series of pulses (0 to +5 V) tends to charge the line. The result is a high level at the receive end (see Figure 12.7).

A solution to this charging problem is to convert the one-sided waveform to a wave swinging plus and minus. The RS232C waveform converts the one-sided TTL signal (Figure 12.8(a) to a plus and minus swing and also inverts the waveform as shown in Figure 12.8(b). The voltage swing must be between ±3 and ±25 V. Generally, this voltage is selected based on what may already be available in the terminal. The normal state of an RS232C transmit line is a negative voltage when no transmission is in progress.

FIGURE 12.7   TTL Waveform Distortion on a Line

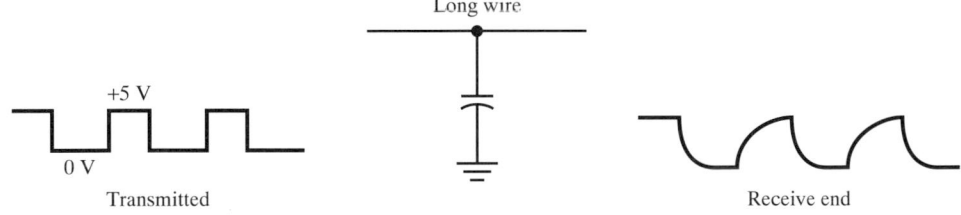

FIGURE 12.8   Serial Data Waveforms (a) ASCII "A" in TTL and (b) RS232C "A"

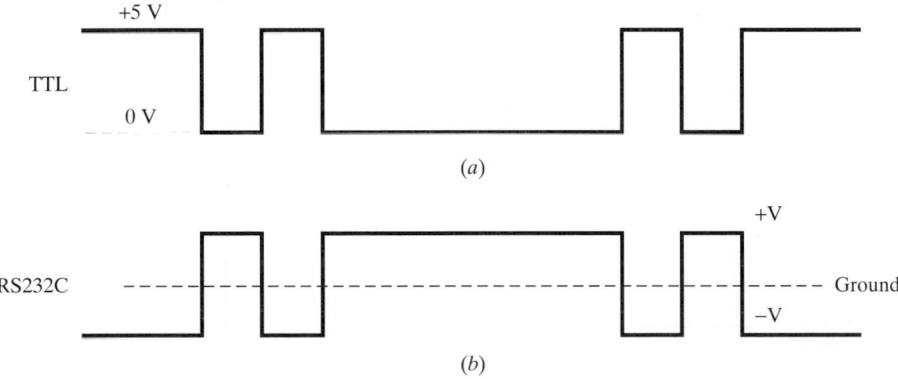

12.7 INTERFACING

Many applications involving a data terminal and a computer need only three wires: transmit, receive, and ground. In other applications, a computer may be too busy to receive data from a terminal at any random moment. In these cases it is necessary to establish a procedure or protocol for transmitting and receiving RS232C information. This is an electronic agreement or request to send information to the computer. If the computer is available and can receive information, then a Clear to Send signal is issued. Figure 12.9 demonstrates.

In the connections for *handshaking* between the terminal and a computer, the four signals involved are:

CTS	Clear to Send	DSR	Data Set Ready
RTS	Request to Send	DTR	Data Terminal Ready

Data Set Ready and Data Terminal Ready are used to determine essential conditions—whether the terminal or computer is turned on, whether a printer is out of paper, and so on.

The use of protocol requires four additional wires between a terminal and a computer. In noncritical applications it represents a high cost in additional cable and this handshaking may be dispensed with. Table 12.11 shows the standard connections for a DB25 connector. This standard is a part of the EIA RS232C document.

Cable Length

The length of an RS232C cable is limited by the baud rate. Longer lengths of line require a lower baud rate. This results from the effect of capacitance on the waveform at increased data rates. The EIA standard limits the capacitance of a line to 2500 pF (see Table 12.12).

When a pulse that represents a bit is transmitted down a line, the edges of the pulse begin to round off. The probability of correctly reading a bit deteriorates as the line length increases. Table 12.13 lists recommended line lengths for given data rates. These figures vary widely depending on the type of wire in use and other considerations. In a particular application there may be no problem in transmitting data at 9600 baud over lengths of 1000 ft. This is because receiver circuitry always looks in the center of each bit position and does its best to correctly distinguish between a zero and a one.

FIGURE 12.9 Connections for Handshaking between the Terminal and a Computer

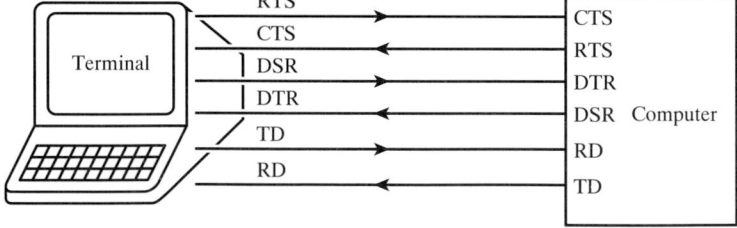

TABLE 12.11 EIA Standard DB25 Connector Pin Assignments

PIN NUMBER	CIRCUIT	DESCRIPTION
1	AA	Protective ground
2	BA	Transmitted data
3	BB	Received data
4	CA	Request to send
5	CB	Clear to send
6	CC	Data set ready
7	AB	Signal ground (common return)
8	CF	Received line signal detector
9	—	(Reserved for data set testing)
10	—	(Reserved for data set testing)
11		Unassigned
12	SCF	Secondary received line signal detector
13	SCB	Secondary clear to send
14	SBA	Secondary transmitted data
15	DB	Transmission signal element timing (DCE source)
16	SBB	Secondary received data
17	DD	Receiver signal element timing (DCE source)
18		Unassigned
19	SCA	Secondary request to send
20	CD	Data terminal ready
21	CG	Signal quality detector
22	CE	Ring indicator
23	CH/CI	Data signal rate selector (DTE/DCE source)
24	DA	Transmit signal element timing (DTE source)
25		Unassigned

Simplex, Half Duplex, and Full Duplex

In the serial transmission of data between two points, three kinds of (data) links may be made. A ***data link*** is the connection between the terminal and the computer. A link that allows the transmission of data in only one direction is referred to as a ***simplex*** data link. A two-way data link is called a *duplex* data link. In a ***full duplex*** data link, data may be transmitted in both directions at the same time; for this arrangement, a terminal may transmit to a computer and simultaneously receive information from the computer. Most modern computer systems operate in full duplex. A ***half duplex*** data link is also a two-way link, but data may not be transmitted in both directions at the same time. One unit must be finished transmitting before the other one can start.

TABLE 12.12 RS232C Parameters

PARAMETER	VALUE FOR RS232C
Line length (recommended maximum—may be exceeded with proper design)	50 ft
Input Z	3–7kΩ
	2500 pF
Maximum frequency	20K baud
Transition time (time in undefined area between 1 and 0) tr = 10–90%	4% of bit period or 1 ms
dV/dt (wave shaping)	30 V/ms
Mark (data 1)	−3 V
Space (data 0)	+3 V
Output Z	3–7kΩ
Open-circuit output voltage, V_o	$3\,V < |V_o| < 25\,V$
V_t = loaded V_o	$5\,V < |V_o| < 15\,V$
	3–7kΩ load
Short-circuit current	500 mA
Power-off leakage	$> 300\Omega$
(V_o applied to unpowered device)	$2\,V < |V_o| < 25\,V$
	V_o applied
Minimum receiver input for proper V_o	$> \pm\,3\,V$

TABLE 12.13 Recommended Line Lengths for Given Data Rates

BAUD RATE	MAXIMUM CABLE LENGTH (FT)
300	4,000
1,200	1,000
2,400	500
4,800	250
9,600	125
19,200	65

Converting between TTL and RS232C

The conversion from TTL to the RS232C signal and back can be done with a simple transistor or with an integrated circuit like the MAX232CPE designed for that purpose. The MAX chip generates its own $+/−10\,V$ supply using four electrolytic capacitors, from a single $+5\,V$ supply. Figure 12.10 illustrates.

Current Loop Transmission

The RS232C standard involves transmitting voltage levels down a line. An older method still in use transmits a pulse of constant current instead. A 20- or 60-mA pulse is used to represent a binary 1 or mark. The 20-mA current is commonly used over the 60-mA current. A

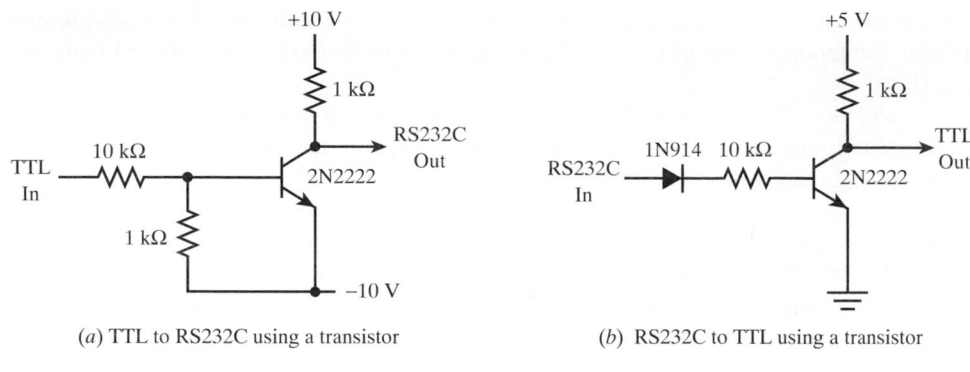

(a) TTL to RS232C using a transistor   (b) RS232C to TTL using a transistor

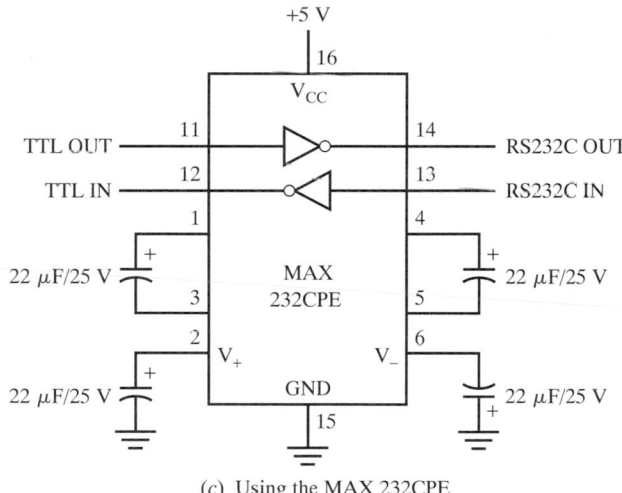

(c) Using the MAX 232CPE

FIGURE 12.10 Converting between TTL and RS232C

zero or space is represented using no current. This method was popular when many Telctype machines were used to receive the same information, as in news services like the Associated Press. The machines could be connected in series over a large area extending hundreds of miles. As long as the current pulse was constant, all machines received the data through a decoding relay. This was a half duplex data link. This method of transmission was quite slow (110 baud) because of the long cable lengths, but nonetheless very reliable. Today this type of information is directed to individual radio and television stations directly by satellite or via the Internet. How things have changed!

12.8 THE RS422 STANDARD

The RS232C standard specifies conditions for serial data transmission over relatively short distances using what is referred to as a single-sided or unbalanced line. Each signal has a single wire and the return is a ground conductor. Single-ended systems are prone to noise pickup, which becomes more severe with increasing distance. Telephone signals have been transmitted over a "pair" of wires for many years. Neither of the wires is ground; in fact, grounding of one of the wires introduces a large amount of AC (60-Hz) hum into the signal.

A system that transmits signals over two wires is said to be "balanced" when there is no ground present. When one conductor is positive, the other is negative. Since neither wire has a ground reference, noise is induced equally in each wire of the pair. The potential between the wires due to the noise then is zero. A balanced wire system can cover larger distances with great immunity to induced noise.

The EIA RS422 standard defines the balanced version of RS232C. A pair of wires is needed for each signal. The added wire in each case becomes the return line for the signal. These signals may include:

Transmitted Data	TxD
Transmitted Data Return	TxD′
Received Data	RxD
Received Data Return	RxD′
Request to Send	RTS
Request to Send Return	RTS′
Clear to Send	CTS
Clear to Send Return	CTS′
Data Set Ready	DSR
Data Set Ready Return	DSR′
Data Terminal Ready	DTR
Data Terminal Ready Return	DTR′
Signal Ground (one wire)	

One disadvantage of RS422 is the increased number of wires required. However, when distance and speed are essential with noise immunity a concern, RS422 defines an answer.

12.9 DIGITAL DATA COMPRESSION/ERROR CORRECTION

When large amounts of data must be stored on a disk, or transmitted to another point, it is desirable to keep the storage space, or the transmission time, to a minimum. This is where digital data compression comes in.

After the transmitted data is received, it would be nice to know if it got to its destination correctly. By adding a few additional parity bits, the received data can be checked for correctness (errors detected). In this section we will examine two simple methods for performing data compression and error detection.

Huffman Coding

Figure 12.11 shows the steps involved in compressing data using a technique called *Huffman coding*. A large block of data consists of the ASCII characters "A" through "E." The number of times each letter occurs in the data block is represented as a percentage

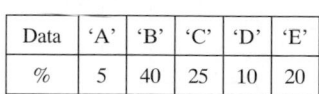

Data	'A'	'B'	'C'	'D'	'E'
%	5	40	25	10	20

(*a*) Original data and percentages

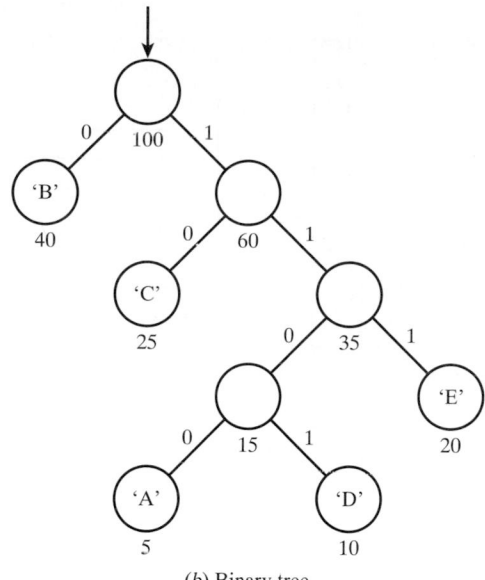

Data	Sequence	Weight
'A'	1 1 0 0	4(.05) = 0.20
'B'	0	1(.40) = 0.40
'C'	1 0	2(.25) = 0.50
'D'	1 1 0 1	4(.10) = 0.40
'E'	1 1 1	3(.20) = <u>0.60</u>

2.10 bits/data item

(*c*) Unique sequences and compression factor

(*b*) Binary tree

FIGURE 12.11 Digital Data Compression

in Figure 12.11(a). Using the Huffman coding technique, a *binary tree* is constructed that contains all of the data items arranged so that smaller percentages are on the left and larger percentages on the right. This is illustrated in Figure 12.11(b).

When the binary tree is *traversed* (walked through), unique binary sequences for each data item are found. Furthermore, the most common data item (the "B") has the shortest bit pattern, and the least common (the "A" and "D" characters) have the longest. On average, it requires only 2.1 bits to store a character. If you consider an 8-bit pattern for each original data item, this gives a compression ratio of 8/2.1, or 3.8. This means that 380,000 bytes of data will compress to 100,000 bytes, with no information lost.

It would be interesting to explore the operation of a digital circuit that takes a serial input and decodes the compressed Huffman sequence. A state machine having state sequences matching the Huffman patterns could be used. One of five outputs will go active as each sequence is recognized. This is left for you to work out as a Study Question.

Digital Error Correction

To handle digital error detection and correction, consider the logic in Figure 12.12. As Figure 12.12(a) shows, a 7-bit number will be transmitted with three additional even parity bits. Note that different combinations of data bits are used with each XOR gate.

In Figure 12.12(b), the 10-bit data/parity word is received with an error in data bit 4, which has changed from a 1 to a 0. The error detection XOR gates use the received data *and* the parity bits to generate a three-bit error code that identifies the incorrect bit (if any). In this example, the error code (E_2 is the MSB) is 100, meaning bit 4 is in error. A code of 000 means no error.

There are many other types of error-correcting codes in which parity continues to be a factor. Once again, a basic logic function plays an important role in a sophisticated operation.

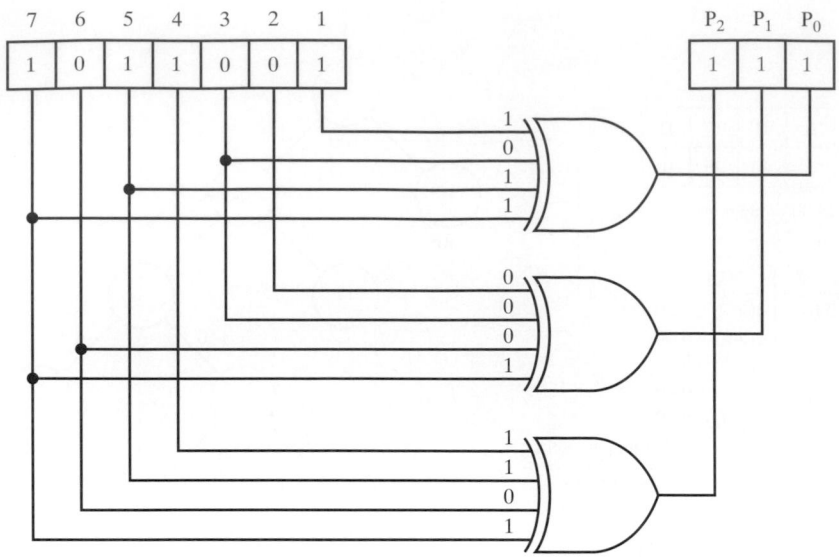

(*a*) Generating parity bits for transmission

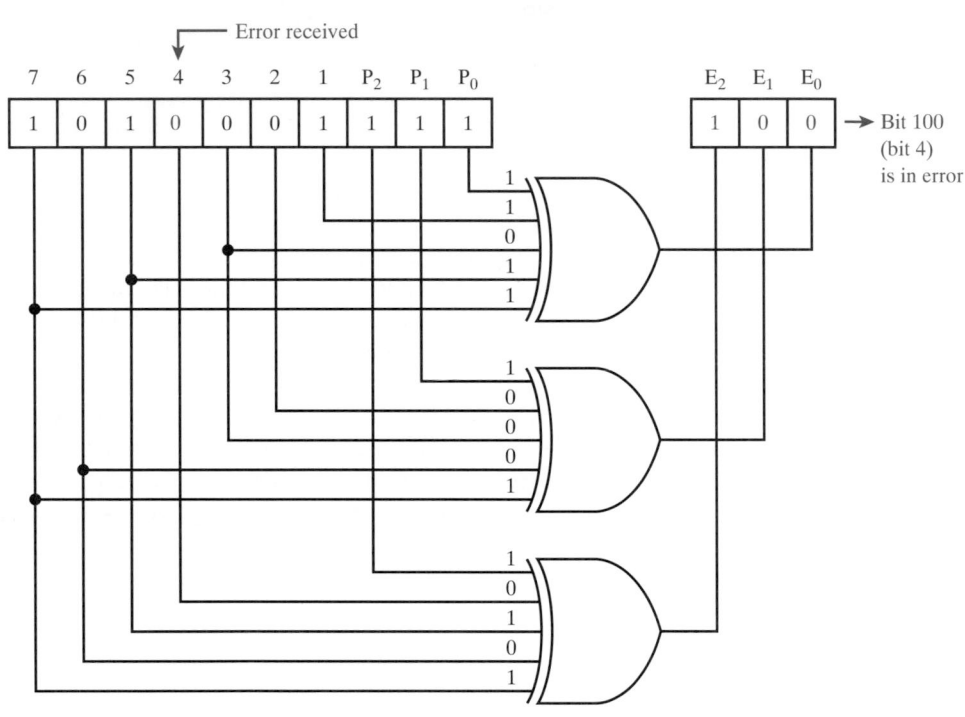

(*b*) Generating error bits from received data

FIGURE 12.12   Digital Error Detection Using XOR Gates

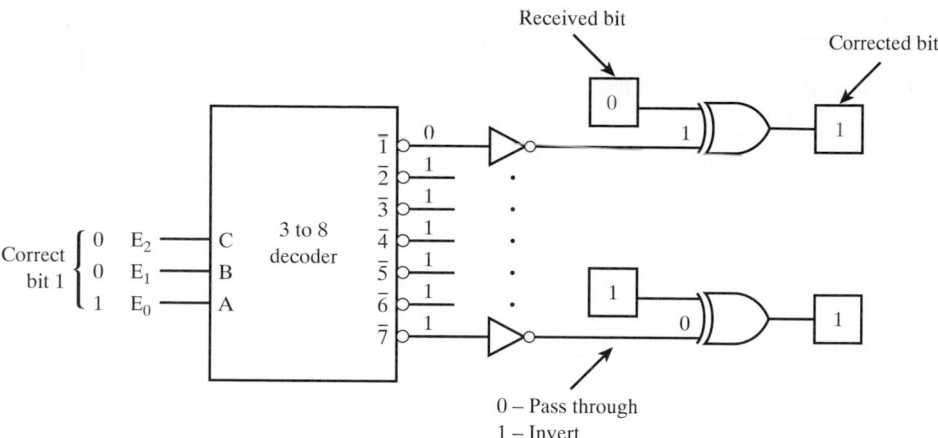

FIGURE 12.13 **Error Correction (Bit 1 Is in Error)**

To correct a bit that is wrong, we need only invert it. The circuit in Figure 12.13 uses the three-bit error code to select one of seven XOR gates and use it as an inverter. If no bits are in error, the error code is 000 and no XOR gates are selected by the decoder.

Adding one more parity bit to the pattern allows us to correct up to 15 bits. Try and determine how this is done by looking for patterns in the XOR inputs of Figure 12.12.

12.10 MODEMS

Frequently data between a terminal and a computer must be communicated over standard telephone lines. In this case, the RS232C waveform is converted to pairs of tones. The device that performs this conversion is called a *modem.* Modem stands for modulator-demodulator. The zero and one information is then transmitted using one frequency to represent a zero and another to represent a one.

Two pairs of frequencies in use—one pair to transmit and one pair to receive—are listed in Table 12.14. The use of four tones allows the simultaneous transmission and reception of data over a standard two-way (duplex) telephone line.

You will notice that two groups of four tones are shown, *originate* and *answer.* In the originate group, we transmit ones and zeros at 1270 and 1070 Hz, respectively. We receive ones and zeros at 2225 and 2025 Hz. This is the opposite of the answer group. This is because for the user, who *calls* the computer, we assign originate frequencies, and

TABLE 12.14 **Pairs of Frequencies**

	FREQUENCIES (Hz)	
DATA	ORIGINATE	ANSWER
Transmit "0"	1070	2025
Transmit "1"	1270	2225
Receive "0"	2025	1070
Receive "1"	2225	1270

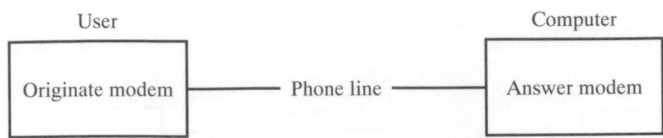

FIGURE 12.14 **Originate and Answer Modems**

to the computer, which *responds,* we assign answer frequencies. Thus two modems (see Figure 12.14) are necessary. Most computer installations (at schools, banks, airports) employ modems. With this kind of connection, the user's transmit frequencies are the same as the computer's receive frequencies, and vice versa.

Sometimes the modem is wired directly to a telephone line and sometimes an acoustic coupler is used. The acoustic coupler uses the telephone's own mouthpiece and earpiece to couple the tones to the telephone line. Modem cards for the personal computer offer very fast transmission speeds (over 56K baud) over ordinary phone lines. A mixture of hardware and software perform data compression and expansion on-the-fly.

Modems are an inexpensive solution to low-speed data transmission. New modems, called *cable modems,* can be rented from your local cable provider. They are capable of millions of bits/second transmission speeds. Cable modems can provide fast Internet service directly to your home. Check with your local cable company about availability.

12.11 LANs

The current popularity of *Local Area Networks (LANs)* is a result of our increasing need to use and exchange information. A LAN may connect computers in an office, on the entire floor of a department store, or across all the buildings of a college campus. LANs, and computer networks in general, provide a means to share system resources, such as hard drives and printers. Individual computers with network interface cards, and other devices, are called *nodes.* A LAN provides a way of connecting many nodes together.

One very common LAN technology is called *Ethernet.* Developed jointly by Digital Equipment Corporation, Intel, and Xerox in 1980, the Ethernet provides a method of transmitting digital data at high speeds (10 Megabits/second originally, with 100 Megabits becoming the standard). Ethernet is a *baseband* system, meaning that one information carrier is present on the wire. *Broadband* systems (such as cable TV) use many signal carriers on the same wire. The Ethernet uses a technique called *Manchester encoding* to transmit the 0s and 1s. An example of Manchester encoded data is shown in Figure 12.15. Phase transitions are used to represent the digital data, with a one-to-zero transition signifying a logic zero, and a zero-to-one transition indicating a logic one. Every bit causes a transition. The Manchester waveform contains both data and a clock mixed together.

FIGURE 12.15 **Manchester Encoding**

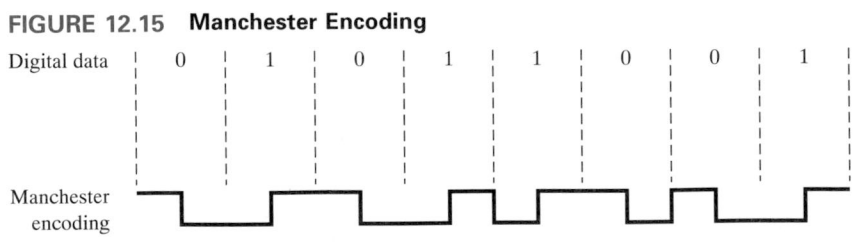

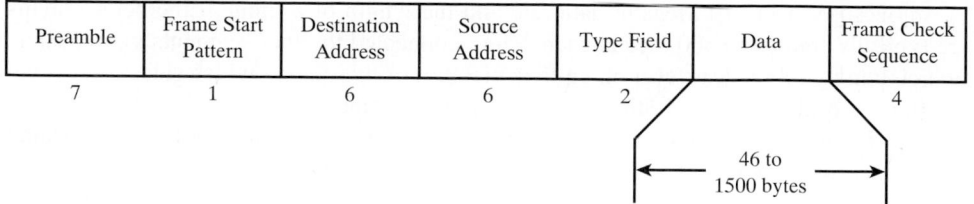

Preamble	Frame Start Pattern	Destination Address	Source Address	Type Field	Data	Frame Check Sequence
7	1	6	6	2		4

46 to 1500 bytes

FIGURE 12.16 **Ethernet Frame**

Ethernet is also a *broadcast* network, meaning all nodes receive the same information, much like a large number of radios all tuned into the same station. The information transmitted over the Ethernet is in the form of a *frame*. The format of a frame is shown in Figure 12.16.

Ethernet LANs are set up similar to the network diagram in Figure 12.17. Signals are carried over coaxial cable (also called 10baseT), to hubs, which serve as concentrators, connecting many nodes together at one point. A typical hub may supply from 8 to 16 nodes. Eight-conductor twisted-pair (10baseT) wires run from the hubs to their associated networked computers. 10baseT wires use RJ45 connectors similar to modular telephone connectors that allow network connections to be made quickly and effectively. 10base2 connections require the coaxial cable to be cut and carefully crimped into a BNC connector. Many networks have shut down due to bad crimps on the BNC connectors.

FIGURE 12.17 **Typical LAN**

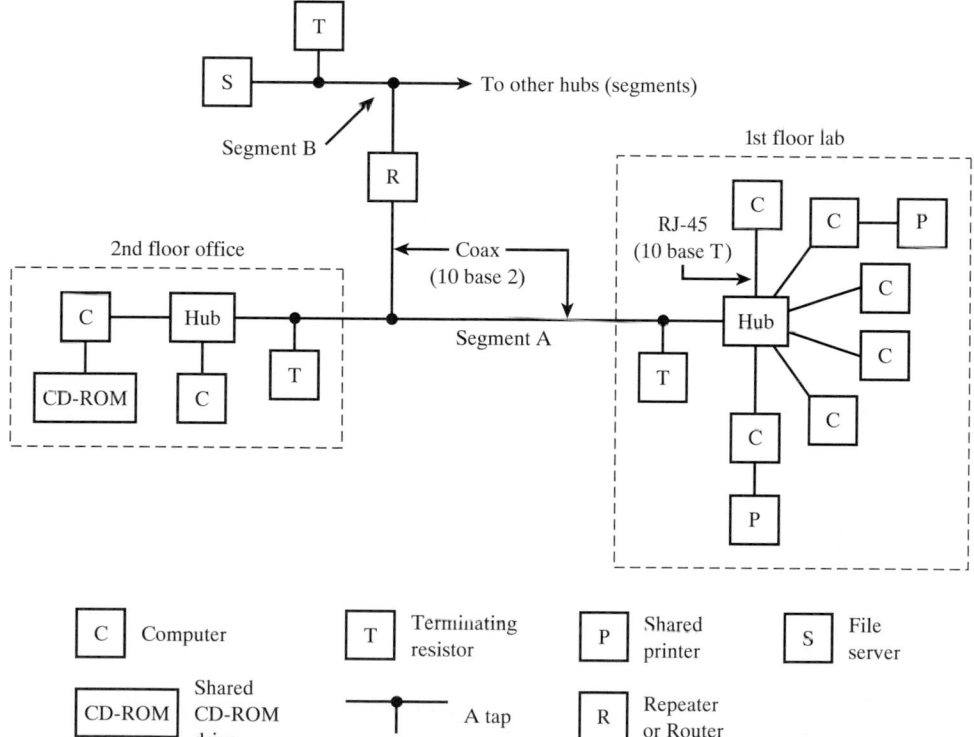

Because of the 10 Megabit data rate and the length of a frame, Ethernet segments are typically limited to 500 meters each, being connected to other segments via repeaters (which blindly rebroadcast everything they pick up) or by routers (which selectively route the frames to their correct destination over a predetermined path).

Every node in a LAN has a unique binary address that is included in every frame transmitted. Frames are addressed to a node with a particular destination address. When a frame is broadcast, however, chances are good that another node may also begin broadcasting its own frame. This is like trying to listen to two radio stations at the same time. The specific name for this event is called a *collision*. When a collision occurs, all Ethernet devices that are transmitting immediately stop and wait a random period of time before retransmitting. This helps to avoid repeat collisions. The entire scheme is called *Carrier Sense Multiple Access with Collision Detection (CSMA/CD)*. The asynchronous nature of the Ethernet makes CSMA/CD a necessity. Other network systems, such as token-ring, avoid the possibility of collisions by controlling the direction and flow of information.

Overall, the Ethernet is a very sophisticated system for providing fast, reliable serial data transmission.

12.12 PRACTICAL APPLICATIONS

In this section we will examine a number of ways parallel communication and serial communication can be put to use.

Decoding ASCII Control Characters

Certain control characters require decoding for many applications. For instance, frequently the BEL or Control G code is used to sound or alert an operator. For decoding, we need only design a simple black box. This can be done all at once or in two stages. Figure 12.18 shows both methods. The second uses the idea that control characters have both b_5 and b_6 at zero. Method 2 uses a separate gate to decode the control $(\overline{\text{CTRL}})$ function. This is useful if other control characters must be decoded—having two wires decoded saves on the number of input gates required.

FIGURE 12.18 Decoding Control-G (0000111 B)

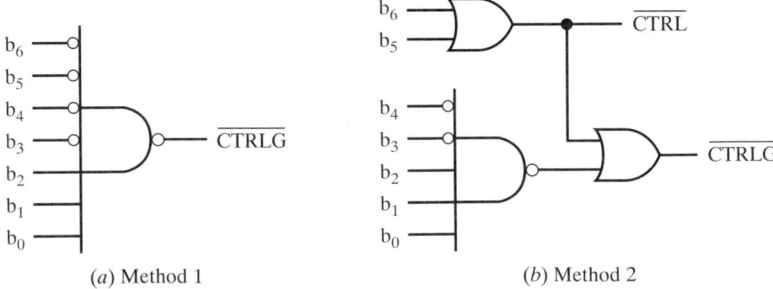

(a) Method 1   (b) Method 2

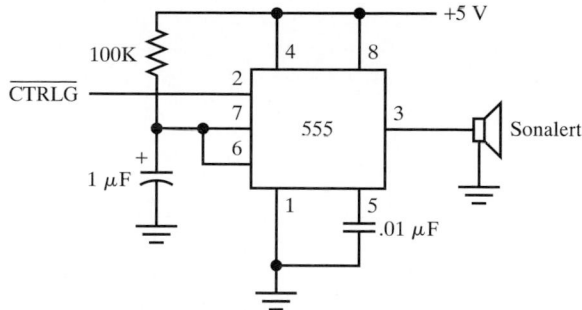

FIGURE 12.19 **Control G Sounding Circuit**

Once the Control G (BEL) signal has been decoded, a sounding device can be activated. This signal can be used to drive a 555 timer connected as a single shot (pulse stretcher) driving a Sonalert. The Sonalert produces a piercing tone when a DC level is applied. Figure 12.19 shows the circuitry.

The 555 timer is an integrated circuit capable of producing pulses to be used in oscillator applications; it also can serve as a one shot in this type of application where a very narrow pulse is widened to allow a 100-ms tone at the Sonalert.

Generating an RS232C Waveform Using a Digital Black Box

To further our understanding of the principles of data transmission, let us generate the ASCII code for the letter A at 300 baud. This signal can then be connected to a data terminal for display. Since we will drive the black box with a binary counter, the character will repeat over and over again.

The code in ASCII for "A" is 1000001. Since there are an even number of ones, an even-parity system for error checking would make the parity bit a zero, thus keeping the number of ones an even number.

The standard method of data transmission begins with a start bit, followed by the seven data bits and the parity bit, ending with two stop bits. The start bit is always a zero. The stop bits are always logic 1s. The normal state of the transmission line when data is not transmitted is a one level. Figure 12.20 shows the correct coding for the complete transmission of the letter "A."

FIGURE 12.20 **Coding for Complete Transmission of the Letter "A"**

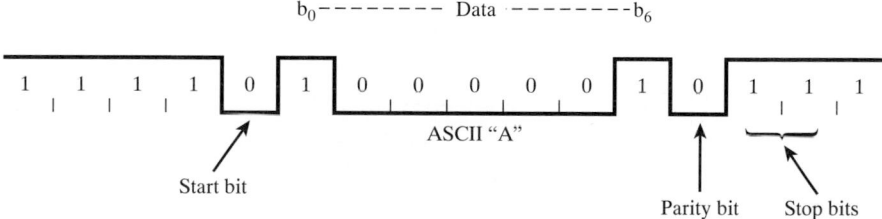

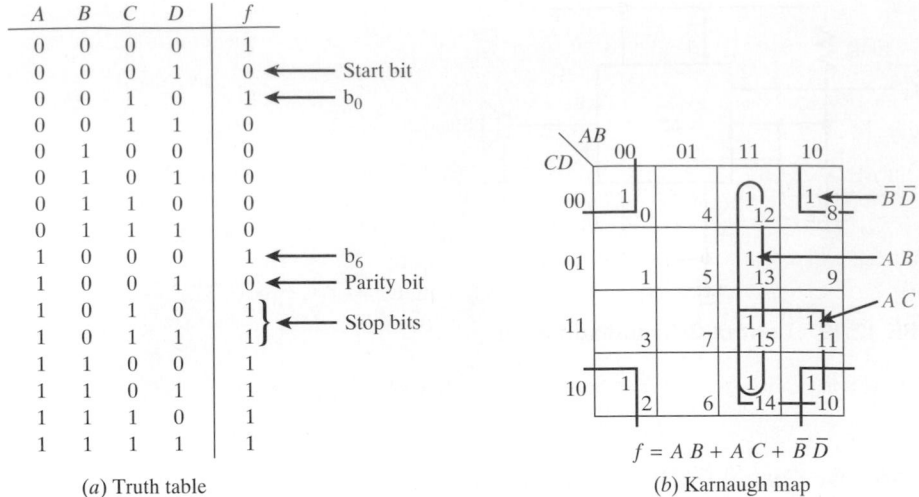

A	B	C	D	f	
0	0	0	0	1	
0	0	0	1	0	← Start bit
0	0	1	0	1	← b_0
0	0	1	1	0	
0	1	0	0	0	
0	1	0	1	0	
0	1	1	0	0	
0	1	1	1	0	
1	0	0	0	1	← b_6
1	0	0	1	0	← Parity bit
1	0	1	0	1	}
1	0	1	1	1	← Stop bits
1	1	0	0	1	
1	1	0	1	1	
1	1	1	0	1	
1	1	1	1	1	

(*a*) Truth table

$$f = A\,B + A\,C + \bar{B}\,\bar{D}$$

(*b*) Karnaugh map

FIGURE 12.21   Design of Digital Black Box to Transmit an "A"

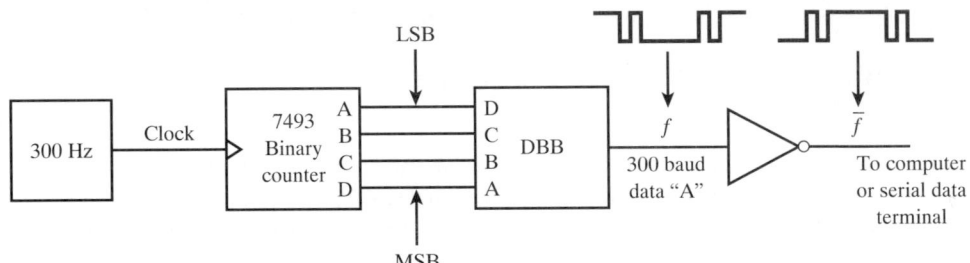

FIGURE 12.22   Connection of DBB to Transmit an "A"

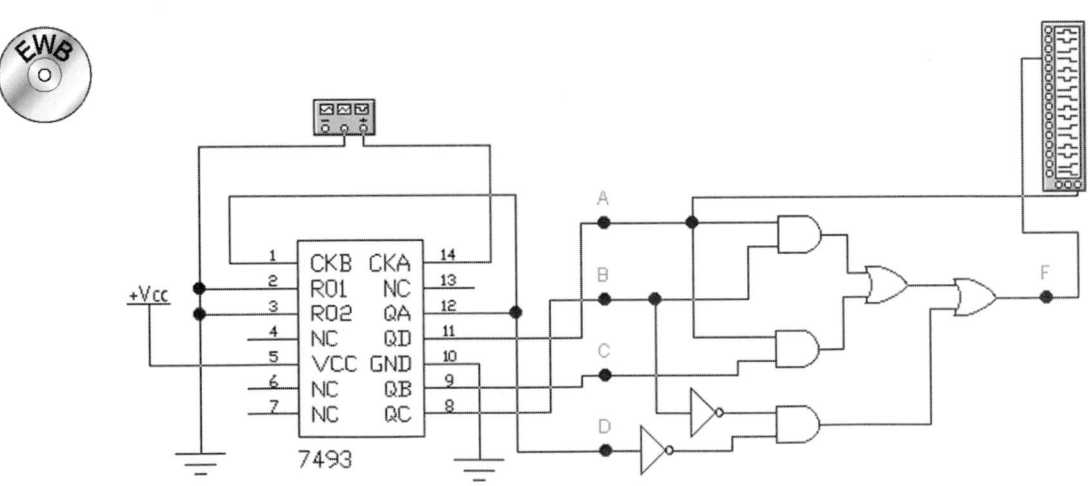

FIGURE 12.23   Electronics Workbench File ASCIIA

We will need a digital black box to produce this pattern of bits at the correct rate to transmit an "A" to the computer or receiver unit. The design begins with a truth table and is simplified with a Karnaugh map, as shown in Figure 12.21.

The DBB when constructed will follow the truth table and when driven by a binary counter will produce the correct waveform. As shown in Figure 12.22, the counter must be driven by exactly 300 Hz to produce 300 baud for the data terminal. In addition, the logic must be inverted to send $\bar{f}$ to the computer, since the RS232C waveform is the opposite of the TTL waveform. When properly "tuned," the terminal should show repetitive A's on the printout, because the counter keeps recycling. Figure 12.23 shows the Electronics Workbench file ASCIIA, which implements the logic equation from Figure 12.21(b). View the waveform with the logic analyzer to verify that it looks like the ASCII "A" waveform.

Large-Scale Integrated Receiver/Transmitter Chips

Digital black box design can be very helpful in experimenting with and using the RS232C standard. However, it is useful only in the transmission of a particular ASCII character. The use of a parallel load shift register greatly simplifies the conversion of an ASCII code in parallel to a series of binary bits for data transmission. The code is loaded in parallel and then clocked out at the desired baud rate.

In practice, and in microcomputer systems, this transmission and reception is accomplished in a specialized receiver/transmitter chip called a **UART** (for universal asynchronous receiver transmitter). Several are available and in common use. Each contains a pair of separate shift registers. One is for transmitting an ASCII character (parallel to serial) and one is for receiving an ASCII character (serial to parallel). Figure 12.24(a) illustrates.

In the example, the receiver is receiving an ASCII "A" in 11-bit transmission code. The transmitter is sending an ASCII "J." This illustrates the separateness of the receiver and transmitter sections. A character can be received at the same time as a different character is being transmitted. The UART chip works with TTL levels and not the plus/minus swing of RS232C. That conversion must be done external to the chip. Both the transmitter and receiver sections have separate clock inputs (RxC = Receiver clock, TxC = Transmitter clock). These clock inputs determine the baud rate at which data is received or transmitted. Depending on the type of UART used, the clock frequency is either 16 or 64 times the desired baud rate. For example, if the desired baud rate is 300, the clock might be 16 × 300 or 4800 Hz.

FIGURE 12.24(a)   **Receiver/Transmitter Chip**

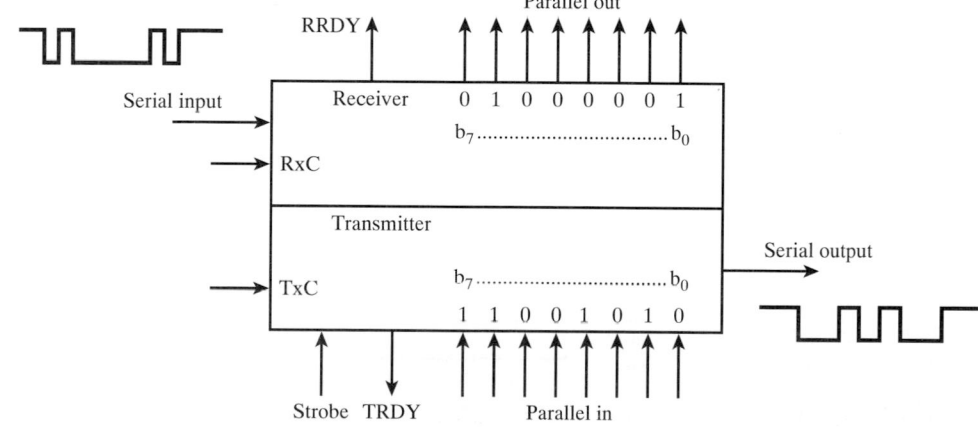

Two status signals called *flags* are also available. These are the Receiver Ready Flag (RRDY) and the Transmitter Ready Flag (TRDY). When a character has been completely received, the RRDY flag is active (high in most applications). When the transmitter is available for use, the TRDY flag is active (high in most applications).

In Figure 12.24(b) the transmitter half of a simple UART is shown. A 12-bit shift register is parallel loaded with the ASCII code to transmit, along with hard-wired start (0) and stop (1) bits. Bits are clocked out at a rate determined by TxC. When a strobe pulse clocks the D flip-flop, the $\overline{Q}$ output goes low, allowing the 4-bit counter to begin counting. After 12 bits have been clocked out, the output of the NAND gate goes low, clearing the flip-flop, which in turn clears the counter. When the counter output goes to 0000 the OR gate allows the shift register to be loaded with new data.

If the transmitter is idle, the shift register output is high due to the hard-wired logic one on the least significant input.

Figure 12.24(c) illustrates the receiver half of the UART. Unlike the transmitter, the receiver uses an RxC clock that is 16 times faster than the incoming bit rate. This allows the bit stream to be sampled in the middle of each bit for higher accuracy.

A falling edge on the serial input clocks a one through the first D flip-flop, allowing both counters to begin counting. After the first eight RxC pulses, the second counter gets clocked once. This causes the start bit to get loaded into the 12-bit receiving shift register. Then every 16 RxC pulses the second counter is clocked again and another bit is loaded into the shift register. After 12 bits have been received, the NAND gate resets the first flip-flop (stopping the counters) and sets the second flip-flop, indicating that a character has

FIGURE 12.24(b)   UART Transmitter Logic

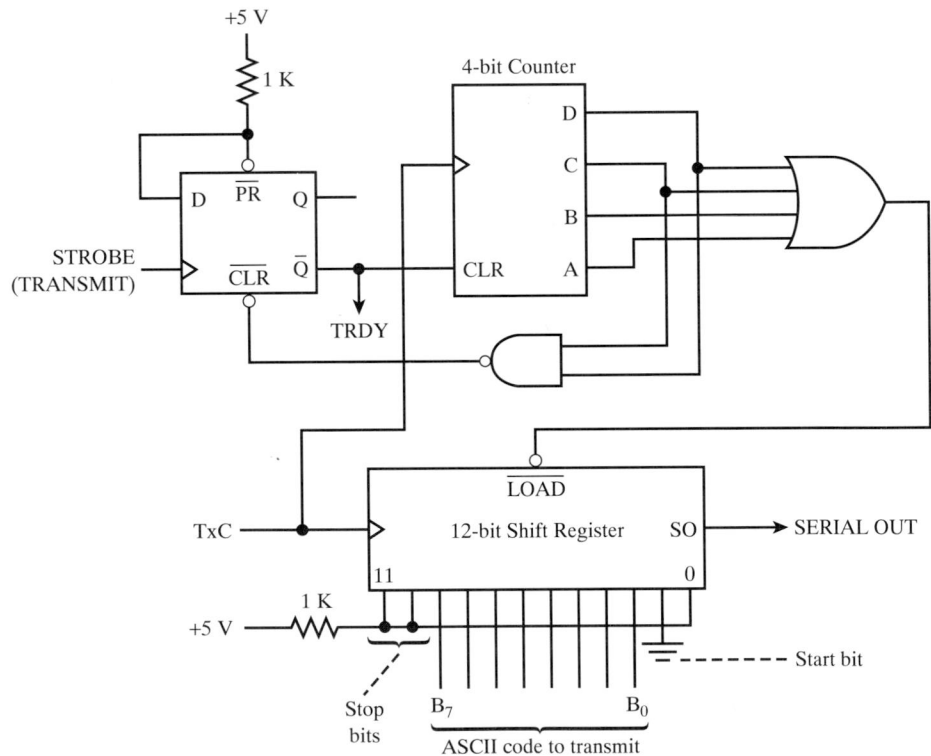

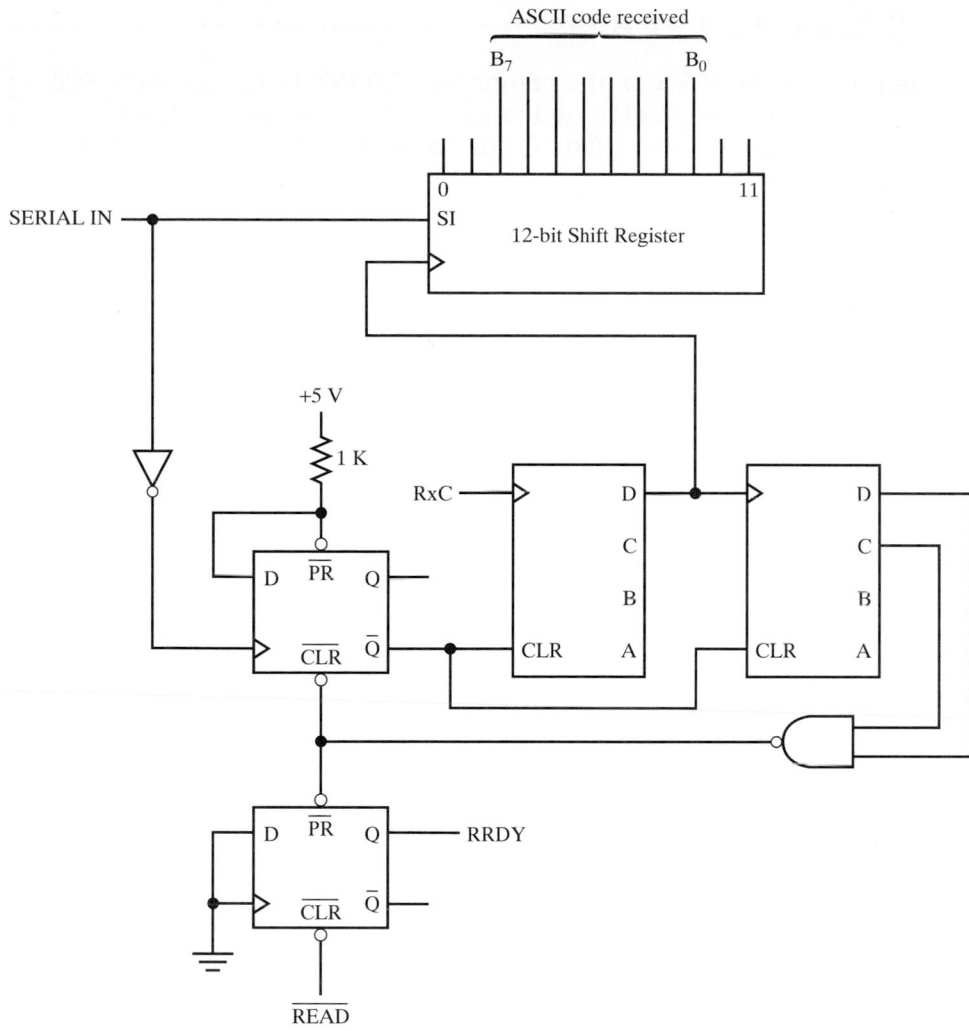

FIGURE 12.24(c) UART Receiver Logic

been received. A latch on the output of the shift register can be used to buffer the received character, to allow reception of a second character to begin immediately.

For flexibility, UARTs are programmable. This means that certain aspects of the 11-bit transmission code may be changed. These involve the number of data bits, the number of stop bits, and the type of parity. The choices are: 5, 6, 7, or 8 data bits; 1, $1\frac{1}{2}$, or 2 stop bits; and odd, even, mark, space, or no parity. Typically, 8 data bits, 1 stop bit, and no parity are selected.

This allows a 10-bit transmission code after the addition of the standard start bit. At a given baud rate, character information transmits faster in 10 bits than 11 bits. At 300 baud:

10 bits: $300/10 = 30$ characters/sec

11 bits: $300/11 = 27.27$ characters/sec

The use of a UART greatly reduces the circuitry needed for data transmission and simplifies design.

C/C++ HELPER

The program ACODES on the companion CD-ROM displays the ASCII code of a character entered by the user in both binary and hexadecimal. The program exits when "Q" or "q" is entered (after displaying its ASCII code).

```c
//ASCII Codes
#include <stdio.h>

main()
{
    unsigned char cin,p;
    do
    {
        printf("Character? ");
        cin = getche();
        printf("\n%c is an ASCII %02XH, or ",cin,cin);
        for(p = 0x80; p != 0; p >>= 1)
            p & cin ? printf("1") : printf("0");
        printf(" B\n");
    } while ('Q' != toupper(cin) );
}
```

A sample execution of ACODES looks like this:

```
Character? A
A is an ASCII 41H, or 01000001 B
Character? a
a is an ASCII 61H, or 01100001 B
Character? 0
0 is an ASCII 30H, or 00110000 B
Character? 7
7 is an ASCII 37H, or 00110111 B
Character? *
* is an ASCII 2AH, or 00101010 B
Character? /
/ is an ASCII 2FH, or 00101111 B
Character? q
q is an ASCII 71H, or 01110001 B
```

12.13 TROUBLESHOOTING TECHNIQUES

There never seems to be an ASCII chart around when you need one. Even memorizing certain key codes, such as "A", "a", and "0" to use with the simple counting method, is restrictive and time-consuming. One way to handle a missing ASCII chart is to let the codes themselves reveal their patterns. To see how this is done, look over the ACODES C/C++ Helper.

SUMMARY

In this chapter we have learned about methods of transmitting parallel and serial data. The ASCII code was discussed extensively and a means of transmitting the code serially was described in detail. This method relies on the RS232C standard, which is widely used in the most modern data terminals. The transmission of data over telephone lines and across a LAN (via Ethernet) was also covered.

STUDY QUESTIONS

General

1. What are the differences between parallel and serial data communication?

2. What type of connection does a printer typically use? What about a mouse?

3. Write down the 8-bit ASCII code for the following characters using an even-parity bit:
   (a) F
   (b) J
   (c) M
   (d) n
   (e) e
   (f) R
   (g) V
   (h) Control-G
   (i) Control-T
   (j) <CR> (carriage return)

4. Sketch the 11-bit transmission code (even parity) for each of the characters from Question 3.

5. At 9600 baud, determine the maximum number of characters per second that can be transmitted using the 11-bit transmission code.

6. If only one stop bit is used, a character may be transmitted more quickly, since only 10 bits are required per character. Repeat Question 5 using a 10-bit transmission code.

7. At 9600 and 19,200 baud, what are the times per bit required?

8. What is the maximum recommended length of cable for data transmitted at 9600 baud?

9. Design a parallel input circuit to decode the following control functions:
   (a) BS
   (b) CR
   (c) Any control character (use NOR logic)
   (d) ESC

10. Design a digital black box that will continuously generate an ASCII * (asterisk).

11. Can a single-parity bit be used to detect the change of more than one bit? Why or why not?

12. Why do UARTs need clocks 16 or 64 times faster than the baud rate being used?

13. Fill in Table 12.15.

TABLE 12.15

ASCII	BINARY	HEXADECIMAL	DECIMAL	P_{even}
L				
m				
!				
ESC				
LF				
H				
#				
6				
EOL				

14. Sketch the 10-bit RS232C waveform for each of the following ASCII characters (1 stop bit).
    (a) 'X'
    (b) 'P'
    (c) '7'

15. Fill in Table 12.16 (10 bits = 1 character).

TABLE 12.16

BAUD RATE	TIME/BIT	TIME/CHARACTER	MAXIMUM NUMBER OF CHARACTERS/SEC
1200			
9600			
19,200			
22,800			
33,600			

16. Draw a circuit to latch an LED in the on state when a control X character is received.

17. Design a digital black box to transmit a string of Ws.

18. Decode each of the RS232C waveforms shown in Figure 12.25. Use a ruler.

(*a*)

(*b*)

(*c*)

FIGURE 12.25   For Question 12.18

19. Can a ROM (or a PROM) be used to convert ASCII to EBCDIC? If so, explain how.

20. Design a circuit with a minimum of TTL chips to decode the first eight control characters (ASCII 00–07). (*Hint:* Use a three-line to eight-line chip.)

21. Repeat Question 20 but decode the first 16 control characters.

22. An exclusive OR gate can be used to compute parity. Can you design a circuit that uses these gates to compute the odd-parity bit for an 8-bit input?

23. The state diagram shown in Figure 12.26 shows how a three-pattern Huffman sequence might be decoded. The patterns are:

    A   0   (recognized in state 1)
    B   10 (recognized in state 3)
    C   11 (recognized in state 4)

    Design the machine (assume it starts in state 0). Take the appropriate output low when in states 1, 3, or 4.

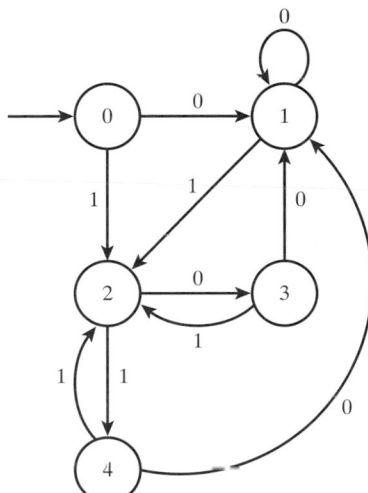

(*a*) Huffman-based state machine

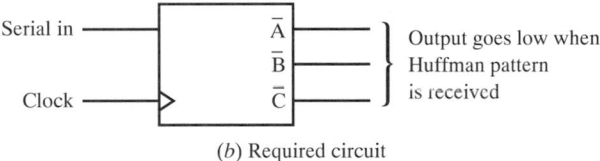

(*b*) Required circuit

FIGURE 12.26   For Question 12.23

24. What are the three parity bits for each of these 7-bit data values (use Figure 12.12)?
    (a) 1011110
    (b) 0111001

25. If bit six is received in error in Question 24, what are the three error bits?

26. What XOR logic is required to error detect a 12-bit data word?

27. What does "modem" stand for?

28. What is a cable modem?

29. What is a LAN?

30. What is Ethernet?

31. Show how the data sequence 10100110 is Manchester encoded.

32. What is a collision?

Electronics Workbench

33. Verify your solutions to Questions 10, 16, and 23.

34. Use Electronics Workbench to design and test the transmitter and receiver portions of the UART in Figure 12.24.

Programmable Logic

35. Write the PAL equations for the ASCII 'A' digital black box.

36. How could the transmitter logic of the UART in Figure 12.24 be placed on a PAL?

TROUBLESHOOTING TECHNIQUES

INSTRUCTIONAL OBJECTIVES

This chapter continues coverage of the broad and elusive subject of troubleshooting. In particular, in this chapter we are concerned with the problems of microprocessor-based circuitry both in the design process and in the maintenance situation. It is the ob-jective of this chapter to start you on a logical path to troubleshooting any type of equipment including microcomputers. To be successful, however, you must add experience to this knowledge.

SELF-EVALUATION QUESTIONS

Keep the following questions in mind and try to answer them when you have completed the chapter:

1. What is the function of a logic analyzer?
2. In what different formats can the data acquired by a logic analyzer be displayed?
3. How can a logic analyzer be triggered?
4. When servicing a microcomputer, what should be checked initially?
5. Is there a common fault of microcomputers?
6. What is signature analysis?

13.1 INTRODUCTION

The high speed of microcomputers combined with the parallel transmission of multiline data makes the servicing of these devices a specialized process. A two- or four-channel real-time oscilloscope becomes an inadequate tool for observing conditions on an 8-, 16-, or 32-bit data bus; they are just not suited for this technology. New equipment and techniques have been developed to answer the new needs created by microcomputers. Logic analyzers are at the top of this list, and their functions are described in this chapter, along with troubleshooting techniques for microcomputers.

13.2 LOGIC ANALYZERS

The analysis of a non-microcomputer-based digital logic is easily carried out with traditional forms of test equipment like the logic probe or oscilloscope. Typically, two or four channels of display on an oscilloscope are enough for equipment debug or servicing if a microprocessor is not involved. When a microprocessor *is* involved, however, traditional test equipment cannot be used effectively because there are far too many signals to look at that are all time related to one another. Since a microcomputer is programmed to follow a specific sequence of activities, and since it operates at its maximum speed, the many digital signals present cannot be captured with a four-channel instrument. An 8-bit microprocessor would have 8 data lines, 16 address lines, and at least 5 lines on the control bus. Other signals including INT, INTA, Sync, and the basic clocks are also present. To observe many of these lines simultaneously requires a high-speed multichannel instrument with a large internal memory for storing data for later display and review. Such an instrument is the ***logic analyzer.*** Figure 13.1 shows a screen shot of the logic analyzer built into Electronics Workbench.

What Do Logic Analyzers Do?

A logic analyzer generates a display on an oscilloscope-type screen that shows the state of its inputs over a period of time. Such a display may be a timing diagram (like that shown in Figure 13.1); hex, binary, or octal format; mnemonic disassembly; or memory map of data taken and stored in the logic analyzer's memory.

The rate at which the logic analyzer takes data is usually determined by a clock, either internal or external. The amount of data taken depends on the available memory. If a logic analyzer has a 4K-bit memory (4096 bits) and uses 16 data channels, each channel could take up to 256 data points ($16 \times 256 = 4096$). These data points would be taken 16 at a time (one per channel). This allows 256 (16-bit) samples. The state of each line at the time the sample is taken is stored in the logic analyzer's memory. Each sample (16 bits wide) may be displayed in timing diagram form, as a number system (binary, octal, or hex), or as a memory map. If it is desired to sample 8 channels instead of 16, there are still 4K bits of memory available. Twice as many samples can be taken ($4096 \div 8 = 512$). If only four channels are needed, another doubling of samples can occur ($4096 \div 4 = 1024$). The sample interval (i.e., the time between samples) is again determined by a clock (internal or external). If an external clock is

(*a*) Icon

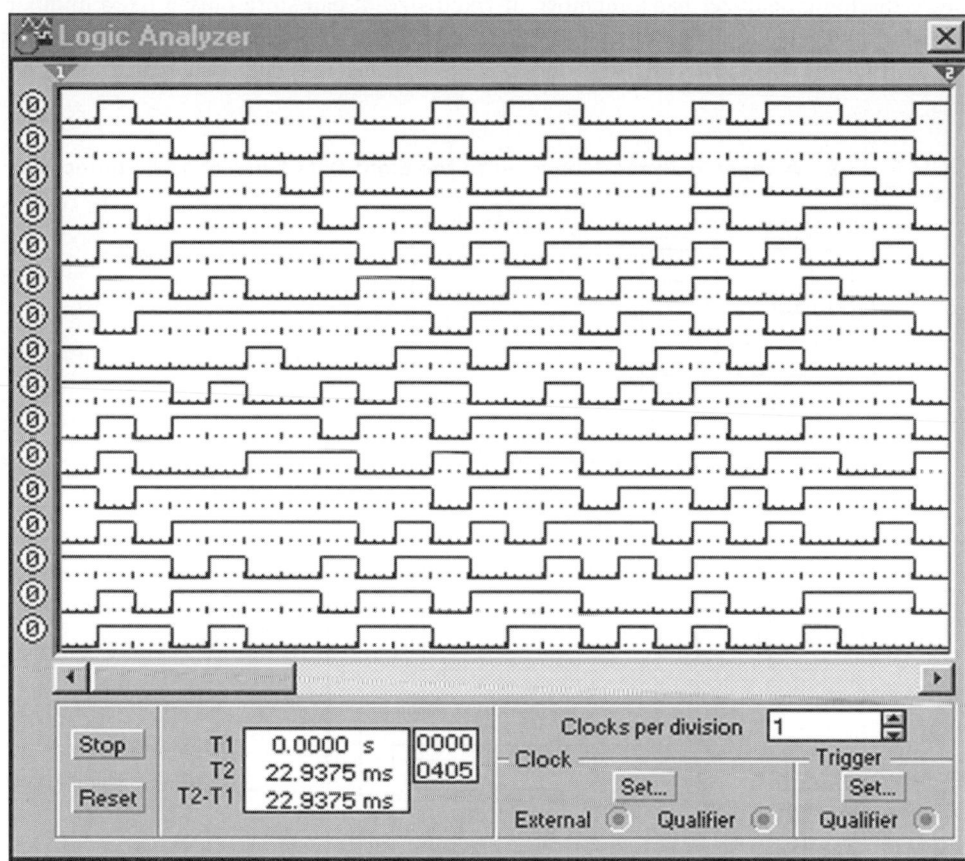

(*b*) Detailed panel

FIGURE 13.1   **Electronics Workbench Logic Analyzer**

used, and if the rate is equal to the rate at which data changes on an 8-bit data bus, the analyzer will "catch" a sequence of a microcomputer's program. This use of a logic analyzer can determine whether a microprocessor under development is functioning correctly. It can be used to isolate sections of bad memory, address or data lines that are hung up (stuck high or low) or to show control line malfunctions. Figure 13.2 shows a logic analyzer connected to the eight data lines of a microprocessor under test.

Starting a Data Sample

Since the logic analyzer has a memory of fixed size, it can store only a fixed number of data points. If data is taken at 1-μsec intervals, the 16-channel, 256-sample memory is filled in 256 μsec. It is important to begin taking this data at a time chosen by

FIGURE 13.2   A Logic Analyzer Connected to the Eight Data Lines of a Microprocessor under Test

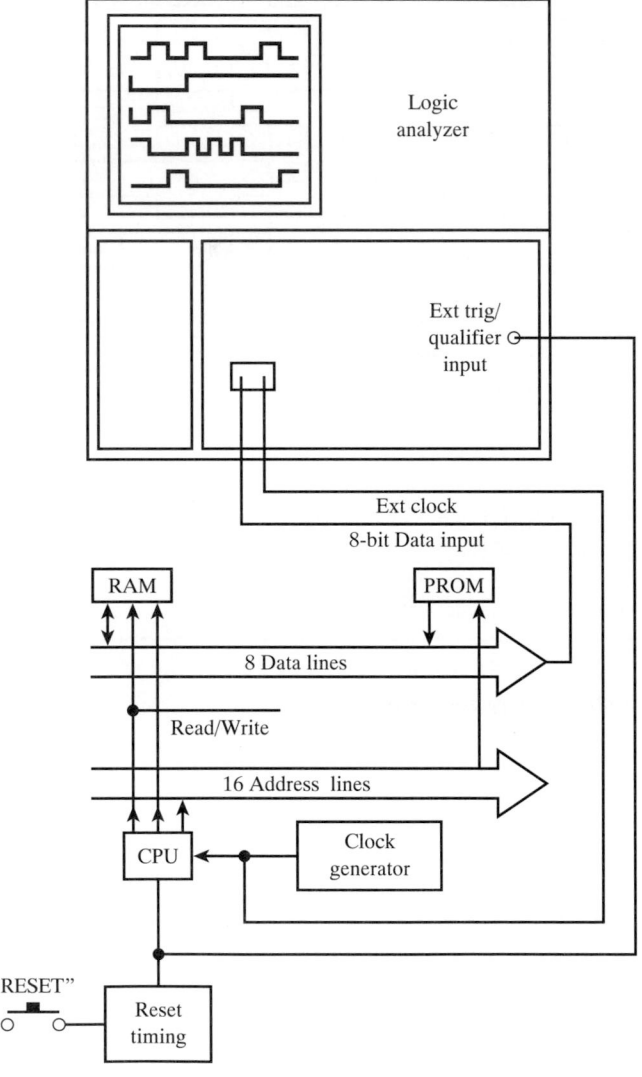

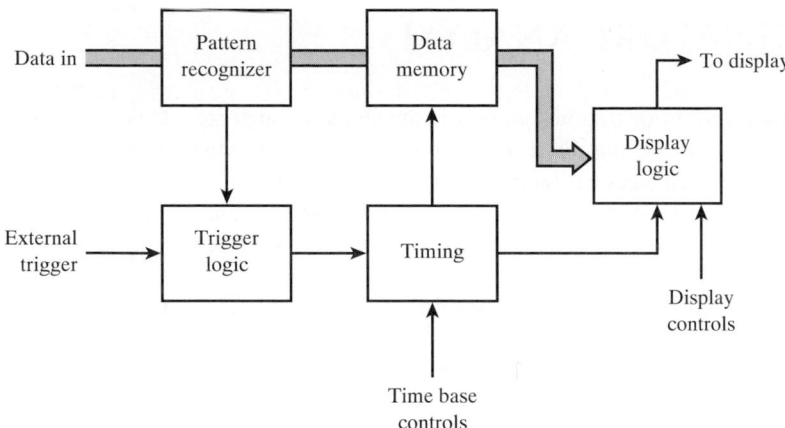

FIGURE 13.3   **Block Diagram of a Typical Logic Analyzer**

the operator. For example, if the data is to be taken after hardware reset of the microprocessor, the entire startup sequence can be observed and debugged. Alternately, it may be desirable to trigger on a particular code or instruction, or perhaps on the 400th occurrence of an instruction. There are many different ways in which to trigger the logic analyzer to begin taking data. In any event, word recognition is a common means of starting the sample process. When a specific data pattern occurs, the analyzer begins taking data. Figure 13.3 shows the circuitry of a typical logic analyzer.

In addition to word recognition, a separate "qualifier" input may be available to add an extra condition that must be met before the analyzer will begin its sample. This is an input that must be high or low (operator selectable) to meet the conditions of triggering.

If a logic analyzer is equipped with a personality module, it can take the acquired data and display it in the mnemonic form of the microprocessor in question. A separate personality module is required for each microprocessor, since different instruction sets are involved. This feature is particularly valuable in troubleshooting both hardware and software problems.

Troubleshooting Using a Microprocessor

In Figure 13.2 a logic analyzer is connected to a malfunctioning microprocessor by way of eight probes to the data bus of the errant microprocessor and the trigger of the analyzer to the reset line of the processor. This causes the analyzer to take data during the initialization sequence. In this way, the very first instruction followed by the CPU can be monitored to determine whether any data lines are stuck or inoperable. Further monitoring of the control or address lines can also help determine problems in high-speed microprocessor systems.

Since there are so many high-speed signals present in the simplest (slowest) microprocessors, it is not practical to do complete troubleshooting using a conventional oscilloscope.

13.3 SIGNATURE ANALYSIS

Another tool of microprocessor troubleshooting is the signature analyzer. This instrument is used to examine waveforms in a piece of errant equipment and reduce them to a number called a signature, which can be referenced on a schematic drawing. Test points within the microcomputer circuitry can be identified by the manufacturer and a signature noted on the print. When troubleshooting becomes necessary, the signature at progressive points in the circuit can be checked and a bad section of the computer located. The signature of a waveform is a hexadecimal number (usually four characters) that is the result of reducing the waveform to an easily readable form.

Looking for the problem in a circuit is a matter of checking the digital signature at key nodes in the circuit against those noted on the drawing. The signature is generated and tested when a special program is run on the computer having the problem. In this way a unique signature can be generated, which in turn becomes a test of the circuitry.

A manufacturer can take certain steps during the design phase to allow for the use of a signature analyzer later, when troubleshooting is required for the equipment. Many original equipment manufacturers (OEMs) are now taking such steps, to increase the value of their equipment to prospective buyers. Signature analysis represents yet another means of troubleshooting the microprocessor-based system.

13.4 TROUBLESHOOTING MICROPROCESSOR-BASED SYSTEMS

The repair or servicing of microprocessor-based equipment requires an understanding of these systems and also a knowledge of basic troubleshooting techniques. Any approach to a problem that does not incorporate a logical technique and a similar understanding is probably doomed to failure. That is not to say that a bit of luck will not shorten the time required to locate and repair a problem. It is important to keep in mind, however, that there is a reason for every system failure. It is best to locate the cause of each failure before replacing a damaged part and putting the unit back into service. To put in a new part before determining why the first one failed is to guarantee that the whole unit will fail again.

Debugging Versus Servicing

There are two types of work that are related, but require different approaches:

1. Failure in units that are new, just built, and never worked before.
2. Failure in units that once worked but failed after long use.

The first type is the worst, and its remedies fall into the "debug" category. That is, we have a product that we desire to make functional. The second type is easier to repair because the unit was working and had already proved both the original design and the construction. Each category has its own types of problem.

1. Never worked
   - Miswired (calls for laborious tracing or wiring).
   - Design flaw—can never work.
   - Chips in the wrong places.

- Other components inserted wrong.
- Electrolytic capacitors inserted backward.
- Diodes inserted backward.
- Transistors inserted incorrectly.
- Defective components.*
- Chips installed backward.
- Power supply problems.
- Ground problems.

2. Worked but failed:

- Component failure due to age or overstress.
- Damaged by accident (e.g., dropped a screwdriver into it, misconnected the power supply, plugged a chip in backward possibly with power on, pulled a card from an edge connector with power on).

A Few Warnings

First of all, do not attempt any extensive repair without a schematic of the unit. It is also a good idea to have the manufacturer's service manual and suggestions.

Second, find out whether you are the first person to look at the unit. If a technically illiterate do-gooder has already touched the project, your job will be ten times as difficult because, for example, you will have to determine what additional damage has been done by nonlogical techniques. (You probably will not be able to talk to the previous "repairman.")

Third, ascertain the symptoms of the problem. How did the unit fail? Talk to any witnesses. Try to get a good description of the unit's problem before touching the job. Symptoms frequently lead the trained troubleshooter to a correct diagnosis. Did the unit fail at power-up? Did the unit fail after extensive use? Was the area in which it was operating excessively hot or cold? Was something spilled into the unit? Was something dropped onto the circuit board? Was a thunderstorm in progress when the unit failed (indicating a powerline surge or damaged I/O lines)?

Before continuing with troubleshooting techniques, we present an important idea. The microprocessor represents a significant advance from traditional circuitry. It is high speed, data oriented, and heavily programmed. Efficient repair of this type of equipment usually requires specialized instruments like logic analyzers.

Troubleshooting Begins with Observation

The repair of micro-based equipment can be a highly organized enterprise. If you develop a system of locating problems, the unit can frequently be repaired quickly. Be forewarned that hit-and-miss techniques are the result of desperation and take a long time to succeed. As you proceed, write down the areas you have worked on, the chips you have changed, and so on.

Obvious Physical Damage

Look the unit over carefully. Check the fuses. If one is blown, you may decide to change it and see if that solves the problem. Occasionally, application of AC power occurs at the peak of the AC wave. This creates a large start-up current, and it is not unusual for a fuse to blow. In that case, the problem was not worth mentioning. However, if the fuse

---

*Do not assume that a new component is operational.

blows again, there is a heavy load on the supply and there is work ahead to find the problem.

Other physical damage includes burned or damaged components or printed circuit lands. Resistors may be discolored, indicating excessive heating, or capacitors may have material oozing out. An integrated circuit may be incinerated. Physical damage is a welcome sight to the repairman. The problem area has been located! When there is no physical damage, you must become a technical Sherlock Holmes. Study the symptoms; observe; deduce. Don't overlook something simple like a broken wire. Wires leading from edge connectors to switches or panels may have been broken as cabinet doors were opened, and so on.

Look for solder bridges, be suspicious of rewiring and component changes, and check the location of all parts. An IC inserted backwards or into the wrong socket can easily complicate the problem.

A simple but often overlooked problem is the corrosion of edge connectors. These components can be effectively cleaned (polished) with a pencil eraser. Never apply solder to gold contacts in an attempt to improve the conductivity of edge connectors.

Power Supply Problems

The power supply is a likely suspect in any problem. Micro-based systems require low voltage ($+5$, $+12$, -12 V, etc.) but at substantial current levels. Systems that have a video display also have a high-voltage section to supply the anode of the CRT. The two basic types of power supply in use today are the standard supply and the switching supply. The switching power supply weighs less because it has a smaller power transformer due to the use of a higher frequency. In the switcher, the AC line is rectified to produce 170 V DC. This drives a multivibrator, which drives a small step-down transformer. The low-voltage alternating current produced is rectified and regulated. Since the multivibrator operates at a frequency greater than the AC line (> 20 kHz), a smaller transformer is required. These supplies are also more compact.

Power supplies contain filter capacitors (electrolytics) that through age may deteriorate, and the usual symptom is internal electrical leakage. The ideal high resistance between the plates decreases, causing heating and then chemical leakage. If a component of the main circuit fails, a heavy load may be placed on the supply. A look at areas of the main circuit that require heavy current may reveal the fault. Other high-current areas include vertical and horizontal sweep circuitry in video displays. Audio sections require more current than the average IC, so it is wise to check these also.

Clock Circuit Problems

All microprocessor-based systems require a precision crystal-controlled clock. Obviously, if the clock fails, the micro cannot run. An oscilloscope should be used to check the clock. If it is a two-phase nonoverlapping clock, the two phases must not overlap.

Other common problems include crystals that quit, clock chips that quit, and related components that also quit.

Intermittent Problems

Locating intermittent problems can be quite annoying. The unit works for a while and then develops some peculiarity, usually a heating problem related to component aging. Frequently such problems can be located with the aid of freeze mist sprayed slowly and judiciously on the circuit board. If a cold IC resolves the problem, you have probably found the culprit. However, if a timing problem exists, changing the temperature of an IC may only correct the difficulty synthetically. Thus timing problems are much more difficult to locate.

Occasionally a cracked land (almost microscopic) or a poorly plated-through hole shows up. These two problems are very difficult to locate, and fortunately they do not occur too often. Careful troubleshooting (described later) and the use of an ohmmeter can help.

Components Change Value with Time

As a piece of equipment ages, its reliability generally decreases. This is due more than anything else to aging of components caused by heating. When each component is originally selected with a sufficient margin from maximum values, the equipment can run nearly forever. However, a product is likely to fail either at the beginning of its career or well down the road of a useful life. A bathtub curve (see Figure 13.4) demonstrates the typical reliability of equipment containing electronic components. This curve is subject to a certain amount of variation and depends on the selection of components during the design and construction phase, but it gives an idea of when to expect failures due to aging.

Components undergo certain changes as they age or "burn in," tending to settle into specific values. Resistors, particularly the composition variety, can open up with use and not show any sign of physical change. Capacitors can open up, although electrolytic (power supply) capacitors tend to short or to draw excessive current.

Resistors that fail are usually in high-current areas: Whenever a component is operated at a level close to its rating, the probability of failure increases greatly. For instance, a 0.5-W resistor operated at 0.5 W has a significantly shorter life than a 0.5-W resistor used at 0.25 W. Capacitors, however, are best used near their rated value. The selection of components in the design phase is important to the life of the product. Often, the price of a component figures into the decision. Such decisions become economic tradeoffs by the original designer. When you replace a component, be sure to use at least the original rating.

Integrated circuits fail for a number of reasons. If an IC is forced to drive currents close to its maximum rating, it will probably fail at some time. ICs manufactured under nonclean conditions contain contaminants that can eat the interconnect wires inside the package. There is nothing you can do about improper production methods, but that is why there are date codes on each IC manufactured. Quality assurance personnel in many companies open a sample of ICs purchased to determine the conditions under which they were

FIGURE 13.4   A Bathtub Curve of Equipment Failure Versus Early/Late Stage of Product Lifetime

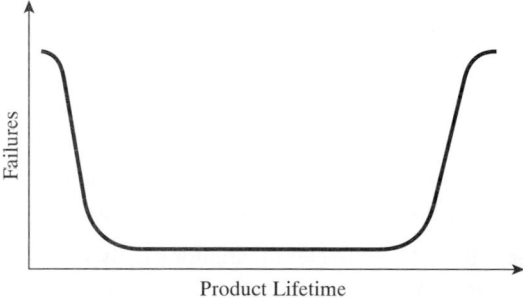

manufactured. You would be quite surprised at the things found inside. Such contaminants are the cause of failures after a year or more. Not the kind of problem that could be tolerated on a Voyager spacecraft passing Pluto!

Static Problems

The high voltages associated with static electricity can have a very adverse effect on electronic circuitry. Be sure to check the environment in which a unit that failed is used. Is there a carpet on the floor? Are people allowed to generate static near the machines by taking off their heavy winter nylon coats in the area?

Although all ICs can suffer from static problems, CMOS devices (most microprocessor chips and support chips) suffer the worst damage. Static that enters from a person's fingers through a keyboard can destroy the CMOS keyboard encoder chip commonly used. Static can find its way into the main circuit board and, for this reason, any CMOS chips should be carefully handled and individuals using the equipment should take proper precautions to ensure that they are "static-free." They are also suspect in any equipment failure.

Note that if a printed circuit board is to be moved, it should be carried in a static bag, placed on conductive foam, or wrapped in aluminum foil. If you receive a board for repair in a paper bag, we suggest you send it back untouched. This is the work of a technical illiterate. Well, perhaps you could look at the problem, but be sure to educate the "carrier."

13.5 A LOGICAL APPROACH TO TROUBLESHOOTING

It is absolutely essential to have available a set of correct schematics for the unit to be repaired. Certain surface troubleshooting is possible without them, but for problems greater than a blown fuse, the drawings are needed. A suggested sequence of events for troubleshooting a nonworking circuit is summarized here. If you choose a hit-and-miss method, you are probably desperate. Learn to use a logical approach to the problem.

1. Have repair equipment on hand to help you.
2. Have the schematics handy for ready reference.
3. Look for obvious physical damage first.
4. Check the power supply, then the microprocessor clock, then all other signals.
5. Compare the symptoms with the circuitry to determine the problem area.
6. Use a logical approach and necessary equipment to find the problem.
7. Don't get angry. Be patient, and use a logical approach. It takes time and experience to become a good troubleshooter in any area. The special nature of the microprocessor makes patience and a logical approach even more important. Study the ideas of this section and work toward proficiency at servicing microcomputer-related products.

Desirable Equipment

We list a few items in order of need. The first two are a minimum and can handle 75% of the problems encountered in troubleshooting microprocessor-based equipment.

1. A digital multimeter (DMM), consisting of a voltmeter to monitor supply voltages, an ohmmeter to check resistances, and an ammeter to monitor supply current.

2. A laboratory oscilloscope (10 MHz or better) to observe digital waveforms and measure pulse widths and periods.

3. A logic probe to catch high-speed glitches (spikes) that may occur, a logic clip to monitor individual ICs, and a logic pen to insert pulses.

4. A logic analyzer to observe the high-speed waveforms on both data and address buses.

Getting Started

The first steps for troubleshooting a microprocessor-based system have already been mentioned: look for obvious damage, and check the power supply. Do the chips actually have power when the supply is loaded down?

Next, check the clock for the microprocessor to be sure it is functioning. Look at areas of high current. In a video circuit, the sweep circuitry requires high currents; in audio sections, the power output circuitry will use larger currents. Is the power supply loaded down? Is it supplying so much current that its output voltage has dropped? If so, you have a shorted IC, a bad capacitor, or damage to another component. Is the CPU running? Check the CPU power and clock signals, and then use the oscilloscope or logic probe to see whether there is bus activity. Are there missing signals? Do RAM/ROM areas have enable signals? Are the chip enables getting used on occasion?

Finding the Problem with Power Applied

If there is no physical damage, you will have to apply power and observe the operation of the circuitry. First, connect the DMM set as an ohmmeter to the power supply lands on the board to determine whether any shorts exist. $5–10 \, \Omega$ for a dozen or more ICs is not unusual. Look for a direct short circuit. When you apply power, don't waste a lot of time after power is on. Get to it. Measure the voltages immediately, and be observant. Look for heating/smoking effects. Use your fingers to find hot spots. Monitor the supply current with the DMM set as an ammeter.

If everything seems normal in a unit that nevertheless will not function correctly, note the symptoms. What is it doing that is not right? Think it through. What areas must be causing a problem? In a microprocessor-based system, static electricity damage may mean a memory problem. Stored-memory (ROM, PROM, EPROM) damage means that the micro does not have a valid program to run and may be running garbage. Check activity at each ROM location. It may be necessary to dump the contents of the ROMs to determine whether they are damaged. You need a development system to check the contents of the ROMs. Is a bit stuck high or low? If so, you've found it!

Doing the Repair

Use caution when replacing burned components. First find out why the component burned up. What else went that may have caused the failure? Never replace components with power applied to the board.

If you must remove a component that is soldered in, use a good solder sucker to remove the part. If you use too much heat in detaching the part, the lands will become damaged and you will have created both a mess and a nightmare. Use the right tool for the job.

When selecting replacement components, the best practice is to use a perfect match. Frequently, the original part can be found. Swapping TTL parts or LS parts can be risky. They are not all the same and may have different characteristics (e.g., 7400 vs. 74LS00).

Desperate? Have a working board from which to swap parts. Change only suspect parts, and write down any swaps so you can go back to at least one working unit.

Power Supply Problems

When a power supply problem is evident, use an ohmmeter to check the rectifiers. A conventional power supply may experience a rectifier failure due to high current, whereas the switcher type may incur such a failure because of high voltage. In either case, rectifiers will fail under heavy load conditions. Are the filter capacitors shorted? Was the supply itself shorted? Check filter capacitors for bulging (or oozing) and overheating. A capacitor should never run warm. If aged, it can develop leakage and become warm due to power losses.

Repairing the Power Supply

When replacing parts in the power supply, use components rated higher than original parts if possible. Apply power to the supply without the microprocessor-based circuitry connected. In other words, run the supply with no load first. If all seems to be in order, connect a resistive load and draw-rated current for a while before reconnecting the main circuitry to the supply. The main circuitry should be checked with an ohmmeter first anyway to rule out any short-circuit problems.

If the supply shows no damage but is not producing one or more voltages as required, check the raw DC source, then the individual regulators. You are bound to find a problem.

A Final Note

The repair of any electronic circuit requires a certain amount of expertise, and ability improves with practice. Record problems for ready reference in the future. Keep a notebook or card catalog of all repair jobs; the information will be useful when similar problems come along.

SUMMARY

In this chapter we presented the tools of troubleshooting microprocessor-based equipment and also discussed troubleshooting in general from a pragmatic viewpoint. Study of these techniques and types of equipment will give you a good start when confronted with an errant microcomputer. Real expertise can come only from experience, and that clearly means that you need to spend time using the logic analyzer to really appreciate its power. The more experience you have with this tool, the more valuable it will become to you.

STUDY QUESTIONS

General

1. A logic analyzer has a 4096-bit memory. If eight channels of data are used, how many data samples can be stored per channel?

2. If the logic analyzer of Question 1 uses 16 channels, how many data samples can be stored per channel?

3. If the sample interval is 100 ns, how long can a 4096-bit, 16-channel analyzer take data?

4. What is the value of a logic analyzer?

5. List some simple things to observe as you begin to look for problems in an errant computer. Where should you begin? Be sure to discuss the basics that every electronic/digital system requires.

6. Discuss the use of a signature analyzer in the repair of a troublesome microprocessor-based system.

7. Get manufacturers' literature on a logic analyzer and review it. What features does it contain?

8. Suppose that in the construction of a microprocessor-based system a pair of data lines was accidentally shorted together, perhaps during soldering or wire wrapping. With so many wires to look at, finding such a problem is no small task. How would this problem show up on a logic analyzer? How would the logic analyzer show a data or address wire that was stuck either high or low? How would you see this?

9. If a pair of wires was shorted together on either a data bus or an address bus, why would an ordinary oscilloscope be difficult to use in finding the problem? Why would a logic analyzer be a much better instrument?

10. You can learn a great deal about a microprocessor system if you already have an idea of what "normal" signals look like on address, data, and control lines. Take an ordinary oscilloscope, ground it correctly to the microcomputer, and observe the signals on all the lines on an operating microprocessor chip. Let the microprocessor run a simple program and carefully look at these typical signals. The shapes and amplitudes may surprise you; sketch them for future reference. Also note the relative times involved.

11. Observe the effect of using the RESET line in a microcomputer. What happens to all the signals when you use the RESET button or line? Do they all go high, low, or perhaps float? Consult the manual on the microprocessor involved to answer this question and then observe some of them in a working microprocessor.

12. Study the schematics of several microcomputers. Identify the various parts and determine their relationship. How are the various memory chips decoded to specific memory areas? How are the I/O areas established? Are the address and data buses buffered? How are status signals obtained?

13. Effective troubleshooting is based on general knowledge, experience, and continued practice. What experiences have you had in repairing electronic and microprocessor-based equipment? What can you do to get more experience in this area? (One answer is to study and monitor working systems.)

14. A simple but major problem in getting a computer working is the basic communication between the terminal and the computer. If RS232C is used, use an oscilloscope to observe the transmit and receive lines of the system and also the control lines (DTR, RTS, CTS, etc). Be sure to note normal levels and the data rates involved. Where are the control signals supposed to be in a working system? (Is DTR high or low in normal operation? RTS? CTS?)

Electronics Workbench

15. Search the web for information on logic analyzers and troubleshooting equipment. How many companies and products do you find?

16. Use the logic analyzer in Electronics Workbench to view the waveforms of the circuit file **TESTCKT** in Figure 13.5.

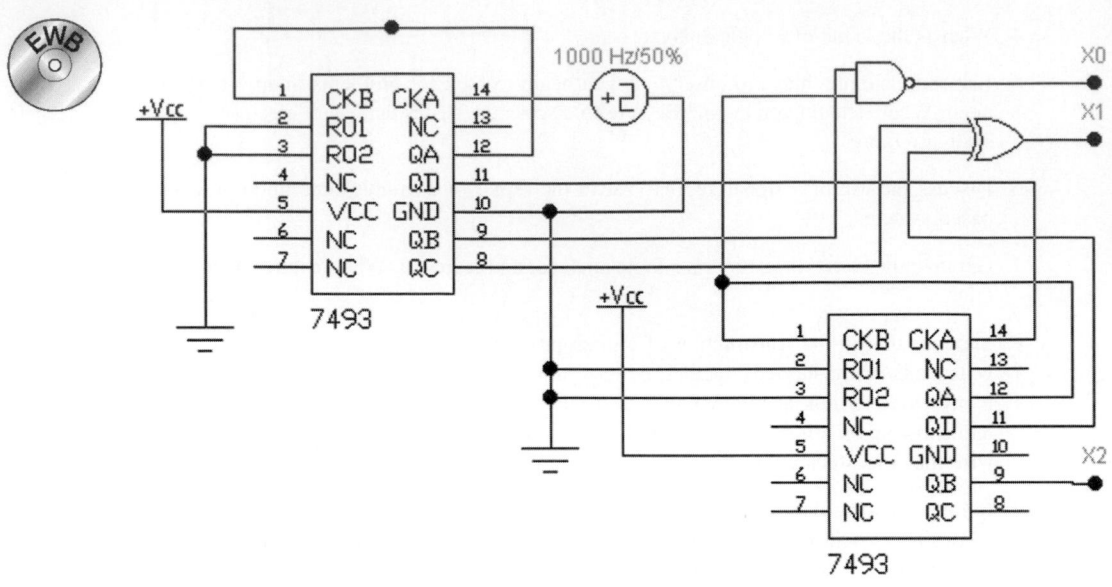

FIGURE 13.5 For Question 13.16

ORGANIZATION OF COMPUTERS

INSTRUCTIONAL OBJECTIVES

When finished with this chapter you should be able to:

1. Identify:
   (a) The main parts of a computer system.
   (b) The different buses in a computer system.
2. Explain:
   (a) Interrupts.
   (b) Interfacing techniques.
   (c) How an instruction is fetched and decoded.

3. Describe where parallel and serial I/O are used in a computer.
4. Explain how the intended application guides the design of a microcomputer.

SELF-EVALUATION QUESTIONS

Keep the following questions in mind and try to answer them when you have completed the chapter:

1. What are the basic parts of a computer and what are their functions?
2. What is a bidirectional bus?

3. How are different instructions represented in binary?
4. What is an interrupt?
5. What are the basic components of the personal computer motherboard?

14.1   INTRODUCTION

The use of computers has transformed the way in which we live. Although we all come in contact with these machines many times a day, we often fail to realize it when a computer is at work performing some useful function for us. Computers were once thought of as large, expensive, and complicated machines. We regarded these machines as being the heart of corporations, governments, or schools and universities. However, large, *general-purpose computers* are no longer the only kind. Now the word "computer" is also applied to smaller machines that are less complicated, less expensive, and more dedicated to a specific task. It is this type of computer, the "micro," that this chapter is really all about. Microcomputers tend to be dedicated to a specific application or purpose and are produced in far greater quantities than are the large, general-purpose machines. They touch our everyday lives in ways we seldom suspect, for microcomputers serve in *control*-type applications. They are designed into different types of equipment, making the equipment perform its function. The primary effect of the microcomputer has been to replace circuit design using standard TTL logic or CMOS logic with a few microcomputer LSI circuits. In this way, a single microprocessor can be programmed to control thousands of applications. Using the same components for many uses has eliminated the costly design of a different TTL circuit for every application.

The cost of a *dedicated microcomputer* has become so small that these devices are finding routine use in appliances and equipment of many types. The first micro was used in the hand-held calculator. It was a 4-bit device intended to perform 4-bit (BCD) arithmetic. At the time (1971) this micro was not thought of as a computer. However, it started the revolution, and when the 8-bit micro (Intel 8008) was created in 1972, it shocked the industry. Table 14.1 lists many different microprocessors, their relative size, and their speed.

TABLE 14.1   Different Microprocessors

PROCESSOR	BITS	SPEED[a]	MAXIMUM MEMORY (BYTES)
4004	4	500 kHz	8K
8008	8	500 kHz	16K
8080A	8	2 MHz	65,536
8085	8	4 MHz	65,536
6800	8	1 MHz	65,536
8086	16	5 MHz	2^{20}
68000	16	12.5 MHz	2^{24}
80386-Pentium	32	>400 MHz[b]	2^{32}

[a]Speed is subject to change by manufacturers.
[b]Current speed of the Pentium II at publication.

14.2 THE OPERATION OF A COMPUTER

All computers have a common basis and as such have the same structure. There is a real difference between a *computer* and a *processor:* the computer is the whole machine; the processor is a small but very important part. Every computer contains five important elements:

1. The ***Processor*** or ***Central Processing Unit (CPU):*** To add, subtract, multiply, divide, shift, AND, OR, XOR, interpret instructions, and make decisions. (In a microcomputer this is sometimes called the MPU.) A floating-point unit (FPU) may be included with the CPU, or externally interfaced. The FPU is actually a dedicated microprocessor designed to perform complex math operations at high speeds.

2. The ***clock:*** To order functions in a desired sequence; the timing section.

3. *The memory section:* To store the program or programs, and data.

4 and 5. *The input and output (I/O) sections:* To give the computer a way of communicating with the outside world.

Figure 14.1 shows the five components of any computer, large or small. The five components are connected by wires or printed circuit paths that are kept as short as possible. Since electrical signals travel at a finite velocity, distance tends to slow things down. The velocity of a pulse is roughly two-thirds of the speed of light or $\frac{2}{3}$ of $(300 \times 10^6$ m/sec). This works out to 1.5 ns/ft. In modern computers, distances down to a few thousandths of an inch remain our enemy! The designer must work carefully to reduce any unnecessary length in signal paths between the elements of a computer. In a large computer, memory is placed as close to the processor as is physically possible. The I/O section, which may involve slow data terminals, can be placed further away without noticeable effect. The program (or sequence of instructions) executed by the computer is stored in the memory section. The CPU reads the instructions one at a time from the memory. Every CPU has a special-purpose register that is used to point to the current instruction address in memory. If our program is stored in memory, we need a way to point to each location that contains

FIGURE 14.1 Computer Block Diagram

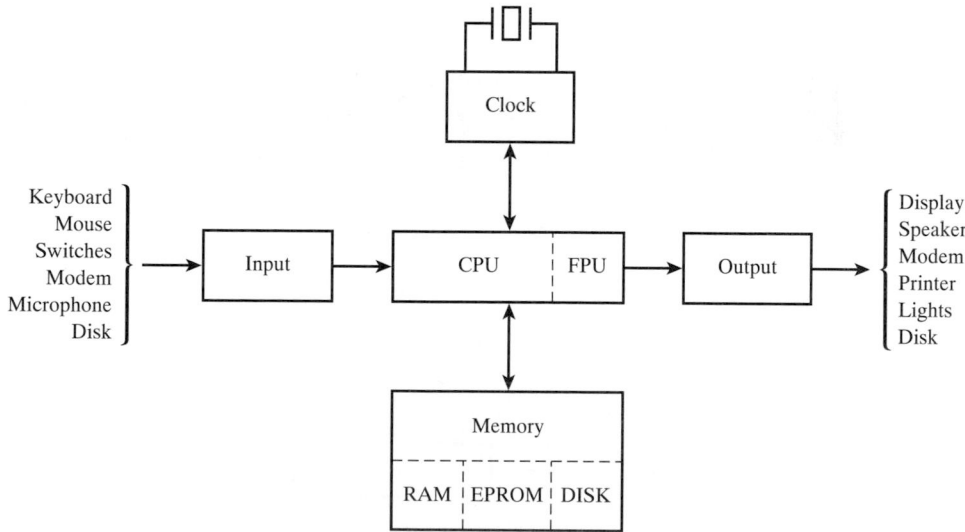

an instruction. This special-purpose register is called either the *program counter* or the *instruction pointer.* This register is a part of the CPU and drives the address bus.

A computer follows a very rigid sequence of operations. These are as follows:

1. The CPU places an address on the bus and reads an instruction from memory on the data bus. This is called ***instruction fetch.***

2. The CPU takes the instruction from the data bus (from memory) and *decodes* it. As this is done, the CPU finds out whether more information is needed to execute that particular instruction. If so, it gets the additional information from memory.

3. The CPU *executes* the instruction.

Thus the basic sequence is: FETCH an instruction, DECODE it to determine its meaning, and EXECUTE it. Let us look at how hardware in a simple microprocessor might be used in the fetch and decode stages.

Instruction Fetch

The CPU loads its program counter (PC or IP) with the address of the instruction in memory. This address is transferred to the memory address register (MAR). The address appears on the address bus and memory picks it up. The address is decoded and the data from the desired memory location is placed on the data bus. The CPU picks up the data from the data bus and stores it in the memory data register (MDR). It is then sent to the instruction register (IR) for decoding. The program counter is incremented by one to point to the next memory location in preparation for the next item in the program. This is illustrated in Figure 14.2.

FIGURE 14.2 Instruction Fetch Hardware

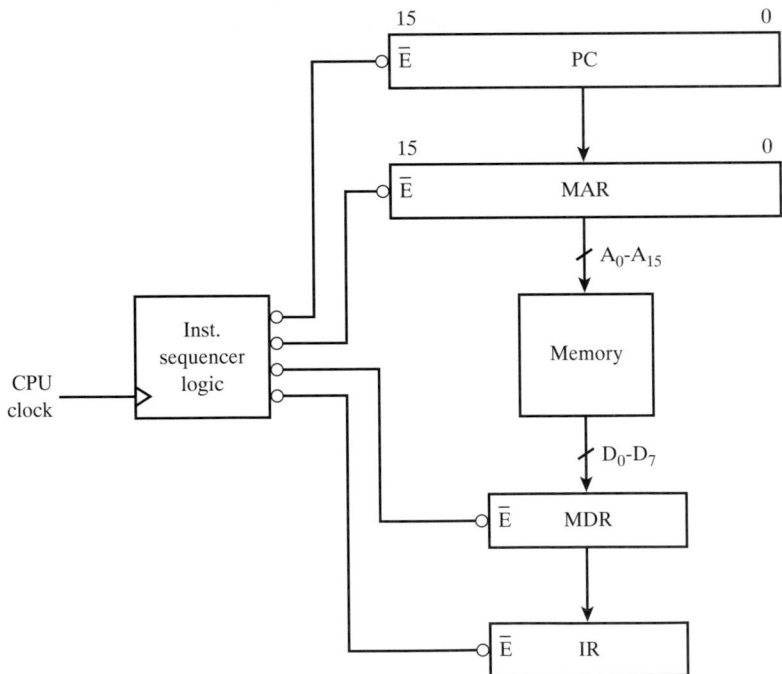

Decode Cycle

The CPU decodes the instruction and determines whether more information is required to perform the instruction. Figure 14.3 shows how the instruction in the instruction register is decoded. Bit patterns assigned by the microprocessor designers are used to control several decoders. If more information is required for the instruction, it will be in the next memory location, which the program counter (PC or IP) is now pointing to. The CPU

FIGURE 14.3 **Instruction Decoding**

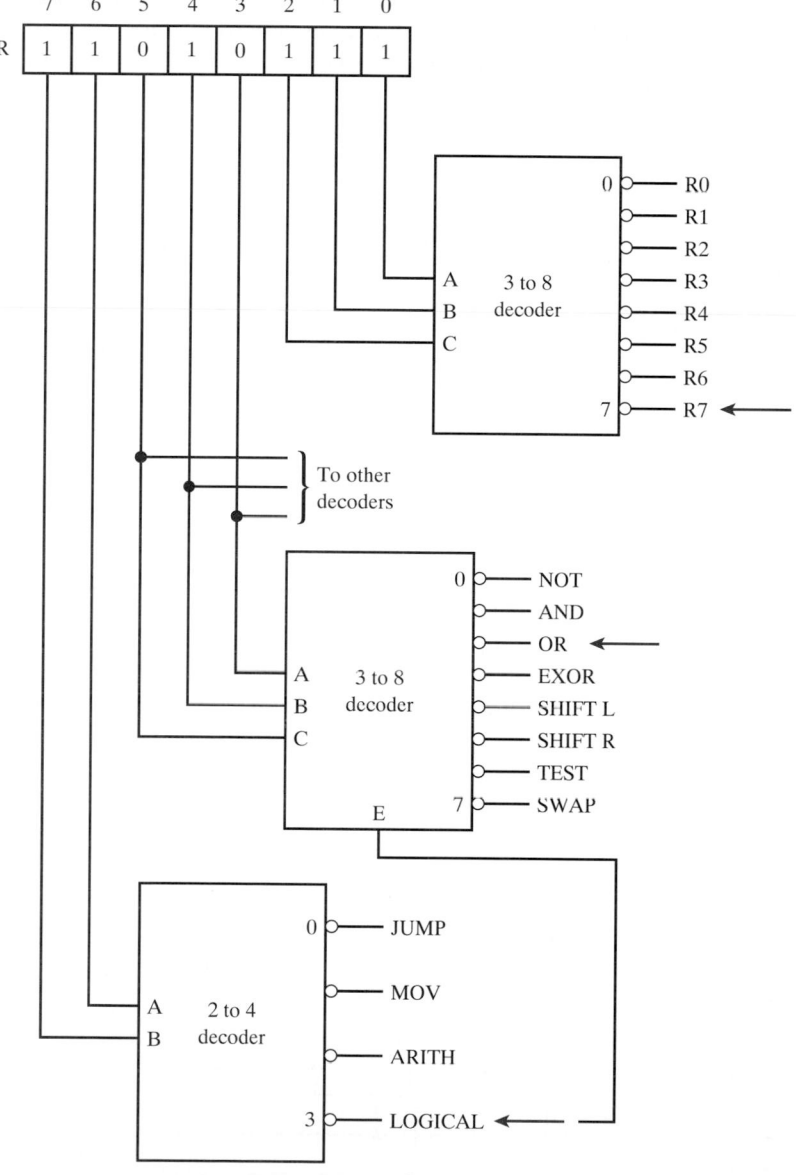

may have to get one, two, or more additional pieces of information from memory to complete the execute cycle. In any event, the information is sent to the CPU from memory on the data bus when an address is provided by the program counter.

Execute Cycle

In the execute cycle, hardware for the decoded instruction is activated and the CPU executes the instruction. Each instruction has a unique binary pattern. The instruction shown in Figure 14.3 involves the logical OR of register R7.

The Interrupt

When a program is running on a computer, it is performing some task. It may be doing calculations or routinely controlling or operating a piece of equipment — that is, the computer is operating normally, which it usually does over long periods of time.

Infrequently, however, events occur that require a computer's immediate attention. Most computers are able to be *interrupted* from the program they are currently running. When a program or computer is interrupted, the computer is usually sent to a program stored in memory at a special location. This program, called an interrupt routine, handles the condition that causes the interrupt.

As an example, suppose a computer is routinely controlling the temperature and humidity in a large building. The computer also routinely monitors lighting in an attempt to save electrical energy (e.g., by turning off lights in unused areas). These are the main jobs of the computer and its program.

However, all the building doors and windows are wired for security purposes. In addition, heat and smoke detectors are placed around the building. If an emergency (fire, burglary) occurs, the computer is interrupted from its main program. At the time of the interrupt, the main program becomes unimportant. The interrupt causes the computer to run a special program that dials a telephone line to alert police and fire departments. The computer can also operate a tape-recorded message or, better yet, operate a commonly available speech synthesizer to exclaim, "Help!"

Computer systems that have many interrupts often *prioritize* them in such a way that, if two or more interrupts arrive at the same time, the one with the higher priority is selected. Figure 14.4 shows one way to implement a simple eight-level prioritized interrupt. The priority encoder outputs the binary value of the highest active-low input. The $\overline{EN}$ output goes low to indicate at least one input is active. The output of the priority encoder is latched so the microprocessor will not miss it during a busy execution cycle.

FIGURE 14.4 Event Interrupt Priority Controller

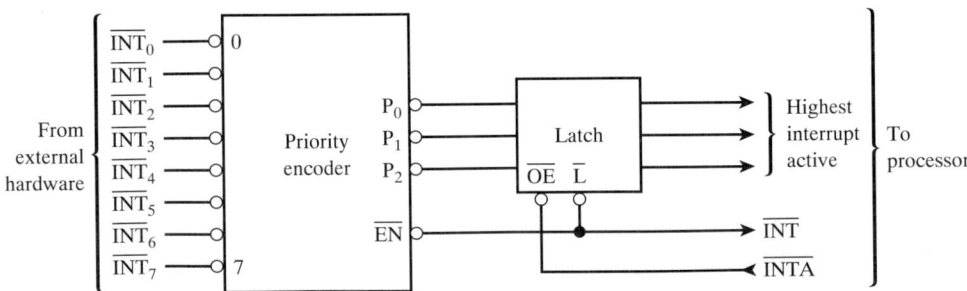

14.3 THE CENTRAL PROCESSING UNIT (CPU)

The CPU (or MPU) is the brain of the computer. It is here that instructions are interpreted and decisions are made. Arithmetic operations, logical operations, data moves, shifting, and compares are performed by the CPU. The CPU obtains instructions one at a time from a sequence of instructions stored in memory. The instruction is decoded to determine its meaning. If more information is required, the CPU gets it from memory. The CPU then performs the task required of it. The instructions of all computers may include the following categories:

1. Move data from one place to another.
2. Input or output.
3. Arithmetic operation.
4. Logical operation.
5. Jump to an address.
6. Compare data.

The CPU may contain an internal instruction *pipeline,* a set of hardware stages with latches in between. The pipeline operates in a manner similar to that of an assembly line. Several instructions may exist in the pipeline at the same time, each in a different stage of execution. The CPU may also contain a *floating-point unit,* dedicated hardware designed to perform high speed math calculations.

14.4 THE MEMORY

There are many types of memory used by computers. All computers have some RAM and most have some ROM. Some computers have slower memories in the form of disk drives and even slower memories like magnetic tape. For the time being, we will concentrate on RAM and ROM, both solid state memories.

Two types of information are stored in the memory (RAM or ROM): the program itself and data required by the program. The program can consist of many main programs operating separately or main programs and subroutines that are used by the main programs. In reality, only one program can operate at a time, since there is only one CPU to perform the tasks each instruction requires. The data used by a program can take on many forms. It could be ASCII data or numerical data, possibly BCD information.

When a computer is first turned on, the RAM type of memory comes up in a random way. A program must be loaded in RAM before a computer can run it. Either it is entered manually or it is loaded from a slower type of memory like a disk storage drive.

Many programmers place programs in a ROM that does not lose information when power is turned off. When the computer is turned on, a program exists in ROM that can be immediately run, saving work for the operator of the computer.

The power of a computer is usually determined by two parameters related to memory:

1. The number of memory locations available.
2. The number of bits a location can store.

As they say, the bigger the better! The data stored in memory can be passed back and forth between the CPU and memory. Data can be read from memory by the CPU, or the

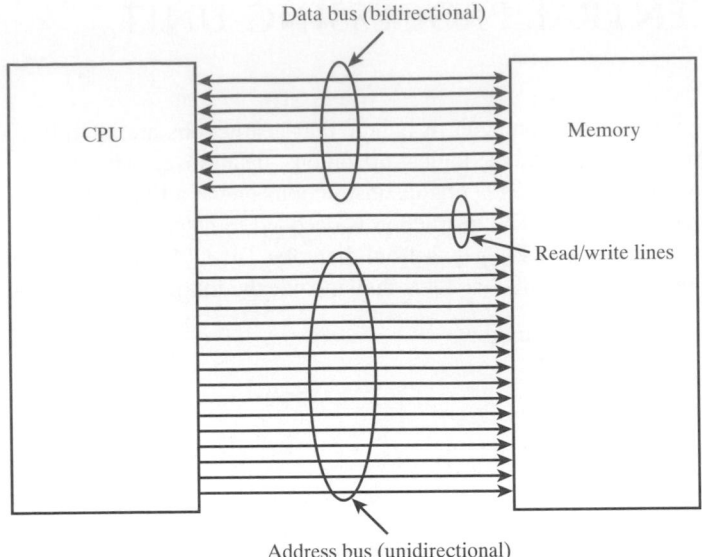

FIGURE 14.5 **Address Bus and Data Bus**

CPU can send data to memory. This is accomplished in the following way. The CPU must first decide which memory address is to be used. Then, if a write into memory is involved, the CPU sends the data to the desired location. If a read from memory is involved, the memory sends the data from the desired memory location to the CPU. The CPU and the memory communicate by two sets of signal paths. These are the ***data bus*** and the ***address bus.*** A "bus" is a group of wires having a common purpose.

The address wires or lines run in one direction (unidirectional). The CPU sends an address down the bus to the memory. The data lines are bidirectional between the CPU and the memory, since data can be passed in either direction. Figure 14.5 shows the address bus and data bus between CPU and memory.

The memory must be able to decode the address information to determine which memory location is involved. Two additional lines, the "read" and "write" lines, are shown between the CPU and memory. These are used to determine which way data flows on the data bus (CPU to memory or memory to CPU). The bus structure allows high-speed *parallel* communication between the CPU and memory.

14.5 THE INPUT/OUTPUT SECTIONS

Input or output from the processor to the outside world can be done in several ways. I/O is usually accomplished with serial or parallel type information. Data terminals usually are serial, and the I/O section may contain a UART to provide serial data. Switches, indicator lamps, and other devices to be controlled frequently require parallel information. This also is available in the I/O section. The components of the I/O section usually are also connected to the address and data buses, and two control signals are used. These are I/O Read and I/O Write. Figure 14.6 shows how simple 8-bit input and output ports can be made.

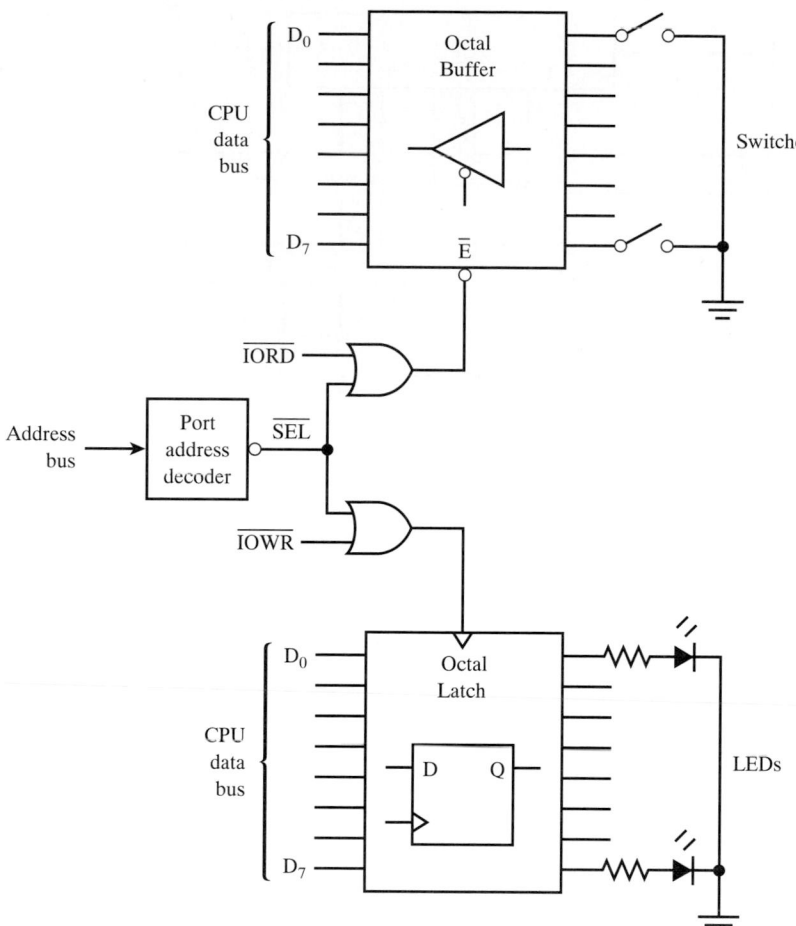

FIGURE 14.6 Simple I/O Port

The System Bus

Three buses exist in a computer system—the address bus, the data bus, and the control bus (see Figure 14.7). The **control bus** includes the memory read and write lines, the I/O read and write lines, and a reset line. Other control-type signals may be included. Figure 14.8 shows how some typical control signals are decoded. Some microprocessors have their own set of bus control chips that handle all traffic on the buses. The three buses together are referred to as the system bus, and most components of a computer system communicate with one another over the system bus. In addition, many computers have a **status bus** that indicates the particular operation the computer is doing at a given moment. This information may be of interest to an operator or someone trying to debug a program.

As you know, many wires are needed in a computer to interconnect the parts. Lengths must be kept short to reduce the time required to communicate. In most computer systems, these signal paths or buses must be carefully designed so that they can operate at high frequencies. Another problem is the cross-talk between adjacent signal paths; that is, the induction by a signal on one wire of a similar signal in nearby wires. Also, when many devices share a common bus, the circuit doing the driving must be capable of driving

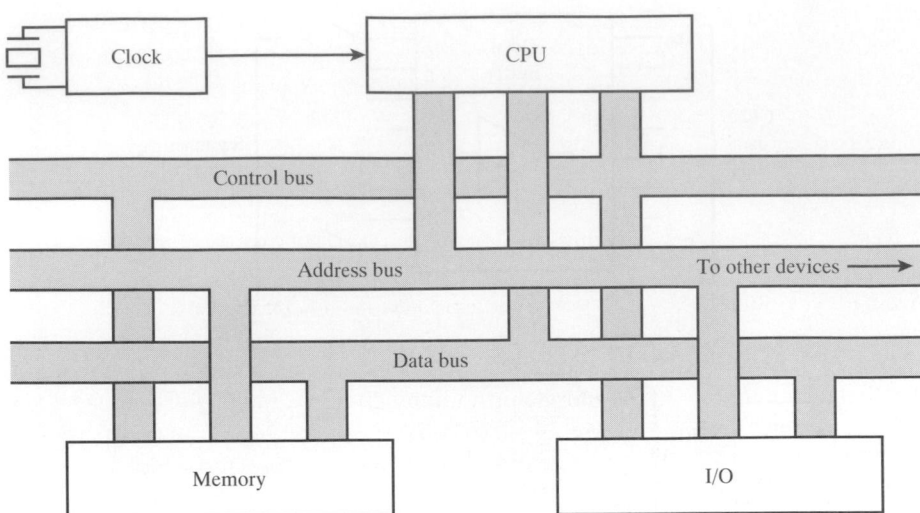

FIGURE 14.7   A Computer and Its Bus Structure

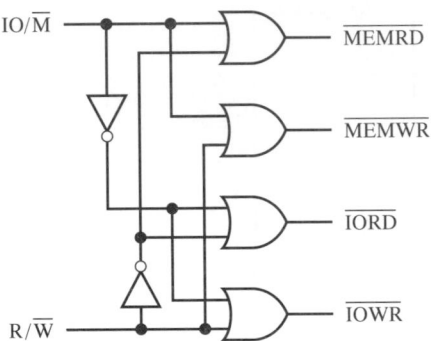

FIGURE 14.8   Decoding Control Signals

many other circuits. Hence, bus drivers may be needed to supply the necessary drive current. Since the CPU is prone to issuing address information, it may first drive a bus driver to produce the necessary current to drive large memory and I/O sections.

One final word: the address bus is generally unidirectional, the data bus is bidirectional, and the components of the control bus may be unidirectional or bidirectional.

14.6   THE PERSONAL COMPUTER

All of the material in this chapter, up to this point, has dealt with general microprocessor-based systems. In this section, we will see how a specific microprocessor-based system, the PC, uses many of the hardware features already described. Although the PC has been around for many years and has evolved into a powerful machine containing very advanced technology, such as CD-ROM drives, it began as a much simpler machine constructed around the 8/16-bit Intel 8088 microprocessor. The 8088 came out in the late 1970s and offered a higher level of computing power than the 8-bit processors of the time. When IBM chose the 8088 for use in its new PC, it paved the way for worldwide acceptance of the new processor. Many companies began copying the architecture of the PC and offered their own compatible 8088-based computer systems. Thus began the PC market.

One reason the PC market grew as fast as it did was due to the usefulness of the features the PC offered. The initial PC contained a keyboard for entering commands and data, a monochrome video display for viewing text and simple graphics, one or two floppy disk drives for storing information and running programs, and a memory large enough for many useful applications. It also came equipped with a software program called DOS, for Disk Operating System, which made it possible to access files on the disk drives and run programs with the use of simple commands.

Most of the electronics within the PC were contained on a single printed circuit board called the motherboard. Memory chips, timing circuitry, interrupt logic, the 8088 microprocessor, and other hardware all resided on the motherboard. Included were a number of expansion slots, plastic connectors with metal fingers into which other circuit boards could be plugged. The PC's system bus was wired to each expansion slot, so any card plugged into an expansion slot had the power of the entire machine available to it. Expansion cards were used to add new features to the basic machine, such as a color video display, a hard disk, or additional memory. Today, hundreds of different expansion cards are available. A small sample of them shows the wide variety of hardware applications:

MODEM/FAX

LAN controller

Data acquisition

Sound/speech synthesis

High-resolution color graphics

Image processing

CD-ROM drive

Hand-held and flatbed scanner

Serial/parallel I/O

Clearly, with the right number and type of expansion cards, the PC can be configured to do just about anything. For our purposes, we will concentrate on the hardware that comes with a base machine, with a few add-ons, namely the hard disk and color-display cards.

Let us now take a detailed look at the inside of the personal computer. Figure 14.9 is the block diagram for a typical PC motherboard. As shown, all communication is through the system bus. The microprocessor may be an 8088 (as found on the original PC), or one of the newer 32-bit processors from Intel, the 80386™, 80486™, or Pentium. A nice feature of the advanced Intel microprocessors is that they all execute programs written for the 8088. So, even if your machine is new, all of your old software will run on it.

If a motherboard contains an 80x86 microprocessor, there is usually a socket provided for an additional chip, the 80x87 Floating-Point Coprocessor. This device is capable of performing mathematical calculations much faster than the processor, and is designed to work parallel with the processor. Motherboards based on the 80486 and Pentium do not contain this socket because the coprocessor is built into the processor itself.

For high-speed data transfers involving memory, the motherboard contains an 8237™ DMA Controller. This device can be easily programmed to move large chunks of data without assistance from the processor.

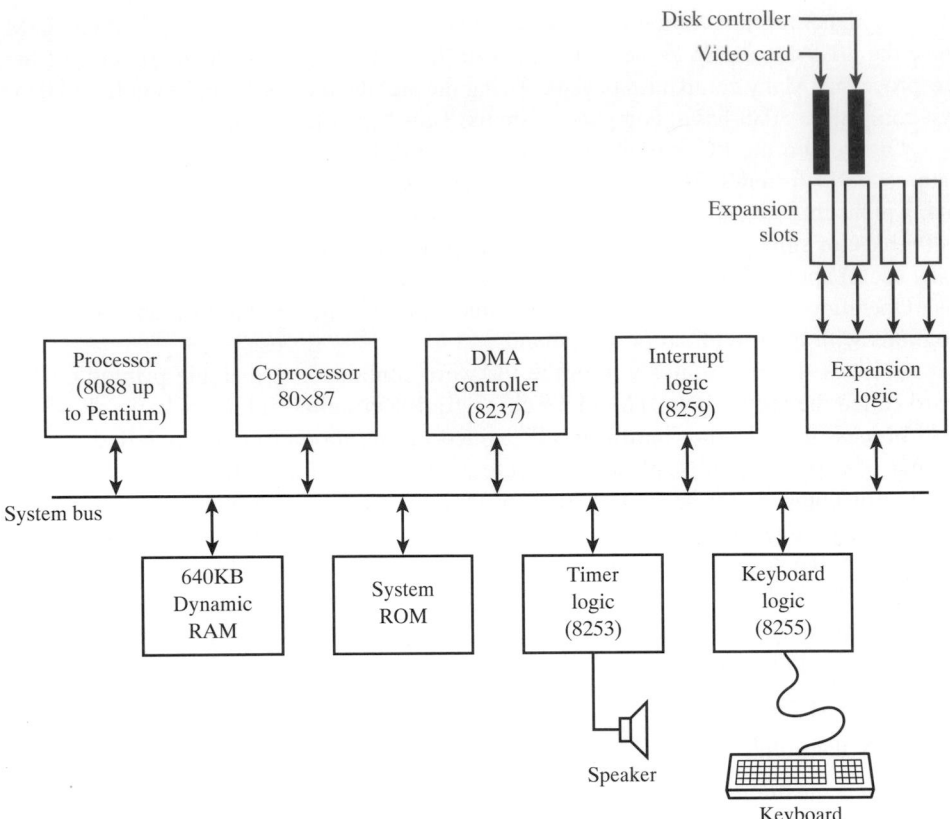

FIGURE 14.9   Block Diagram of a Typical PC Motherboard

The PC has many features that require the use of the interrupt system. An 8259™ Programmable Interrupt Controller is included to handle the interrupts generated by the PC's time-of-day clock, keyboard, serial and parallel I/O devices, and disk drives.

Today, it is rare to find a PC that does not contain at least 640KB of RAM. All application programs written for the many versions of DOS use RAM during their execution. The memory is contained in a handful of dynamic RAM chips. Special timing circuitry is used to ensure that each memory chip gets refreshed and accessed properly.

The motherboard also contains a small amount of ROM as well. This ROM is referred to as system ROM and is used to control the PC when it is turned on. The system ROM is responsible for checking and initializing all peripherals and devices on the motherboard, and for starting up the disk drive to load the operating system.

As mentioned before, the PC maintains a time-of-day clock. This clock is a combination of software and hardware. A special timing device, the 8253™ Programmable Interval Timer, is used to generate timing pulses at regular intervals. These pulses interact with the interrupt logic and DOS to simulate the passage of time. The 8253 also controls the PC's speaker. With proper programming, it is possible to make the speaker beep and generate other sounds.

A parallel I/O device, the 8255 Programmable Peripheral Interface, is used to monitor and read the PC's keyboard and motherboard option switches.

Finally, a group of tri-state buffers are used to drive the system bus signals on the expansion slots. This makes it possible for circuitry on an expansion card to access every device on the motherboard. In the next section we will see a detailed example of a microprocessor-based system.

14.7 A PRACTICAL APPLICATION

In this section we will take a detailed look at how the 8088 microprocessor is interfaced to all the components required to make a complete single-board computer. All of the concepts and techniques we have covered in this textbook should put us in a good position to follow the design of the single-board computer from a digital perspective. Interested readers may even wish to construct the computer system. Read on, and see how your digital expertise has grown.

The Timing Section

The timing section has the main responsibility of providing the CPU with a nicely functioning stable clock. Any type of digital oscillator will work in many cases. It is then necessary to decide on a frequency for the oscillator. Many times this frequency is the operating frequency of the CPU being used. Microprocessors are commonly available with different clock speeds.

One important factor limiting the clock speed is the speed of the memories being used in the system. A 12-MHz CPU might require RAMs or EPROMs with access times less than 100 ns! In our design, we will use a 10-MHz crystal, together with the 8284 clock generator. This is fast enough to provide very quick execution of programs, while at the same time allowing for use of less expensive RAMs with longer access times (200 ns).

The circuit of Figure 14.10 shows a 10-MHz crystal connected to the 8284 clock generator. The output of the 8284 drives two buffers so that any external loading on the CLK signal will not affect its operation. One of the outputs, CPU-CLK, is the master CPU clock signal. Because many other circuits might also require the use of this master clock, we make the CLK signal available too. The CPU therefore gets its own clock signal. It is desirable to separate the clock in this fashion to aid in any digital troubleshooting that may need to be done. By making multiple clocks available, it is easier to trace the cause of a missing clock, should that problem occur.

In addition to the clock, the CPU must be provided with a reset pulse upon application of power. It is very important to properly reset the CPU at power-up to ensure that it begins executing its main program correctly. The 8284 has built-in reset circuitry that uses an external R-C network to generate the power-on reset pulse. The values shown in Figure 14.10 (100K ohms and 10 μF) produce a reset pulse of about 1 ms in duration, long enough to satisfy the hardware reset requirement of the CPU and other system devices.

The CPU Section

Once we have a working timing section, we must design the CPU portion of our system. During the design of this section we answer our question about interrupts, and pose a few more important questions. For instance, do we need to buffer the address and data lines?

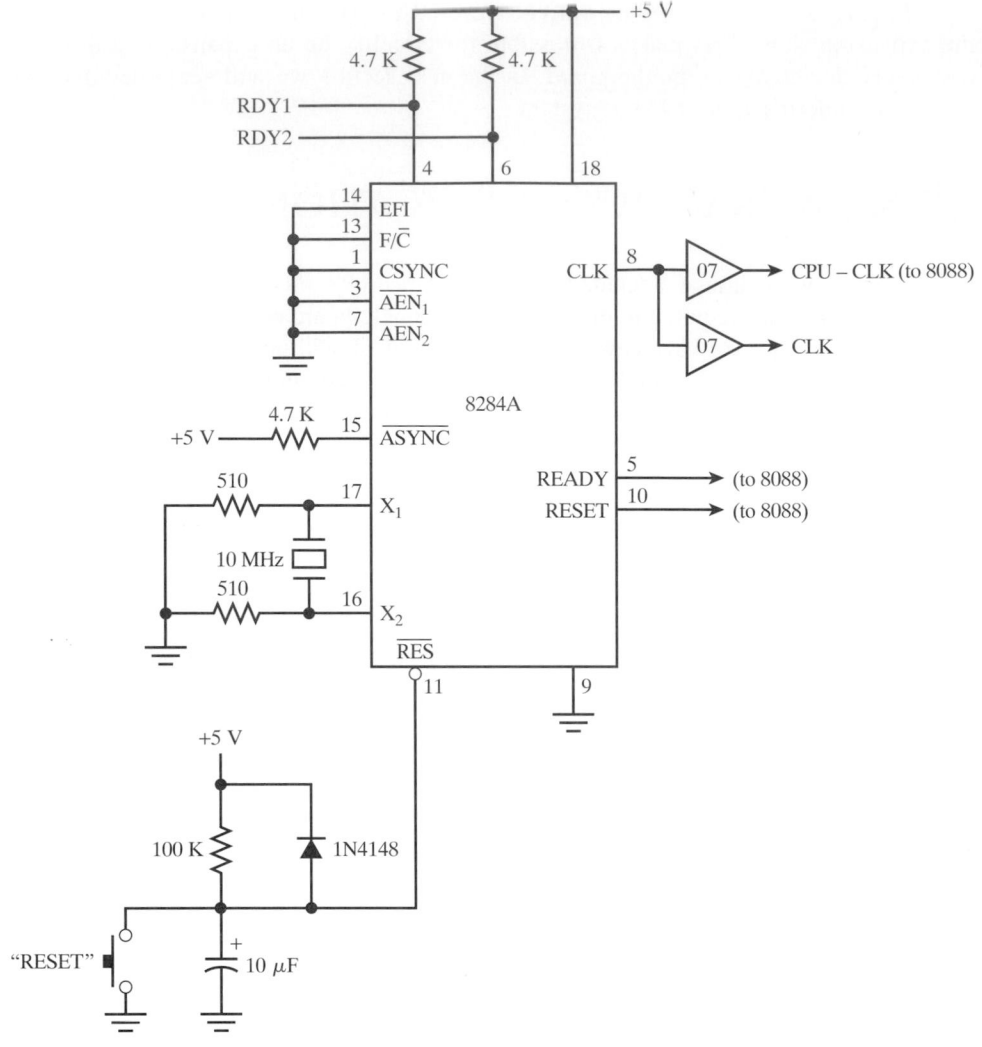

FIGURE 14.10   Clock Generator for Single-Board Computer

Do we want to give bus-granting capability to an external device? Should the system operate in maxmode or minmode? Take a good look at Figure 14.11 before continuing with the reading.

The figure details the connections we must make to the CPU for it to function in our minimal system. On the right side of the CPU we see the data and address lines. These signals are used in both the memory and I/O sections. The CPU is capable of driving only a few devices by itself (one RAM and one EPROM safely). Because the system we are designing will contain RAM, EPROM, and serial *and* parallel I/O, it is best to buffer the address and data lines. As Figure 14.11 shows, a 74LS244 octal buffer is used to drive address lines A_8 through A_{15}. Address lines A_0 through A_7 are multiplexed together with the eight data lines, requiring an 8282 octal latch (together with the ALE signal from the 8288) to demultiplex and drive the lower byte of the address bus. These 16 address lines will allow for 64KB of system memory in our design.

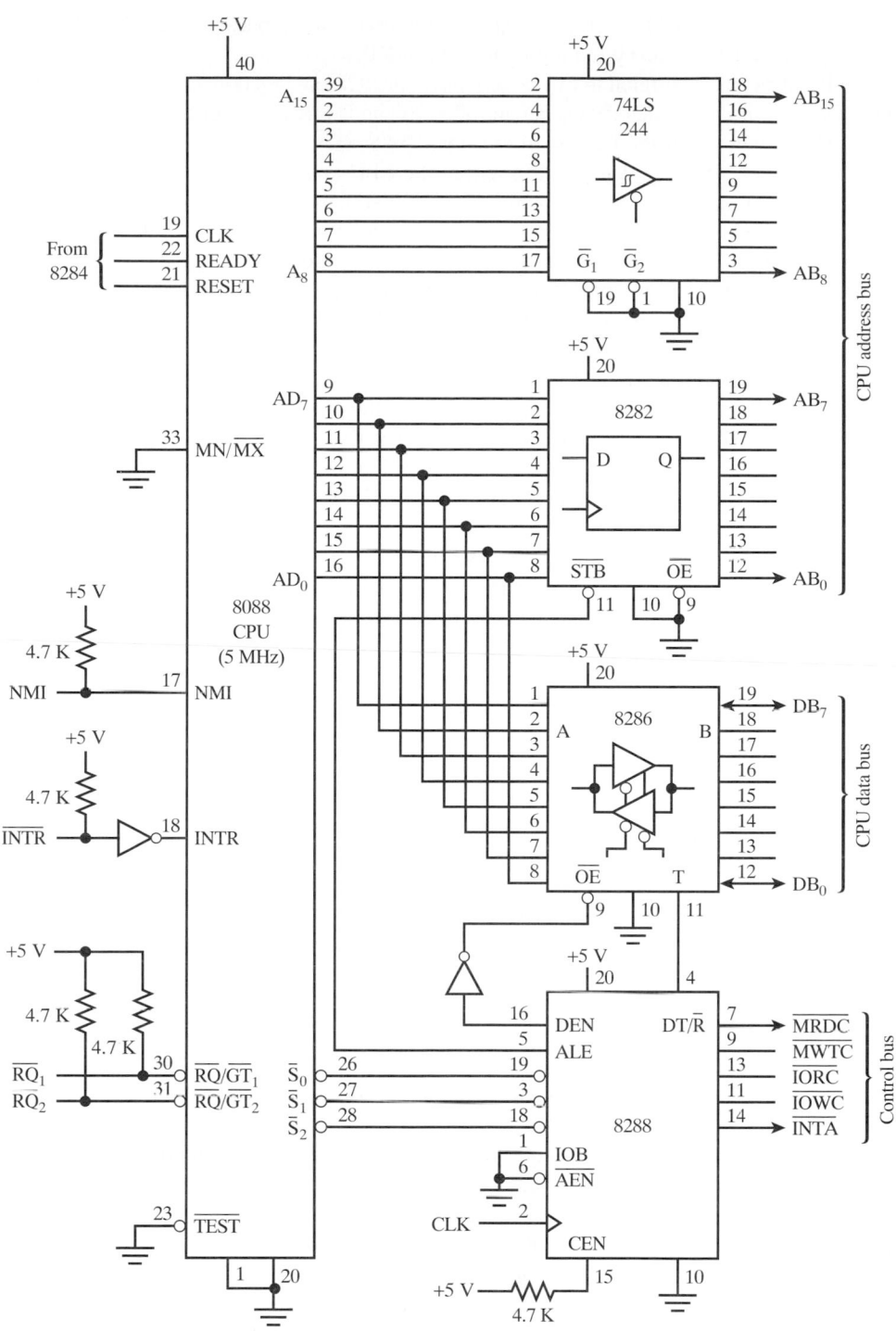

FIGURE 14.11 **CPU Section of Single-Board Computer**

The upper four address lines are not used in this system, because we will not be expanding the system memory requirements past 64KB.

The data bus is buffered in two directions by the 8286 bidirectional line driver/receiver. The direction of data in this device is controlled by the DT/$\overline{\text{R}}$ output of the 8288.

The schematic of the CPU section shows the 8088's MN/$\overline{\text{MX}}$ pin wired to ground. This selects maximum mode operation within the CPU and requires that we use the 8288 bus controller to generate memory and I/O control signals.

The decision to operate the processor in minmode or maxmode depends on a number of factors. If low chip count is necessary, then minmode can be used and the 8288 eliminated. If coprocessor support will be needed in a future expansion of the system, it is best to operate in maxmode from the beginning. The 8288 may also eliminate the need for additional decoding logic in a minimum mode system. Furthermore, bus-granting capability is available only in maxmode. Because no devices in the minimal system use this feature, both $\overline{\text{RQ}}/\overline{\text{GT}}$ inputs are pulled high. The pullup resistors will not prevent us from connecting a DMA device to either input at a future time.

Because our goal is to design a system with a minimum of hardware, extensive interrupt support logic will not be necessary. The processor's two external hardware interrupt inputs should serve our needs adequately.

An inverter is used to make the 8088's high-level INTR interrupt respond to a low-level signal. This technique keeps the INTR input in the inactive state if no devices are connected to $\overline{\text{INTR}}$. NMI is also pulled up to a high level. NMI is edge-sensitive. When no interrupting device is connected, NMI will remain in the high state, and no interrupt will be requested. If we connect a device to NMI in the future, the pullup resistor will not have an adverse effect on any rising-edge NMI signal that is generated.

Technically, although there are five integrated circuits in the CPU section, a *bare-bones* system could get by with only the 8088 running in minmode. But this would most likely require the addition of hardware in the future (to drive the buses and/or possibly switch to maxmode). Any unexpected expansion of hardware is a costly, and sometimes impossible, venture. So, although the minimal system already contains a handful of integrated circuits, choosing maxmode for our project leaves the door open for easy expansion in the future.

The Memory Section

A number of questions must be answered before we get involved in the design of our memory section. For instance, how much EPROM memory is needed? How much RAM? Should we use static or dynamic RAM? Should we use full or partial address decoding? Will we allow DMA operations?

The answer to each of these questions will help specify the required hardware for the memory section. If we first consider what *applications* we will be using, the previous questions will almost answer themselves. Our application at this time is educational. We desire an 8088-based system that will run short-machine language programs. Keeping this point in mind, we will now proceed to find answers to our design questions.

A programmer, through experience, can estimate the required amount of machine code needed to perform a desired task. The software monitor that we will need to control our system will have to be placed in the EPROMs of our memory section. One standard 2764 EPROM will provide us with 8,192 bytes of programmable memory. This is more than enough EPROM to implement our software monitor. We will still have space left over in the EPROM in case we want to add more functions to the monitor in the future.

The amount of RAM required also depends on our application. Because we will be using our system to test only short, educational programs, we can get by with a few hundred bytes or so. Because dynamic RAMs are generally used in very large memory systems (64K, 256K, and more), we will not use them because most of the memory would go to waste. Other reasons exist for not choosing dynamic RAMs at this time. They require complex timing and refresh logic, and will also need to be wired very carefully to prevent messy noise problems from occurring. Even if we use a DRAM controller, we will need some external logic to support the controller, which itself could be a very costly item.

For these reasons, we decide to use static RAM. Even though a few hundred bytes will cover our needs, we will use one 6264 static RAM, thus making our RAM memory 8192 bytes long also. The 6264 is a low-power static RAM, with a pinout almost exactly identical to the 2764 we are using for our EPROM memory. So, by adding only two more integrated circuits (plus a few for control), our memory needs are taken care of.

Figure 14.12 shows how we use a 74LS138 3- to 8-line decoder to perform partial-address decoding for us. Because we are not concerned with future expansion on a large scale, partial-address decoding becomes the cheapest way to generate our addressing signals. Address lines A_{13} through A_{15} are used because they break up the 8088's memory space into convenient ranges (8KB blocks in this case). We completely ignore the state of the upper four address lines (A_{16} through A_{19}). If future expansion beyond the 64KB range is necessary, the upper four address lines must be used to enable the 74LS138.

With A_{13} through A_{15} all low, the 74LS138 decoder will output a 0 on the output connected to the RAM's chip-enable input. With A_0 through A_{12} selecting individual locations within the 6264 RAM, we get an address range of 00000H to 01FFFH. Thus, any time the processor accesses memory in the range 00000H to 01FFFH, the RAM will be enabled. This is a good place for system RAM, because the interrupt vector table must be stored in locations 00000H through 003FFH.

When A_{13} through A_{15} are all high (as they are after a reset causes the initial instruction fetch from FFFF0H), the chip-enable of the 2764 8KB EPROM is pulled low (by the 74LS138). Together with information on the thirteen lower address lines, this maps the EPROM into locations FE000H to FFFFFH. Because partial-address decoding is being used, we can imagine the upper four address lines to be anything we want. This is why we conveniently made them low for the RAM range and high for the EPROM range. Other acceptable RAM ranges are 10000H through 11FFFH, 50000H through 51FFFH, and C0000H through C1FFFH. These are only three more of the 16 possible RAM address ranges, all of which look identical to the processor. EPROM ranges can be found in a similar manner.

In addition to the RAM and EPROM chip-select signals, the 74LS138 also decodes six additional blocks of addresses. Table 14.2 shows the address range associated with each output of the 74LS138. If additional 8KB RAMs or EPROMs need to be added at a later date, the free decode signals can be used to map them into the desired range.

If we were allowing DMA operations, we might not want the 74LS138 to operate in the same way. The enable inputs of the 74LS138 provide us with a way to disable it (all outputs remain high) during a DMA operation, so that an external device may take over the system.

The $\overline{\text{MRDC}}$ and $\overline{\text{MWTC}}$ signals generated in the CPU section are used to control the transfer of data between the processor and memory.

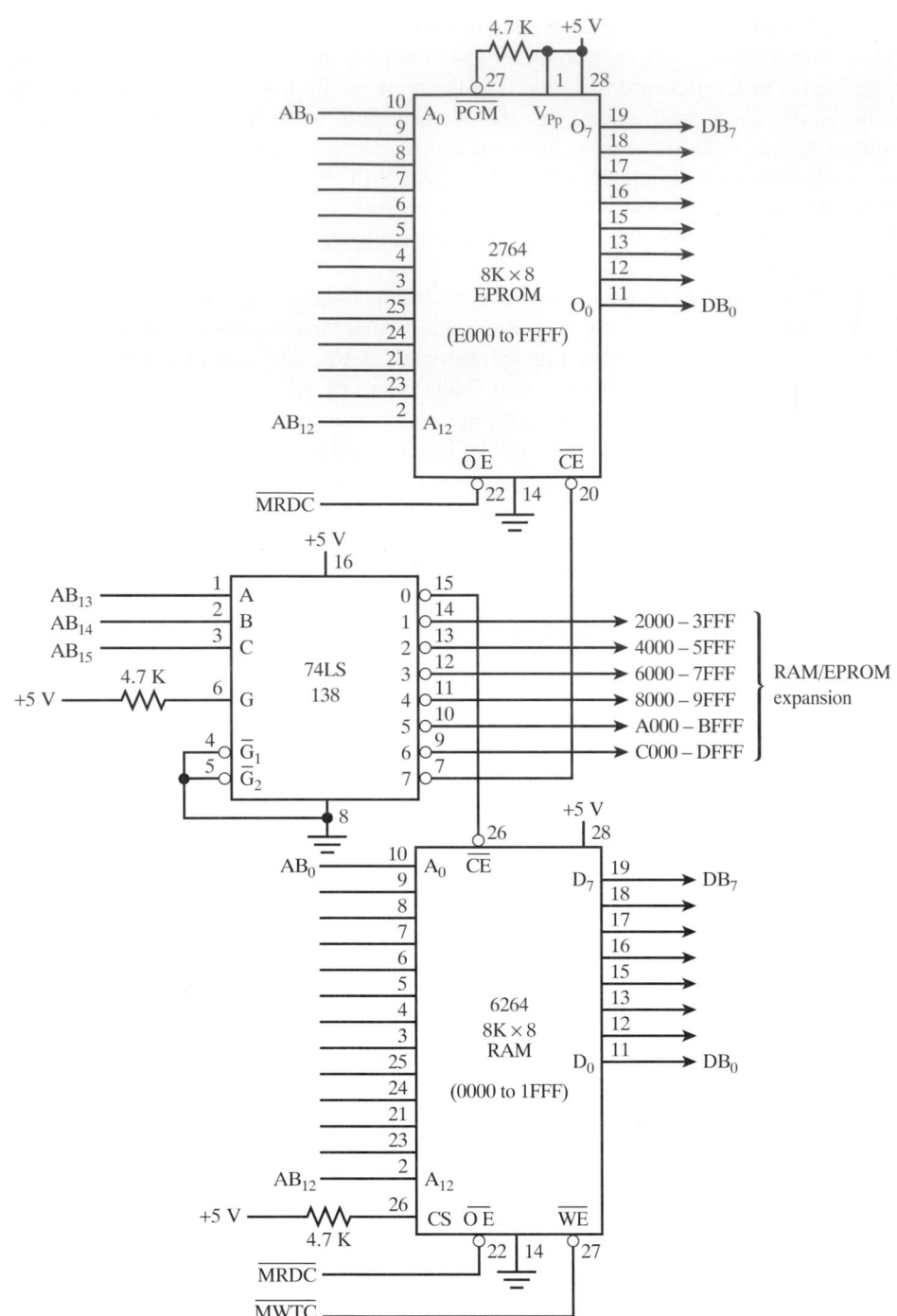

FIGURE 14.12 Memory Circuitry for the Single-Board Computer

TABLE 14.2   Partially Decoded Address Ranges in the Minimal System

74LS138 OUTPUT	DECODED ADDRESS RANGE	USE
0	x0000 to x1FFF	Main RAM
1	x2000 to x3FFF	Free
2	x4000 to x5FFF	Free
3	x6000 to x7FFF	Free
4	x8000 to x9FFF	Free
5	xA000 to xBFFF	Free
6	xC000 to xDFFF	Free
7	xE000 to xFFFF	Main EPROM

x = don't care (can be anything from 0 to F)

The Serial Section

The serial section of our computer will contain all hardware required to communicate with the outside world (via an EIA-compatible data terminal). One question that must be answered concerns the baud rate at which we will be transmitting and receiving. A very acceptable speed is 2400 baud. Speeds higher than this will be too fast to read on the screen, and slower speeds will take too long to read.

Figure 14.13 shows the schematic of the serial section, where an 8251 is used to provide serial communications. The chip-enable input of the 8251 is controlled by output 2 of the 74LS138 port-address decoder. Address lines A_5 through A_7 are used by the 74LS138 to decode eight port-address ranges. The 8251 responds to any I/O accesses to ports 40H through 5FH. The remaining seven groups of port addresses are available for expansion. One of these groups will be used to access an 8255 to provide parallel I/O (as shown in the next section).

The MC14411, together with a 1.8432-MHz crystal, generates the required transmitter and receiver clock frequencies for standard baud rates from 300 to 9600. A DIP switch or jumper can be used to select one of these rates.

The 8251 communicates with the processor via the 8-bit data bus. $\overline{\text{IORC}}$ and $\overline{\text{IOWC}}$, together with A_0, control read and write operations in the 8251. CLK is provided to take care of the 8251's internal activities, and RESET is used to initialize the 8251 at power-up.

Because no modem is connected, the 8251's $\overline{\text{DSR}}$ and $\overline{\text{CTS}}$ inputs can be grounded. This ensures that the 8251 is always ready to communicate.

Serial data enter and leave the 8251 on RxD and TxD. These signals are connected to a MAX232CPE, which converts the 8251's TTL signal levels into RS232-compatible voltage levels, and vice versa. Four $22\text{-}\mu\text{F}$ electrolytic capacitors are used to create a $\pm 10\text{-V}$ swing on the output of the MAX232. This eliminates the need for an external power supply for these two voltages. The serial-in and serial-out signals from the MAX232 can be wired to a DB25 connector or other suitable connector.

Although any port address in the range 40H through 5FH will activate the 8251, the software (via A_0) uses only ports 40H and 41H. Port 40H is the 8251's *data* port, which is used to read and write to the receiver and transmitter. Port 41H is the 8251's *status* port, which is used by the software to determine when it is safe to access the receiver or transmitter.

If one serial channel is not sufficient for your needs, a second one can be added by interfacing a second 8251. One of the free port-address ranges should be used to enable

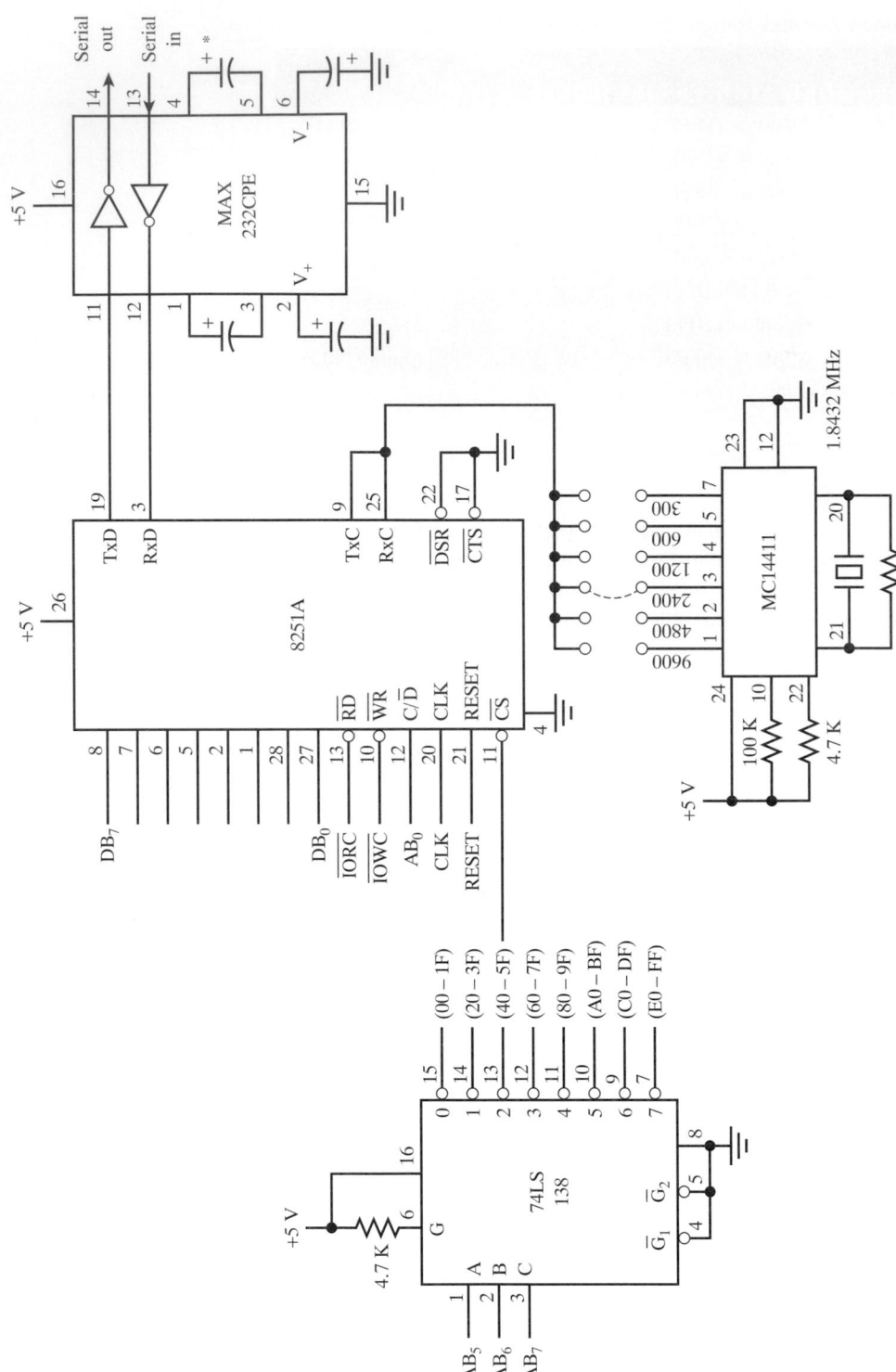

FIGURE 14.13 Serial I/O Circuitry for Minimal System

* All electrolytic caps are 22μF/25V.

the second 8251. Its baud rate clock will be supplied by the MC14411, and the other half of the MAX232 can be used to drive the second set of serial data lines. A second serial channel is useful for downloading machine code into the minimal system's memory (although this can also be done with a single channel), or for echoing data to a printer.

The Parallel Section

If parallel I/O is needed, simple latching and buffering circuitry can be used to add a single I/O port using one of the free port-address decoder ranges. If more than one port is needed, it is best to use a multiport device such as the 8255. The 8255 provides three programmable I/O ports and is easily interfaced with the processor.

Figure 14.14 shows how an 8255 is connected in the minimal system. All of the usual data and I/O signals are connected. Because there are four internal ports in the 8255

FIGURE 14.14 Parallel I/O Circuitry for Minimal System

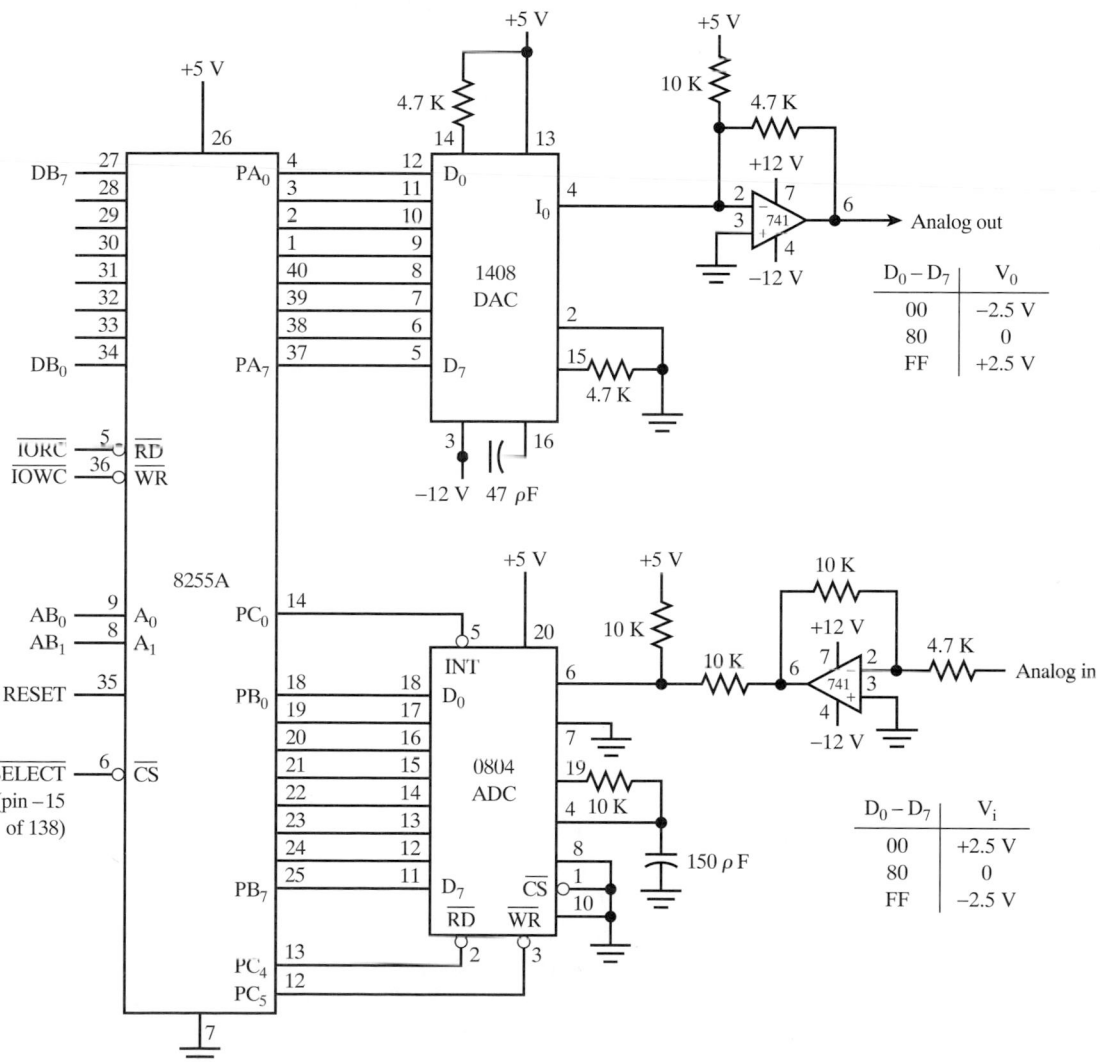

(three for data and one for control), two address lines are required to select one of the four internal ports. A_0 and A_1 are used for this purpose. With the chip-enable input of the 8255 wired to the first output of the 74LS138 port-address decoder, port addresses 00H through 03H can be used to select the 8255.

Figure 14.14 also shows how the 8255 is used to provide the minimal system with analog I/O capability. This additional circuitry may not be needed in many applications. In that case, the 8255 merely provides 24 bits of parallel I/O. When analog I/O is a requirement, the circuit of Figure 14.14 provides an acceptable range of analog input and output voltages. A 1408 8-bit digital-to-analog converter is connected to port A of the 8255 (which must be programmed for output operation). The current output of the 1408 is converted into a ± 2.5-V swing by a 741 op-amp.

Port B of the 8255 is used to read the output of an 0804 8-bit analog-to-digital converter (which must be programmed for input operation). A second 741 is used to adjust the input voltage range of ± 2.5 V to the 0- to 5-V swing needed by the 0804. The 0804 is controlled by 2 bits in the 8255's C port. With the 0804 connected in this way, it is possible to digitize over 8,000 analog samples in 1s (one sample every 125 μs).

Control of the analog circuitry is provided by instructions in the monitor program. Clearly, there is a great deal of cooperation between the software written for a microprocessor and the hardware that supports the software commands. Without a good understanding of the basics of digital logic and binary numbers, the design task would be much more difficult.

14.8 TROUBLESHOOTING TECHNIQUES

Only hands-on experience can truly develop the necessary skills required to work with a digital machine as complex as a computer. Through trial and error, we learn how to diagnose problems and find solutions. For someone who may just have constructed the single-board computer presented in the last section, here are some things to do if the single-board computer does not work when it is turned on:

- ◆ Feel around the board for hot components. A chip that is incorrectly wired or placed backwards in its socket can get very hot. You may even smell the hot component.

- ◆ Make sure all the ICs have power by measuring with a DMM or oscilloscope. Put the probe right on the pin of the IC, not on the socket lead.

- ◆ Use an oscilloscope to examine the CLK output of the 8284. Push the RESET button to verify that the RESET signal is being generated properly.

- ◆ Look at the Address, Data, and Control lines with an oscilloscope or logic analyzer. Activity is a good sign. There may be something as simple as a missing address or data line, or crossed lines. No activity means the processor is not receiving the right information. By examining the logic analyzer traces, you should be able to determine if the memory and I/O address decoders are working properly, as well as the address and data bus drivers.

- ◆ Verify that the EPROM was burned correctly, and is in the right socket and not switched with the RAM. You should be able to connect a logic analyzer to verify that the processor fetches the first instruction from address FFFF0H. You should also be able to see the first instruction byte come out of the EPROM as well.

◆ Examine the TxD output of the 8251. Activity at power on or RESET is a good sign, since the monitor program is designed to output a short greeting to let us know it is alive. If TxD wiggles around, but the serial output of the MAX232 does not, there could be a wiring problem there. Putting the capacitors in backwards is bad for the MAX chip.

◆ Check every connection again from a fresh schematic. Many times a missing connection is found, even when every attempt was made to be careful during construction.

◆ Change all the chips, one by one. Look for bent or missing pins when you remove them.

◆ When all else fails, tell someone else everything you've done and ask whether he or she can suggest anything else.

◆ You could also set the project aside for a while to get your mind off it. The solution to the problem may present itself to you when you least expect it. You may suddenly remember that you wired separate ground paths on each half of the board, but did not connect them together to make a single common ground. Silly things like that are really fun to find.

SUMMARY

In this chapter we studied the basic parts of a computer: clock, CPU, memory, and I/O. We also studied the buses used to control the flow of information in a computer, namely, the data, address, control, and status bus.

We also had a brief introduction to the hardware of the personal computer, and took a detailed look at the design of an actual microprocessor system.

STUDY QUESTIONS

General

1. List six applications of microprocessors found around you that were *not* listed in this chapter.

2. Draw the block diagram of a computer. Label each part and all buses.

3. A microprocessor chip may be 0.2 in. on a side. Calculate the time required for an electrical signal to travel the width of the chip, assuming a speed of $\frac{2}{3}c$ (two-thirds the speed of light).

4. If a memory unit is located 1.5 m from the CPU, calculate the delay in transmitting information from one to the other.

5. What is a floating-point unit?

6. What is an interrupt?

7. Describe the fetch, decode, and execute operation and tell how it is modified if further data is needed by the CPU.

8. Explain how the address bus is used. Explain how the data bus is used. What signals are traveling on these buses?

9. How is the CPU different from the memory section of a computer?

10. How does port I/O differ from I/O to a memory device?

11. What is the purpose of the timing section of a computer?

12. In human terms, when you hear a phone ring we might say that you are experiencing an interrupt. Why is this analogy valid?

13. What is the function of the program counter or the instruction pointer?

14. How are decoders used to interpret an instruction?

15. What instruction is decoded in Figure 14.3 if its pattern is 10101100?

16. Design a circuit that prioritizes four interrupt signals. Use basic logic gates in your design.

17. Look at the chips on a motherboard you have access to. What kind of devices do you find?

18. A dedicated hardware controller has eight instructions that may operate on one of two registers. Four-bits of data are included in the instruction byte, as indicated in Figure

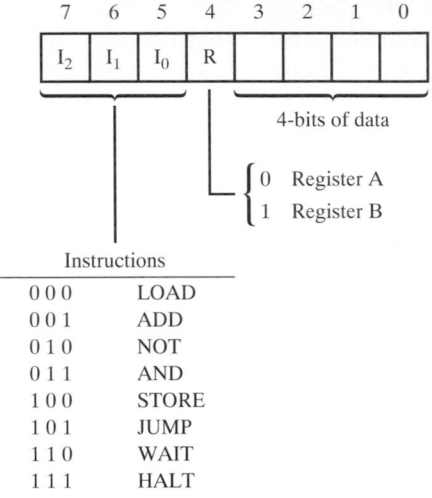

FIGURE 14.15   For Question 14.18

14.15. Show how the instruction byte is decoded, with the appropriate register enabled.

19. Show how two priority encoders (see Figure 14.4) can be used to encode one of sixteen interrupts. Generate a 4-bit interrupt number and an $\overline{\text{INT}}$ signal.

20. What are the hardware differences between an input port and an output port?

21. Determine the truth table for the control signal decoders in Figure 14.8.

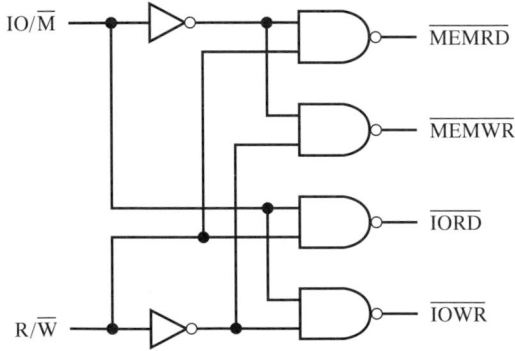

FIGURE 14.16   For Question 14.22

23. What is the difference between the 8282 and the 8286 in the CPU schematic of Figure 14.11? How do they relate to the processor signals AD_0 through AD_7?

24. What changes must be made to the memory address decoder (Figure 14.12) to use 16 KB EPROMs in the single-board computer?

25. What changes must be made so that the port address decoder in Figure 14.13 only decodes port addresses 40H through 47H?

Electronics Workbench

26. What portions of the 80x86 single-board computer can be simulated in Electronics Workbench?

Programmable Logic

27. Write the PAL equations for the control signal decoder in Figure 14.8.

28. How could programmable logic be used in the 80x86 single-board computer?

ANSWERS TO SELECTED ODD-NUMBERED STUDY QUESTIONS

Chapter 1

1.1   (a) 13.625   (e) 3539
        (b) 55.75   (f) 183.875
        (c) 890.75   (g) 255.15625
        (d) 0.114257812   (h) 879.578125

1.3   (a) 22.40625
        (b) 0.00352478
        (c) 19921.546875
        (d) 2874.787109375

1.5   (a) 09A2 H
        (b) FDE7 H
        (c) 0F9F H
        (d) 1FFE H

1.7   (a) 4B H = 1001011B = 113 Q
        (b) 121 H = 100100001B = 441 Q
        (c) FE7 H = 111111100111B = 7747 Q
        (d) 6F0F H = 110111100001111B = 67417 Q

1.9   (a) 11111110
        (c) 01010101
        (e) 11000011
        (g) 10010100

1.11 21 = 00100001   02 = 00000010
      00 = 00000000   C2 = 11000010
      40 = 01000000   03 = 00000011
      DB = 11011011   40 = 01000000
      40 = 01000000   76 = 01110110
      FE = 11111110

1.17 (b) 001000001010.010011001101 B

1.19 (a) 1243.20 Q
      (b) 14232.77 Q
      (c) 6624.0020304 Q

1.23 6, 72, 110, and 272.

Chapter 2

2.1   (a) 66 D = 42 H = 01000010 B
     (b) 169 D = A9 H = 10101001 B
     (c) 44.8 D = 2C.$\overline{CC}$H = 101100.110011001$\overline{100}$ B
     (d) 85.3 D = 55.4$\overline{C}$CH = 1010101.010011001100 B

2.3   −118, 138

2.5   (a) 19 D = 13 H = 10011 B
     (b) −26 D = −1A H = −11010 B
     (c) 8.14D = 8.23D H = 1000.001000111101 B

2.7   (a) Accumulator entries:

$$
\begin{array}{r}
206.4 \\
\times\ \ 36 \\
\hline
7430.4
\end{array}
\qquad
\begin{array}{r}
11001110.011001100\overline{110} \\
\times\ 100100. \\
\hline
?
\end{array}
$$

$$
\begin{array}{r}
1100111001.1001100110 \\
1100111001100.1100110011 \\
\hline
1110100000110.0110011001
\end{array}
$$

(7430.399414062)

     (b) Accumulator entries:

$$
\begin{array}{r}
141.3 \\
\times\ .6 \\
\hline
84.78
\end{array}
\qquad
\begin{array}{r}
10001101.010011001\overline{100} \\
\times\ \ \ \ \ \ .10011001\overline{1001} \\
\hline
?
\end{array}
$$

$$
\begin{array}{r}
1000110.1010011001100 \\
1000.1101010011001 \\
100.0110101001100 \\
.1000110101001 \\
.0100011010100 \\
\hline
1010100.1011100101110
\end{array}
$$

(84.724365234)

2.11 (a)
$$
\begin{array}{r}
0010\ 0111 \\
+\ 1000\ 0111 \\
\hline
1010\ 1110 \\
+\ 0110\ 0110 \\
\hline
1\ 0001\ 0100
\end{array}
$$
(14 Answer)

$$
\begin{array}{r}
9A \\
-\ 13 \\
\hline
87
\end{array}
$$
Calculation of complement

2.13 (b)
$$
\begin{array}{r}
28\ H \\
-\ 79\ H \\
\hline
-\ 51\ H
\end{array}
$$
   −79 H = 10000111 B

$$
\begin{array}{r}
0010\ 1000 \\
+\ 1000\ 0111 \\
\hline
0\ 1010\ 1111 \\
(-)0101\ 0000 \\
+\ \ 1 \\
\hline
(-)0101\ 0001
\end{array}
$$
   Answer

Chapter 3

3.1 The three main logic families are TTL, CMOS, and ECL.

3.3 ECL is faster than TTL because its transistors are not driven into saturation.

3.5 DIP stands for dual inline package.

3.7 A date code of 9711 means the 11th week of 1997.

3.9 Yes.

3.11 TTL output circuitry is called totem-pole because the output transistors look like they are stacked on top of each other.

3.13 Fan out refers to the number of inputs a single output can drive.

3.19 (a) SSI, (b) LSI, (c) MSI, (d) VLSI

3.27 Application-specific integrated circuit.

Chapter 4

4.1 (b) f is a 1 when A is a 1 or B is a 0 or C is a 1
  (d) f is a 1 when you do not have (not($A = 1$ and $B = 1$) or
   ($A = 0$ and $B = 1$))

4.5 (a) $f = \overline{\overline{\overline{A\,B} \cdot \overline{C\,D}}}$
  (c) $f = \overline{\overline{\overline{A\,B} \cdot \overline{A\,\overline{B}} \cdot \overline{\overline{A}\,B\,C}}}$

4.7 $X = \overline{A\,B}$ $Y = \overline{B + C}$

4.17 An open-collector gate lacks an internal pullup resistor on its collector. Adding an external pullup resistor is necessary to make logic one levels.

4.25 Connecting two spare inverters to the inputs of a spare NAND will simulate the OR function.

Chapter 5

5.1 (a) $f = A + BC + \overline{B}\,\overline{C}$
  (c) $f = \overline{B}\,\overline{C} + \overline{A}\,B$

5.3 (a) S/P $f = \overline{A}\,C + A\,\overline{C}$ P/S $f = \overline{\overline{A}\,\overline{C} + A\,C}$
  (c) S/P $f = \overline{A}\,B + A\,\overline{B}\,C$ P/S $f = \overline{\overline{A}\,\overline{B} + A\,B + A\,\overline{C}}$

5.5 S/P $f = \overline{A}\,\overline{B} + A\,B + A\,\overline{C}$ (or $\overline{B}\,\overline{C}$)
  P/S $f = \overline{A}\,B + A\,\overline{B}\,C$

5.7 (a) $f = B\,\overline{C} + A\,\overline{B}\,C$ $\overline{f} = B\,C + \overline{A}\,\overline{B} + \overline{B}\,\overline{C}$
  (b) $f = \overline{A}\,B + A\,\overline{B} + A\,C$ $\overline{f} = \overline{A}\,\overline{B} + A\,B\,\overline{C}$
  (c) $f = \overline{B}$ $\overline{f} = B$

5.13 $\overline{A} + A$ is always true because $0 + 1$ is 1 and $1 + 0$ is also 1.

5.15 $f = BC(\overline{A} + A) + \overline{A}\,\overline{B}(\overline{C} + C)$
   $= BC + \overline{A}\,\overline{B}$

5.17 Before: 4 (2 AND, 1 OR, 1 INV). After: 3 (1 AND, 1 OR, 1 INV).

5.19 The basic technique behind Quine-McCluskey is to compare terms between groups of terms, looking for a one variable difference.

5.23 Spare gates can be used to make other logic functions (via DeMorgan's Theorem), reducing the possibility of having to add a new package to the circuit.

5.25 Karnaugh maps only yield the simplest Boolean equation, not necessarily the simplest function.

Chapter 6

6.1 The flip-flop triggers when a zero-to-one transition occurs on the clock input.

6.3 An asynchronous input causes an effect immediately, whereas a synchronous input has no effect until a clock pulse is applied.

6.7 A flip-flop is set when its Q output is high.

6.9 Clock the 8-bit shift register once every 30 ms to obtain a 240 ms delay.

6.13 The propagation delay of a flip-flop guarantees that only one state change is possible on the output each clock pulse. This causes bits in a shift register to only shift one bit position with each clock pulse.

6.21 (a) 000, 100, 010, 001, back to 000
(b) 0000, 1000, 1100, 0110, 0011, 0001, back to 0000

Chapter 7

7.1 Six flip-flops are required to count from 0 to 63.

7.3 The maximum frequency is $1/(32 \text{ ns}) = 31.25$ MHz.

7.7 The count goes backwards from 1111 to 0000.

7.13 (a) $10(10) = 100$ (d) $16(16) = 256$
(b) $10(16) = 160$ (e) $10(10)(10) = 1000$
(c) $10(2) = 20$ (f) $16^4 = 65536$

7.15 Ripple is eliminated in a synchronous counter by clocking all flip-flops at the same time.

Chapter 8

8.1 The synchronous part of a synchronous circuit are the storage elements that change state only on the edge of the input clock.

8.3 A truth table shows what happens to Q after a clock pulse. An excitation table shows what J and K levels are needed to get a certain behavior at the Q output.

8.5 Three flip-flops are required to keep track of seven state numbers.

8.13 "Don't cares" in a truth table allow Boolean reduction by helping to make groups of 1s that may be circled.

8.17 There are a total of 10 inputs to the state machine. This will require a ROM with 2^{10}, or 1024, locations.

8.21 No, 10011010 is not divisible by 5 (machine ends in state 4).

8.25 Digital clocks, microprocessors, UARTs, pinball machine, telephone.

Chapter 9

9.1 Reduce chip count, increase functional flexibility.

9.3 Specify the design, choose a PAL, write the equations and create the circuit file, assemble your design, program and test the PAL.

9.5 The equations are:

$$O_0 = I_0I_2 + \bar{I}_1$$
$$O_1 = I_1\bar{I}_2 + \bar{I}_0$$

9.7 The first line of the PAL specification identifies the PAL being used in the design.

9.9 A registered PAL contains flip-flops in its output stage.

9.13 Gate delay is specified in a VHDL statement like this:

```
Sum <= (A xor B) after 4 ns;
```

9.15 A registered PAL should be used to make a counter.

Chapter 10

10.1 $2^5 = 32$ $2^8 = 256$

10.3 The circuit in Figure 10.48(b) is preferred because it uses one chip instead of three.

10.21 Carry lookahead adders use additional logic to perform additions of individual bits in parallel. Ripple adders add two bits at a time, taking much longer to generate the same sum as the carry lookahead adder.

10.23 A successive approximation register is an A/D converter that has a fixed conversion time and uses special internal hardware to determine the binary result beginning with the MSB.

Chapter 11

11.1 (a) $2^{15} = 32,768$
     (b) $32,768 \times 8 = 262,144$

11.3 (a) Yes
     (b) No

11.5 (a) 16,384 bits
     (b) 256K bits

11.7 $2^{12} = 4096$ (Twelve address lines are required.)

11.9 8192

11.11 8K = 8192 = 2000 H
      Therefore, if memory starts at 2000 H, an 8K memory will end at 3FFF H.

11.13 (a) $2^{14} = 16384$
          $2^{13} = 8192$
          For a 12K memory, you need 14 address lines.
      (b) 12K = 4K + 8K = 4096 + 8192 = 12288 = 3000 H
          Starting at 0, a 12K memory ends at 2FFF H
      (c) At 5000 H, memory ends at 7FFF H

11.15 Word processing
Image processing
Very large information systems
Speech analysis/recognition

Multiuser systems
Large data bases
Pattern recognition
Weather analysis

Chapter 12

12.3 (a) 11000110
(c) 01001101
(e) 01100101
(g) 01010110
(h) 00010100

12.5 $1800/11 = 163.63\overline{63}$ characters per second

12.7 $104.17 \, \mu s$, $52.08 \, \mu s$

12.17 S/P: $f = \overline{A + \overline{B}\,C + D}$
P/S: $f = \overline{A}\,C\,\overline{D} + \overline{A}\,B\,\overline{D}$

12.19 Yes, by using the ASCII character code as the address to the ROM. The data stored in any location is the equivalent EBCDIC code.

12.25 Both error patterns should be 110 (for bit 6).

12.27 MODEM stands for modulator/demodulator.

12.29 A LAN is a local area network, a collection of computers that share files and resources.

Chapter 13

13.1 512

13.3 $25.6 \, \mu sec$

13.5 Check the power supply for correct DC level, then AC ripple. Check to be sure that the clock is running correctly. See if there is activity on the address and data lines. Check the level of the ready, hold, and interrupt lines. Look for obvious physical damage (burned parts or lands). Check the I/O devices to be sure they are operational.

13.9 By examining data on data lines, it would be easy to find two bits that were always the same, indicating that they are shorted together. It would also be easy to find a data line that is always tied low or always tied high. Observing the address lines may turn up the same type of problem.

Chapter 14

14.3 $c = 300 \times 10^6 \, \text{m/s} \times 39.370 \, \text{in./m} = 1.1811 \times 10^{10} \, \text{in./s}$
delay $= 0.2/(1.1811 \times 10^{10} \times \frac{2}{3}) = 25.4 \, \text{ps}$

14.5 A floating-point unit is a dedicated math processor, typically contained on-chip with its host processor.

14.7 The fetch cycle reads an instruction from memory. During the decode cycle the processor decodes the instruction and may discover that additional data is required from memory. The execute cycle completes the instruction, possibly modifying memory in the process.

14.9 The CPU fetches, decodes, and executes instructions. The memory only stores instructions and data.

14.11 The timing section provides synchronization for all other sections of a microcomputer.

14.13 The program counter contains the address used to fetch the next instruction.

14.15 Arithmetic, shift R, R4.

14.21 Truth table for decoder circuit in Figure 14.8:

IO/M̄	R/W̄	MEMRD	MEMWR	ĪORD	ĪOWR
0	0	1	0	1	1
0	1	0	1	1	1
1	0	1	1	1	0
1	1	1	1	0	1

14.23 The 8282 latches the lower half of the address bus. The 8286 is a bidirectional buffer for the data bus.

POWERS OF 2 AND 16

Hexadecimal Columns

6		5		4		3		2		1	
HEX	**DEC**	**HEX**	**DEC**	**HEX**	**DEC**	**HEX**	**DEC**	**HEX**	**DEC**	**HEX**	**DEC**
0	0	0	0	0	0	0	0	0	0	0	0
1	1,048,576	1	65,536	1	4,096	1	256	1	16	1	1
2	2,097,152	2	131,072	2	8,192	2	512	2	32	2	2
3	3,145,728	3	196,608	3	12,288	3	768	3	48	3	3
4	4,194,304	4	262,144	4	16,384	4	1,024	4	64	4	4
5	5,242,880	5	327,680	5	20,480	5	1,280	5	80	5	5
6	6,291,456	6	393,216	6	24,576	6	1,536	6	96	6	6
7	7,340,032	7	458,752	7	28,672	7	1,792	7	112	7	7
8	8,388,608	8	524,288	8	32,768	8	2,048	8	128	8	8
9	9,437,184	9	589,824	9	36,864	9	2,304	9	144	9	9
A	10,485,760	A	655,360	A	40,960	A	2,560	A	160	A	10
B	11,534,336	B	720,896	B	45,056	B	2,816	B	176	B	11
C	12,582,912	C	786,432	C	49,152	C	3,072	C	192	C	12
D	13,631,488	D	851,968	D	53,248	D	3,328	D	208	D	13
E	14,680,064	E	917,504	E	57,344	E	3,584	E	224	E	14
F	15,728,640	F	983,040	F	61,440	F	3,840	F	240	F	15

Powers of 2

2^n	n	
1	0	$2^0 = 16^0$
2	1	$2^4 = 16^1$
4	2	$2^8 = 16^2$
8	3	$2^{12} = 16^3$
16	4	$2^{16} = 16^4$
32	5	$2^{20} = 16^5$
64	6	$2^{24} = 16^6$
128	7	$2^{28} = 16^7$
256	8	$2^{32} = 16^8$
512	9	$2^{36} = 16^9$
1,024	10	$2^{40} = 16^{10}$
2,048	11	$2^{44} = 16^{11}$
4,096	12	$2^{48} = 16^{12}$
8,192	13	$2^{52} = 16^{13}$
16,384	14	$2^{52} = 16^{13}$
32,768	15	$2^{60} = 16^{15}$
65,536	16	
131,072	17	
262,144	18	
524,288	19	
1,048,576	20	
2,097,152	21	
4,194,304	22	
8,388,608	23	
16,777,216	24	

Powers of 16

16^n	n
1	0
16	1
256	2
4,096	3
65,536	4
1,048,576	5
16,777,216	6
268,435,456	7
4,294,967,296	8
68,719,476,736	9
1,099,511,627,776	10
17,592,186,044,416	11
281,474,976,710,656	12
4,503,599,627,370,496	13
72,057,594,037,927,936	14
1,152,921,504,606,846,976	15

APPENDIX B

RULES OF BOOLEAN ALGEBRA

$A + 0 = A$
$A + 1 = 1$
$A + A = A$
$A + \overline{A} = 1$
$A(B + C) = AB + AC$
$(AB)C = A(BC)$
$AB = BA$

$A \cdot 0 = 0$
$A \cdot 1 = A$
$A \cdot A = A$
$A \cdot \overline{A} = 0$
$(A + B) + C = A + (B + C)$
$A + B = B + A$
$\overline{\overline{A}} = A$

$A \oplus 0 = A$
$A \oplus 1 = \overline{A}$
$A \oplus A = 0$
$A \oplus \overline{A} = 1$

DeMorgan's Theorem

$AB = \overline{\overline{A} + \overline{B}}$ $\qquad$ $\overline{A + B} = \overline{A}\,\overline{B}$

$A + B = \overline{\overline{A}\,\overline{B}}$ $\qquad$ $\overline{AB} = \overline{A} + \overline{B}$

LOGIC GATE SUMMARY

FUNCTION	SYMBOL	BOOLEAN EQUATION	TRUTH TABLE	TTL PACKAGE
Buffer	$A \longrightarrow\!\!\!\rhd\!\!\!\longrightarrow f$	$f = A$	$\begin{array}{c\|c} A & f \\ \hline 0 & 0 \\ 1 & 1 \end{array}$	7407
Inverter	$A \longrightarrow\!\!\!\rhd\!\!\!\circ\!\!\!\longrightarrow f$	$f = \overline{A}$	$\begin{array}{c\|c} A & f \\ \hline 0 & 1 \\ 1 & 0 \end{array}$	7404
AND	$\begin{array}{c} A \\ B \end{array} \!\!\!\rhd\!\!\!\longrightarrow f$	$f = AB$	$\begin{array}{cc\|c} A & B & f \\ \hline 0 & 0 & 0 \\ 0 & 1 & 0 \\ 1 & 0 & 0 \\ 1 & 1 & 1 \end{array}$	7408
NAND	$\begin{array}{c} A \\ B \end{array} \!\!\!\rhd\!\!\!\circ\!\!\!\longrightarrow f$	$f = \overline{AB}$	$\begin{array}{cc\|c} A & B & f \\ \hline 0 & 0 & 1 \\ 0 & 1 & 1 \\ 1 & 0 & 1 \\ 1 & 1 & 0 \end{array}$	7400
OR	$\begin{array}{c} A \\ B \end{array} \!\!\!\rhd\!\!\!\longrightarrow f$	$f = A + B$	$\begin{array}{cc\|c} A & B & f \\ \hline 0 & 0 & 0 \\ 0 & 1 & 1 \\ 1 & 0 & 1 \\ 1 & 1 & 1 \end{array}$	7432
NOR	$\begin{array}{c} A \\ B \end{array} \!\!\!\rhd\!\!\!\circ\!\!\!\longrightarrow f$	$f = \overline{A + B}$	$\begin{array}{cc\|c} A & B & f \\ \hline 0 & 0 & 1 \\ 0 & 1 & 0 \\ 1 & 0 & 0 \\ 1 & 1 & 0 \end{array}$	7402

FUNCTION	SYMBOL	BOOLEAN EQUATION	TRUTH TABLE	TTL PACKAGE
XOR		$f = A \oplus B$ $= A\bar{B} = \bar{A}B$	$\begin{array}{cc\|c} A & B & f \\ \hline 0 & 0 & 0 \\ 0 & 1 & 1 \\ 1 & 0 & 1 \\ 1 & 1 & 0 \end{array}$	7486
XNOR		$f = \overline{A \oplus B}$ $= AB + \bar{A}\,\bar{B}$	$\begin{array}{cc\|c} A & B & f \\ \hline 0 & 0 & 1 \\ 0 & 1 & 0 \\ 1 & 0 & 0 \\ 1 & 1 & 1 \end{array}$	74266

IEEE/IEC STANDARD SYMBOLS

A new form of schematic symbol for logic functions has been standardized by the *International Electrotechnical Commission* and the IEEE. The logical function being performed is found by examining labels and identifiers on the symbol. Figure D.1 shows a small subset of the IEEE/IEC symbols. The cryptic meaning of the symbols inside the "box" in some cases refers to the number of inputs that must be active ("1" for one input, ">1" for at least one input, and "=1" for one-and-only-one input). In other cases, the Boolean operation being performed is indicated ("&" for AND). Inversion is represented by a triangular symbol attached to the input or output.

Devices that have inputs that have a common effect on the outputs, such as counters, decoders, and shift registers, have slightly different notation (*dependency notation*) that indicates what signals depend on each other. Pin numbers and common signal names are only one way the dependencies are represented.

You may agree that it is easier to work with the original symbols. You may also agree that the original symbols, without their truth tables, do not imply much meaning regarding their function. Both sets of symbols have their uses, with the IEEE/IEC symbols gaining more acceptance as time passes.

Function	Original Symbol	IEEE/IEC Symbol

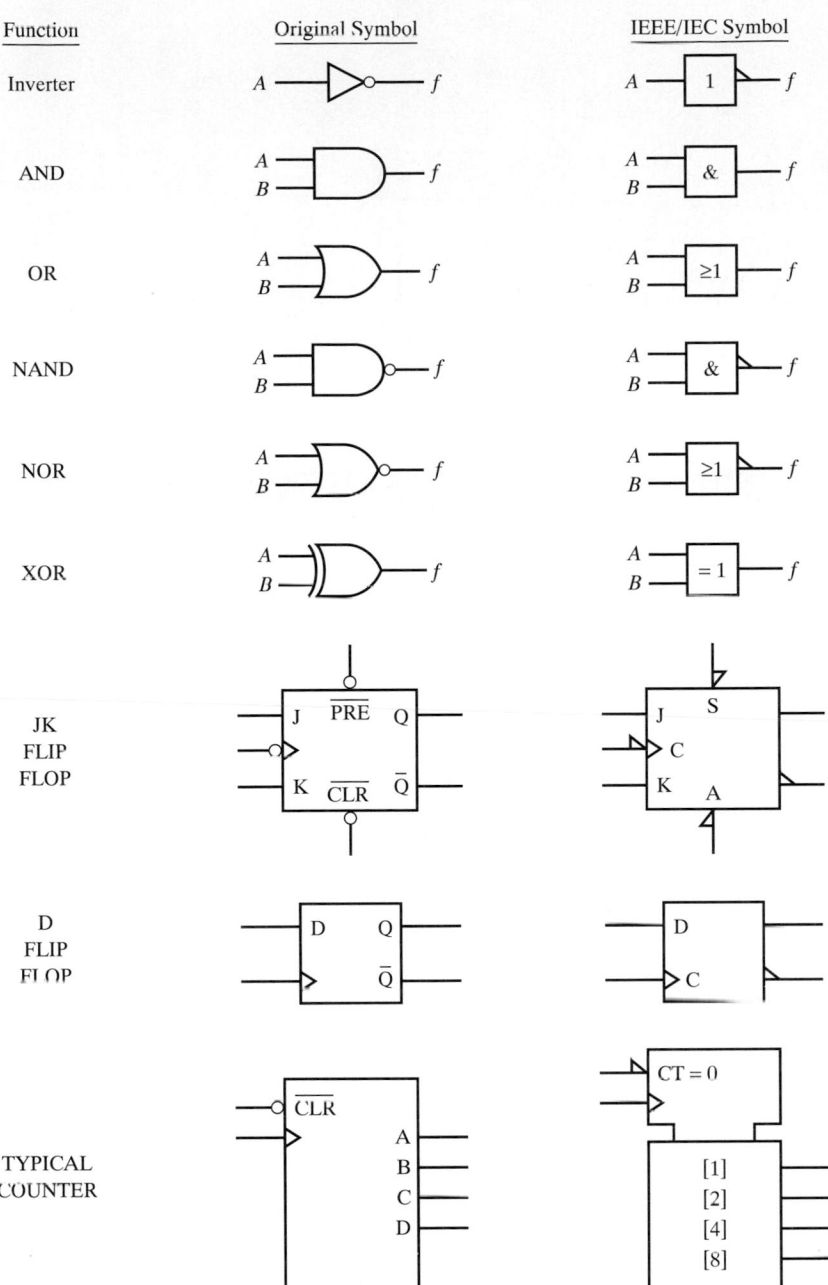

Inverter		
AND		
OR		
NAND		
NOR		
XOR		
JK FLIP FLOP		
D FLIP FLOP		
TYPICAL COUNTER		

FIGURE D.1   IEEE/IEC Symbols

APPENDIX

E

ASCII CHART

HEX	ASCII	HEX	ASCII	HEX	ASCII	HEX	ASCII	HEX	ASCII
00	NUL	1A	SUB	34	4	4E	N	68	h
01	SOH	1B	ESC	35	5	4F	O	69	i
02	STX	1C	FS	36	6	50	P	6A	j
03	ETX	1D	GS	37	7	51	Q	6B	k
04	EOT	1E	RS	38	8	52	R	6C	l
05	ENQ	1F	US	39	9	53	S	6D	m
06	ACK	20	SP	3A	:	54	T	6E	n
07	BEL	21	!	3B	;	55	U	6F	o
08	BS	22	"	3C	<	56	V	70	p
09	HT	23	#	3D	=	57	W	71	q
0A	LF	24	$	3E	>	58	X	72	r
0B	VT	25	%	3F	?	59	Y	73	s
0C	FF	26	&	40	@	5A	Z	74	t
0D	CR	27	'	41	A	5B	[	75	u
0E	SO	28	(	42	B	5C	\	76	v
0F	SI	29	)	43	C	5D	]	77	w
10	DLE	2A	*	44	D	5E	^ (↑)	78	x
11	DC1 (X-ON)	2B	+	45	E	5F	– (←)	79	y
12	DC2 (TAPE)	2C	,	46	F	60	`	7A	z
13	DC3 (X-OFF)	2D	-	47	G	61	a	7B	{
14	DC4	2E	.	48	H	62	b	7C		
15	NAK	2F	/	49	I	63	c	7D	}
16	SYN	30	0	4A	J	64	d		(ALT MODE)
17	ETB	31	1	4B	K	65	e	7E	~
18	CAN	32	2	4C	L	66	f	7F	DEL
19	EM	33	3	4D	M	67	g		(RUB OUT)

ACK	acknowledge	EM	end of medium	NAK	negative acknowledge
BEL	bell	ENQ	enquiry	NUL	null
BS	backspace	EOT	end of transmission	RS	record separator
CAN	cancel	ESC	escape	SI	shift in
CR	carriage return	ETB	end of transmission block	SO	shift out
DC1	device control 1	ETX	end of text	SOH	start of heading
DC2	device control 2	FF	form feed	STX	start of text
DC3	device control 3	FS	file separator	SUB	substitute
DC4	device control 4 (stop)	GS	group separator	SYN	synchronous idle
*DEL	delete	HT	horizontal tabulation	US	unit separator
DLE	data link escape	LF	line feed	VT	vertical tabulation

*not strictly a control character

APPENDIX

F INTEGRATED CIRCUIT REFERENCE

This quick reference contains many of the most common and useful TTL logic functions. Use it during a design to verify that you are using a function that exists, or to estimate package costs.

PART NO.	PINS	FUNCTION	PART NO.	PINS	FUNCTION
74LS00	14	Quad 2 NAND Gate	74LS51	14	Dual 2-Input AND/OR Invert Gate
74LS01	14	Quad 2-Input NAND Gate	74LS54	14	Quad 2-Input AND/OR Invert Gate
74LS02	14	Quad 2 NOR Gate	74LS55	14	Dual 4-Input AND/OR Invert Gate
74LS03	14	Quad 2 NAND Gate	74LS73	14	Dual J-K Flip-Flop with Clear
74LS04	14	Hex Inverter	74LS74	14	Dual D Flip-Flop
74LS05	14	Hex Inverter (Open Collector)	74LS75	16	Quad Latch
74LS08	14	Quad 2-Input AND Gate	74LS76	16	Dual J-K Flip-Flop with Preset and Clear
74LS09	14	Quad 2-Input AND Gate (Open Collector)	74LS78	14	Dual J-K Flip-Flop w/Preset Comm. Clk & Clr
74LS10	14	Triple 3 NAND Gate	74LS83	16	4-Bit Full Adder
74LS11	14	Triple 3-Input AND Gate	74LS85	16	4-Bit Magnitude Comparator
74LS12	14	3-Input NAND Gate (Open Collector)	74LS86	14	Quad Exclusive OR Gate 2-Input
74LS13	14	Dual Schmitt Trigger	74LS90	14	Decade Counter
74LS14	14	Hex Schmitt Trigger	74LS91	14	8-Bit Shift Register
74LS15	14	Triple 3-Input AND Gate	74LS92	14	Divide-by-Twelve Counter
74LS20	14	Dual 4 NAND Gate	74LS93	14	4-Bit Binary Counter
74LS21	14	Dual 4-Input AND Gate	74LS95	14	4-Bit Parallel-Access Shift Register
74LS22	14	Dual 4-Input NAND Gate (Open Collector)	74LS96	16	5-Bit Shift Register
74LS26	14	Quad 2-Input Interface Positive NAND Gate	74LS107	14	Dual J-K Master-Slave Flip-Flop
74LS27	14	Triple 3 NOR Gate	74LS109	16	Dual J-K Positive Edge Flip-Flop
74LS28	14	Quad 2-Input Positive NOR Buffer	74LS112	16	Dual J-K Negative Edge Flip-Flop
74LS30	14	8-Input NAND Gate	74LS113	14	Dual J-K Negative Edge Flip-Flop
74LS32	14	Quad 2-Input Positive OR Gate	74LS114	14	Dual J-K Negative Edge Flip-Flop
74LS33	14	Quad 2-Input NOR Buffer (Open Collector)	74LS122	14	Retriggerable Monostable Multivibrator
74LS37	14	Quad 2-Input NAND Buffer	74LS123	16	Monostable Multivibrator with Clear
74LS38	14	Quad 2-Input NAND Buffer (Open Collector)	74LS125	14	Tri-State Quad Buffer
74LS40	14	Dual 4 NAND Buffer	74LS126	14	Quad Buffer (Tri-State)
74LS42	16	BCD-to-Decimal Decoder	74LS132	14	Quad Schmitt Trigger
74LS47	16	7-Segment Decoder/Driver	74LS133	16	13-Input NAND Gate
74LS48	16	BCD to 7-Segment Decoder/Driver	74LS136	14	Quad Exclusive OR Gate
74LS49	14	BCD to 7-Segment Decoder/Driver	74LS138	16	Expandable 3/8 Decoder

PART NO.	PINS	FUNCTION	PART NO.	PINS	FUNCTION
74LS139	16	Expandable Dual 2/4 Decoder	74LS257	16	Quad 2-Input Multiplexer Tri-State
74LS145	16	BCD-to-Decimal Decoder/Driver	74LS258	16	Quad 2/1 Multiplexer
74LS147	16	10-Line to 4-Line Priority Encoder	74LS259	16	8-Bit Addressable Latch
74LS148	16	8-Line to 3-Line Priority Encoder	74LS260	14	Dual 5-Input NOR Gate
74LS151	16	8-Input Multiplexer	74LS266	14	Quad Ex-NOR Gate
74LS153	16	Dual 4-Input Multiplexer	74LS273	20	8-Bit D Type Register
74LS154	24	Single 4-16 Decoder	74LS279	16	Quad S-R Latches
74LS155	16	Dual 2/4 Demultiplexer	74LS280	14	9-Bit Odd/Even Parity Generator/Checker
74LS156	16	Dual 2/4 Demultiplexer (Open Collector)	74LS283	16	4-Bit Full Adder
74LS157	16	Quad 2/1 Multiplexer	74LS289	16	64-Bit RAM Open Collector
74LS158	16	Quad 2/1 Multiplexer (Inv. Out)	74LS293	14	4-Bit Binary Counter
74LS160	16	Presettable Decade Counter	74LS298	16	Quad 2-Input Multiplexer with Storage
74LS161	16	Presettable Binary Counter	74LS299	20	8-Bit Universal Shift/Storage Register
74LS162	16	Presettable Decade Counter with Clear	74LS322	20	8-Bit Serial/Parallel Reg.
74LS163	16	Presettable Binary Counter with Clear	74LS323	20	8-Bit Univ. Shift/Storage Reg.
74LS164	14	8-Bit Shift Register	74LS347	16	BCD to 7-Segment Decoder
74LS165	16	Parallel Load 8-Bit Serial Shift Register	74LS352	16	Dual 4-Bit Multiplexer (Inv. Out)
74LS166	16	8-Bit Shift Register	74LS353	16	Dual 4-Bit Multiplexer (Inv. Out)
74LS168	16	Synch. Decade Up/Down Counter	74LS364	20	Octal D-Type Transparent Latch
74LS169	16	Synch. Binary Up/Down Counter	74LS365	16	Hex Buffer (Tri-State)
74LS170	16	4x4 Register File	74LS366	16	Hex Inverter (Tri-State)
74LS173	16	Quad D Register (Tri-State)	74LS367	16	Hex Buffer (Tri-State)
74LS174	16	Hex D Flip-Flop with Clear	74LS368	16	Hex Inverter (Tri-State)
74LS175	16	Quad D Flip-Flop	74LS373	20	Octal Transparent Latch
74LS181	24	Arithmetic Logic Unit	74LS374	20	Octal Dual Flip-Flop (Tri-State)
74LS189	16	64-Bit RAM Tri-State	74LS375	16	Quad Latch
73LS190	16	Up/Down Decade Counter	74LS377	20	Octal D Register, Common Enable
74LS191	16	Up/Down Binary Counter	74LS378	16	Hex D Register, Common Enable
74LS192	16	4-Bit Up/Down BCD Counter	74LS379	16	4-Bit Register, Common Enable
74LS193	16	4-Bit Up/Down Binary Counter	74LS386	14	Quad Ex-OR Gate
74LS194	16	4-Bit Bi-Directional Universal Shift Register	74LS388	16	Quad D Reg., Std & 3-State Out.
74LS195	16	4-Bit Parallel-Access Shift Register	74LS390	16	Dual 4-Bit Binary Counter
74LS196	14	Presettable Decade Counter	74LS393	14	Dual 4-Bit Binary Counter
74LS197	14	Preset Binary Counter	74LS399	16	Quad 2-Mux with Q & Q Outputs
74LS221	16	Dual One-Shot	74LS490	16	Dual 4-Bit Decade Counter
74LS240	20	Octal Inverting Bus/Line Driver	74LS533	20	Octal Transparent Latch (Tri-State)
74LS241	20	Octal Bus/Line Driver	74LS534	20	Octal D-Type Flip-Flop (Tri-State)
74LS242	14	Quad Bus Transceiver Inverting	74LS540	20	Octal Buffer/Line Driver (Tri-State)
74LS243	14	Tri-State Quad Transceiver	74LS541	20	Octal Buffer/Line Driver (Tri-State)
74LS244	20	Octal Driver Non-Inverting Tri-State	74LS640	20	Octal Bus Transceiver (Inverting) Tri-State
74LS245	20	Octal Bus Transceiver Non-Inverting	74LS641	20	Octal Bus Transceiver (True) O.C.
74LS247	16	BCD to 7-Segment Decoder/Driver	74LS644	20	Octal Bus Transceiver (True/Inverting) O.C.
74LS248	16	BCD to 7-Segment Decoder/Driver	74LS645	20	Octal Bus Transceiver (True) Tri-State
74LS249	16	BCD to 7-Segment Decoder/Driver	74LS670	16	4x4 Register File (Tri-State)
74LS251	16	Tri-State 8-Channel Multiplexer	74LS688	20	8-Bit Magnitude Comparator
74LS253	16	Dual 4-Input Multiplexer Tri-State			

365

GLOSSARY

Access time. The time required by a memory to produce data after an address has been specified.

Accumulator. Register in a computer that contains the result of math and logical operations.

ACIA. A type of receive/transmit integrated circuit.

Address bus. The parallel path in which the binary memory address from the CPU is sent to the memory section.

Address. A number identifying a location in memory where data are stored.

ALU. Arithmetic logic unit; the portion of a computer that performs math and logical operations.

Analog. A continuously changing signal (i.e., having an infinite number of possible levels).

Analog-to-digital (A/D) converter. A device that converts an analog voltage or current into a corresponding binary number.

AND gate. Logic gate that produces a logic "1" only when all inputs are logic 1s.

Answer. Name of the modem used by a computer to communicate with user whose call it answered.

ASCII. The (7-bit) American Standard Code for Information Interchange.

ASIC. Application-specific integrated circuit.

Asynchronous input. This type of input causes immediate action by itself.

Asynchronous transmission. Serial transmission of data in which a start bit is used to synchronize the receiver clock.

Baseband. Transmission method where only one information carrier is used.

Baud. Rate at which serial information is transmitted.

BCD. Binary coded decimal (also 8421 coded binary).

Bidirectional bus. Any parallel group of wires that can carry signals both ways. A bidirectional data bus can *send* and *receive* information.

Binary counter. A counter where the total number of states equals 2^N, where N is the number of flip-flops in the counter.

Binary. A numbering system based on the number 2.

Bistable flip-flop. Another name for the two-state flip-flop.

Bit. The smallest element of a binary number.

Boolean algebra. Symbolic logic (algebra) used in the design of digital circuits.

Boolean reduction. An algebraic method of reducing the number of logic gates necessary to perform a logic function.

Broadband. Transmission method where many information carriers are used.

Bubble memory. Magnetic bubbles in minor or major loops store binary data.

Buffer. A logic gate with a high fan out used to drive many other gates.

Burning (an EPROM). Programming data into a ROM.

Bus. Parallel paths used for transmitting data at high speeds from any of several sources to any of several destinations.

Byte. A group of 8 bits.

Carry bit. A bit located at MSB + 1 that is set or reset depending on the result of adding the two MSBs.

Carry lookahead adder (CLA). An adder that generates all of the sum and carry bits at the same time.

Cascade. To make a larger counter by connecting smaller counters together, output to input.

Character generator. A special ROM that converts ASCII codes to dot matrix format.

Chip select or chip enable. An input to a memory that connects the data lines to a data bus, hence activates the chip.

Clock. A signal or input line that must be used to cause a flip-flop to toggle, a counter to count, or a shift register to shift.

CMOS. Complementary symmetry metal oxide semiconductor.

Comparator. A device that compares two analog or digital values.

Complement, 1's. Representation of a binary word in which the value of each bit is exchanged.

Complement, 10's. Found by adding 1 to the 9's complement; used most commonly in BCD subtraction.

Complement, 2's. 1's complement plus 1; most commonly used in binary subtraction.

Complement, 9's. Representation of a BCD word in which the original word was subtracted from 9s.

Contact bounce. Noise resulting from the action of closing a mechanical switch.

Control bus. The CPU bus used in controlling all sections of the computer.

Control word. Group of bits that controls the addition to the accumulator during binary multiplication.

Core memory. A magnetic memory shaped like a donut. Magnetization in the clockwise and counterclockwise directions represents a one or a zero, respectively.

Counter. A connection of flip-flops that produces a binary count from a single input of pulses.

CPU. Central processing unit. The element of a computer that performs all mathematical and logical operations.

Current loop. Method of serial transmission in which current pulses, not voltages, are sent on a wire.

Data bus. The parallel path in which information from the CPU is sent to other parts of the computer.

Data link. The connection between a terminal and a computer. It can be direct wire, telephone, satellite, laser, or fiber optic.

Data selector. See *Selector*.

Data terminal. Electronic device used to communicate with a computer. Terminals may either print output on paper (like a typewriter) or display it on a television screen or custom display device.

DB25. DB25P and DB25S are the standard plug and socket connectors used in serial data transmission facilities.

D-C latch. A flip-flop. A 1-bit memory device.

Decade counter. More common term for a modulo-10 counter. The counting sequence is from zero to nine.

Decimal. A number system using 10 digits, beginning with zero and ending with nine.

Decode cycle. The CPU operation in which an instruction fetched from memory is decoded, to guide the CPU to its next cycle.

Decoder. A circuit to select one of several memory chips when an address is specified.

Dedicated microcomputer. A microcomputer programmed to perform only one task.

DeMorgan's theorem. Rules for gate substitution that allow changes in the way a circuit is constructed: $A B = \overline{A} + \overline{B}$ and $\overline{A B} = \overline{A} + \overline{B}$

Demultiplexer. A circuit that converts a binary (parallel) signal to a few lines.

Digital. A signal with only two possible states.

Digital-to-analog (D/A) converter. A device that converts a binary number into a corresponding analog voltage or current.

DIP. Dual inline package.

Disk. A magnetic disk memory used in large computer systems.

Diskette. Flexible disk magnetic memory in $5\frac{1}{4}$ or 8-in.1 diameter format.

DMA. Direct memory access. The process in which an external device takes over the computer's buses and performs operations on the computer's memory.

Don't care. A state denoted by the symbol "X," meaning that the line so designated has no effect. It can be high or low with no effect on the circuit.

Dot matrix. Characters are represented on a display in a grid of dots of 5×7, 7×9, 9×11, or 11×13.

D-type flip-flop. A flip-flop with a data input. Data is clocked into the flip-flop by the clock line (also has asynchronous preset and clear inputs).

Dynamic memory. A memory cell that is a charged or discharged capacitor that must be refreshed to maintain its charge.

EAPROM. Electrically alterable programmable read-only memory. Ultraviolet light is not needed to change data in a memory location.

EBCDIC. The (8-bit) Extended Binary Coded Decimal Interchange Code.

Echo. To transmit back to a display terminal.

ECL. Emitter-coupled logic.

Edge triggering. Most flip-flops toggle on a transition (edge) (zero to one or one to zero). When the pulse travels from one state to the other, however, toggling is initiated (triggered) by the edge.

EEPROM. Electrically Erasable PROM. Can be erased with a signal instead of ultraviolet light.

EIA. The EIA is the national trade association representing a full range of manufacturers in the electronics industry. (For more information, contact the Electronic Industries Association, Type Administration Office, 2001 Eye Street, N.W., Washington, DC 20006.)

Eight-bit ASCII. ASCII code with the addition of a parity bit.

Eleven-bit transmission code. Universal code used to transmit and receive serial data.

Enable. A line that is used to allow a signal to pass to a circuit.

EPROM. Erasable programmable read only memory; programmed by user and erased by ultraviolet light.

Ethernet. Communication method based on baseband transmission of 0s and 1s using CSMA/CD.

Even parity. The number of ones transmitted in a particular word must be even.

Excitation table. A different way of representing the truth table of a flip-flop. Emphasis is placed on what is required to cause a specific transition on the Q output.

Exclusive OR (XOR) gate. A modified OR gate that produces a logic "1" only when one and only one of its inputs is at a logic "1" level.

Execute cycle. The cycle in which the CPU performs an operation on internal registers or an external memory location or I/O device.

Fan out. The number of inputs an output is capable of driving.

FET. Field effect transistor.

Flags. Signals or bits used during handshaking to determine the transmittal or reception of a character.

Flip-flop. A 1-bit memory device that can be either SET or RESET.

Floating-point unit. Dedicated hardware that performs high-speed math.

Full adder. A modified half adder. A full adder adds the carry signal from a previous stage.

Full duplex. Simultaneous transmission of data in two directions.

GAL. Gate array logic.

Gate cost. The number of gates required to perform a DBB function.

Gate delay. See Propagation delay.

General-purpose computer. A computer able to perform a variety of jobs or tasks.

Half adder.   A binary adder used to add two logic levels together and produce a carry and a sum.

Half duplex.   Transmission of data in two directions, but only one direction at a time.

Handshaking.   Signals used by both the computer and the terminal to control the flow of information back and forth between them.

Hardware.   Actual electronic devices (transistors, resistors, capacitors, integrated circuits, LEDs, etc.).

Hexadecimal.   A numbering system based on the number 16.

Huffman code.   A code where the binary patterns are unique and of minimal length.

I/O ports.   Hardware devices that allow eight or more bits of information to be exchanged between the processor and the outside world.

Instruction fetch.   The CPU operation used to fetch a word from memory to be interpreted as an instruction.

Instruction pointer.   See *Program counter*.

Interrupt.   An external signal (condition or set of conditions) causing the computer to immediately halt its current operation and perform a specific task.

Inverter.   A device used to switch logic levels. An inverter converts a logic "0" to a logic "1" and vice versa.

J-K Flip-flop.   A flip-flop with synchronous inputs J and K, and asynchronous inputs, clear and sometimes preset.

LAN.   Local area network.

Latch.   Essentially a flip-flop or binary storage device.

Legal character.   Any character belonging to a particular number system.

Load.   A line that is used to enter a binary number into a register or counter in parallel.

Load/recirculate.   The command whereby a shift register is first loaded with data and the data is then recirculated.

Logic analyzer.   A multichannel instrument that can sample at least eight and sometimes more than 100 channels of data. Each channel can store many samples in a memory for later observation.

Long Word.   Four bytes (32 bits) of information.

LSB.   The least significant bit: the character in the right-most position of a number. It has the smallest weight.

LSI.   Large-scale integration.

Manchester encoding.   A technique used to encode 0s and 1s on the Ethernet. 0s are encoded as rising edges, 1s as falling edges.

Mark.   Indicates a high level in serial transmission.

Memory map.   A display of how different sections of memory are used (e.g., where ROM or RAM is located), and which ranges of memory are empty.

Memory-mapped I/O.   An input-output exchange that uses a decoded memory address instead of a port.

Microcode.   ROM-based data that controls the activity inside a microprocessor.

Modem.   Modulator/demodulator device used in the transmission of serial data over telephone lines.

Modulo-N counter.   A counter that contains N states. Typically, the count goes from zero to $N-1$.

MOS.   Metal oxide semiconductor.

MSB.   The most significant bit: the character in the left-most position of a number. It has the largest weight.

MSI.   Medium-scale integration.

Multiplexer.   Logic gate used to select one of two (one of four, one of eight, etc.) inputs and transfer them to an output. Decoders are very useful when looking at many logic signals.

NAND gate.   AND gate with an inverted output.

Negative level.   Describing a family of integrated circuits that operate on negative voltages and signals.

Negative logic. Logic in which the least positive value is a logical "1."

Nibble. Four bits.

Noise margin. Voltage difference between a TTL minimum output "1" and minimum input "1" $(2.4 - 2.0 = 0.4)$. It is also the difference between an output minimum "1" and an input minimum "1."

NOR gate. An OR gate with an inverted output.

Octal latch. A prepackaged 8-bit D latch.

Octal. A numbering system based on the number 8.

Odd parity. The number of ones transmitted in a particular word must be odd.

Open collector. Output circuitry in which there is no internal load. The collector is brought out to provide pulldown capability.

OR gate. A logic gate that produces a logic "1" when any input is at a logic "1" level.

Originate. Name of the modem used by the person calling a computer over telephone lines.

Package cost. The number of ICs required to perform the DBB function. Usually less than the gate cost.

PAL. Programmable array logic.

Parallel. The presentation of all bits of a binary number on an equal number of wires at the same time.

Parity. Method used to help detect errors (during the transmission or reception of data).

Pipeline. Multistage register-logic hardware for executing instruction sequences in parallel.

PLA. Programmable logic array.

Port. An electronic device in a computer that allows information exchange between the computer and the outside world.

Positive level. Describing a family of integrated circuits that operates on positive voltages and signals.

Positive logic. Most positive voltage signal represents a logic "1."

Processor. See *CPU*.

Product of the sums. Boolean method utilizing all DBB inputs that produce a zero output.

Program counter. The CPU register that contains the current instruction address in memory.

PROM. Read only memory programmed by the user one time only.

Propagation delay. The time needed for a gate's output to change state in response to an input.

Pseudo-random pulse generator. A shift register circuit used to produce random bit sequences.

Pullup resistor. A resistor used to pull an input up to a logic one level.

Qualifier. A condition necessary to cause a logic analyzer to start taking data. Used as an AND with the trigger word.

Quine-McCluskey method. Technique used to minimize a logic function by comparing terms.

Radix. The base of the number system.

RAM. Random-access memory—read/write.

Read/write. Describing a memory that may be read from or written to.

Refresh. Dynamic memories must have the data stored in their small capacitive elements recharged or refreshed. The "Refresh" command triggers such recharging.

Register. A temporary storage device used for one or more bytes.

Registered output. An output that contains a storage element.

RESET. A reset flip-flop has $Q = 0$.

Ring counter. A shift register wired so that its output is fed back to its input, allowing data to circulate in a ring.

Ripple counter. A counter where the output of each flip-flop is used as a clock for the next flip-flop. The output of the counter is not valid until the propagation delay of all flip-flops has passed.

Rise time. In pulse analysis, the amount of time a signal takes to rise from 10% of a high level to 90% of a high level.

ROM. Read-only memory; factory programmed.

RS232C. A standard used in serial data communications.

Sample interval. The time between samples (reciprocal of sample rate).

Sample rate. The rate at which data is taken.

Saturated logic. Form of logic in which the transistors are overdriven, causing reduced speed.

Selector. A logic element that chooses one signal from a group of signals. Same as *decoder*.

Serial. Describing the presentation of one bit at a time on a single wire.

SET. A set flip-flop has Q = 1.

Shift register. A connection of flip-flops used to store long binary numbers and shift them to the left or right as necessary.

Sign bit. Takes the place of the MSB when using signed numbers.

Signature analysis. An electronic method of checking patterns in digital circuits.

Signed number. Number range that includes both positive and negative numbers.

SIMM. Single inline memory module. Standard memory for the PC.

Simplex. Transmission of data in only one direction.

Space. Indicates a low level in serial transmission.

SSI. Small-scale integration.

Start bit. Bit or signal used in serial transmission to indicate the beginning of new information.

State diagram. A graphical way of representing the behavior of a state machine.

State machine. A circuit containing several different states reachable from each other, with state changes happening synchronously.

State transition table. A data table containing all the information from the state diagram.

Static memory. A device in which each memory cell is a flip-flop.

Status bus. CPU outputs showing the present state of operation in a computer.

Step size. The amount of voltage controlled by the LSB of a digital-to-analog converter.

Stop bit. Bit or signal used to indicate the end of a serial transmission.

Successive approximation register (SAR). Special A/D converter with fixed conversion time.

Sum of the products. Boolean method utilizing all DBB inputs that produce a one output.

Switch. Network device that forwards packets to specific destinations.

Synchronous counter. A counter where all the flip-flops are clocked at the same time. The output is valid after a single propagation delay time.

Synchronous logic. A logic circuit that contains memory, and uses its current state and inputs to determine its next state. State changes occur when a clock pulse is received.

Synchronous transmission. Serial transmission of data requiring a separate clock signal.

Synchronous. Describing input that requires a clock pulse to cause activity. Such input cannot act on its own.

T1 line. 24-channel TDM signal running at 1.544 Mbps. Channels are 8 bits wide.

Tape memory. Slow memory for long-term program storage.

Time division multiplexing (TDM). A digital multiplexing technique where multiple data signals are sampled sequentially and transmitted over a single wire, in serial format.

Timing diagram. A drawing of the simultaneous waveforms that exist as a device (counter, shift register, etc.) proceeds through its various states.

Toggle. When a flip-flop alternates from set to reset, it is said to toggle.

Transceiver. A circuit that can be switched to move data in one of two directions.

Trigger word. A pattern that is used to begin data acquisition.

Tri-state logic. Logic gates that are capable of being electronically disconnected from a wire.

Tri-state. To electronically disconnect the output of a gate from a wire.

Truth table. Table containing the key to the logical operation of an integrated circuit. It is a structure that organizes all of the input and output combinations of a digital circuit.

TTL. Transistor-transistor logic.

TV teletype. A device that displays character information on an ordinary television screen.

UART. Universal asynchronous receiver transmitter.

Unsigned number. Number range that begins at zero and does not go negative; only positive values are allowed.

VCO. Voltage-controlled oscillator. The output frequency is proportional to input voltage. VCOs are used when frequency must be a function of voltage.

VHDL. VHSIC Hardware Description Language.

VHSIC. Very-High-Speed Integrated Circuit.

VLSI. Very-large-scale integration.

Weight. The importance of a position in a number.

Wired OR. Function in which many gates are tied together at the output to commonly affect an input or produce a result.

Word. Two bytes (16 bits) of information.

INDEX

373